Secondary metabolism in plant cell cultures

Secondary metabolism in plant cell cultures

Edited by

PHILLIP MORRIS
ALAN H. SCRAGG
ANGELA STAFFORD
MICHAEL W. FOWLER

Wolfson Institute of Biotechnology
University of Sheffield
Sheffield

The right of the
University of Cambridge
to print and sell
all manner of books
was granted by
Henry VIII in 1534.
The University has printed
and published continuously
since 1584.

CAMBRIDGE UNIVERSITY PRESS
Cambridge
London   New York   New Rochelle
Melbourne   Sydney

CAMBRIDGE UNIVERSITY PRESS
Cambridge, New York, Melbourne, Madrid, Cape Town,
Singapore, São Paulo, Delhi, Tokyo, Mexico City

Cambridge University Press
The Edinburgh Building, Cambridge CB2 8RU, UK

Published in the United States of America by Cambridge University Press, New York

www.cambridge.org
Information on this title: www.cambridge.org/9780521279338

First published 1986
First paperback edition 2011

A catalogue record for this publication is available from the British Library

ISBN 978-0-521-32889-0 Hardback
ISBN 978-0-521-27933-8 Paperback

CONTENTS

PREFACE

It is perhaps only a matter of time before someone writes a book, 'Plant Cell Biotechnology - the first century'! Given the, albeit superficial, view that 'biotechnology' has happened since about 1970, it is easy to forget that it is almost one hundred years since Haberlandt first carried out his pioneering experiments on maintaining plant cells in a viable state away from the parent body by bathing them them in simple nutrient solutions. Haberlandt's contribution to plant cell culture is often underestimated but was the key in opening up the area. Since that time, while not without its ups and downs, progress in the technology has been steady, with particular developments taking place over the last decade or so. In the mid 1950's attention was focussed upon the possibility of using plant cells on an industrial scale for natural product synthesis. At the time however, the technology lagged far behind the dream, and both progress and interest flagged with that realisation. The great upsurge of interest in 'biotechnology' in the 1970's carried plant cell culture with it. Rapid developments followed, including techniques for protoplast fusion, the establishment of a wide range of often recalcitrant species in culture, micro screening techniques for natural products including RIA and ELISA systems, process technology in the form of large scale culture, continuous culture systems and immobilised cell technology. Some progress also occurred in developing the knowledge base of cell physiology and natural product biosynthesis, but unfortunately at nothing like the rate necessary to match developments in the technology. It was to some extent a recognition of this imbalance in the development of plant cell biotechnology that lead to the format of the IAPTC Meeting held in Sheffield in the summer of 1985.

The meeting was organised under the auspices of the U.K. 'branch' of the International Association of Plant Cell and Tissue Culture. Previous meetings had focussed on plant propagation, protoplast fusion and related area. No meeting had however been held in the U.K. which focussed upon, and specifically addressed the status of plant cell culture and natural product synthesis. The meeting covered a period of two days in the July of 1985, and spanned a wide range of topics including; natural product synthesis, strain selection and screening, stability and variability, relationships between natural product synthesis and cell and tissue differentiation, and large scale culture and cell immobilisation. It says much for the strength or resilience (!) of plant cell culture in the U.K., that almost all of the presentations in a very full programme came from this country. The range of material presented also is indicative of the very wide variety of approaches to plant cell culture, which can only be to the good of the subject.

The presentations made during the meeting engendered a great deal of excellent discussion. It is always intriguing in such situations to see quite often that views expressed in other places which appear to be totally opposed are not after all so far apart. The bringing together of people with different views in open forum and constructive debate does much towards the intellectual development of the subject, and this was certainly true of the Sheffield meeting. The meeting was good humoured and active and it was especially encouraging to see the younger participants fearlessly engaging the more elderly (and entrenched?) members present. Great sport was had by all!

So where does the subject go from here? That there is widely developed activity in plant cell biotechnology is obvious. The development of the shikonin process by Mitsui and Co. in Japan is also extremely encouraging. It is equally obvious however, that the fundamental base to the subject is woefully inadequate. Already developments are being limited by a lack of information on the synthetic pathways to secondary metabolites, on the properties of key enzymes, of enzyme control systems, and on cellular transport phenomenon. One can only hope that meetings such as that held in Sheffield will bring home to the various funding bodies the need to support a lot more research into basic aspects of plant cell physiology and biochemistry, if the longer term goal of industrial application at a significant level is to be achieved. We can live in hope!

ACKNOWLEDGEMENTS

The organisation of any meeting takes a great deal of time and effort. For the success and smooth operation of the Sheffield IAPTC Meeting, I would wish to express my personal thanks to Dr. Phillip Morris for taking on the job as Meeting Organiser. Thanks are also due to his able assistants, Dr. Alan Scragg, and Dr. Angela Stafford, for their most supportive contributions.

The great pain of coping with manuscripts of all shapes and sizes was cheerfully coped with by our secretaries Jan Stacey and Carol Cavanagh, who provided a relative haven of peace to keep us all going! Maureen Hollingsworth as usual provided excellent slide and photographic material.

In these days of shrinking university resources, financial assistance from external bodies is not just gratefully received, it is almost a necessity to allow meetings to be held. In this context we would like to thank Imperial Tobacco PLC., Cadbury Schweppes PLC., Allied Breweries Ltd., Sanofi (UK) Ltd. and ICI PLC., for their most generous contributions in support of the meeting. We are also grateful to acknowledge help with equipment displays and donations from Gibco Ltd., L.H. Fermentation Ltd., Gelman Ltd., and F.T. Scientific.

M.W. Fowler
Wolfson Institute of Biotechnology.
April 1986.

SECTION 1

SECONDARY PRODUCTS FROM PLANT CELL CULTURES

1. ASPECTS OF ALKALOID PRODUCTION BY PLANT CELL CULTURES

L.A. Anderson,   J.D. Phillipson & M.F. Roberts

Department of Pharmacognosy,
The School of Pharmacy,
University of London,
29-39 Brunswick Square,
London. WC1N 1AX. U.K.

PHARMACEUTICAL USES OF ALKALOIDS
For centuries plants have been an important source of drugs. Many plant extracts and isolated constituents are well established in clinical practice and are likely to remain so for some time until better, cheaper, less toxic or more efficacious alternatives become available.

Analysis of prescriptions dispensed in the U.S.A. has shown that throughout the period 1959-1973, 25% contained natural products (Farnsworth & Morris 1976). More recent analyses have shown that these figures have not changed, the proportion remains about 25% but the relative cost of these items has escalated from $3 billion to $8 billion (Farnsworth 1984).

Of the pharmacologically active principles found in plants, alkaloids are arguably the most important group. Some 30 or so are used therapeutically and they cover a broad spectrum of pharmacological effects (Table 1). Only a few of these 30 alkaloids are now produced by chemical synthesis, notably caffeine and ephedrine. Some alkaloids have further uses, for example, quinine from <u>Cinchona</u> bark, is used extensively in the food industry as a bitter flavouring.

Table 1   Examples of pharmaceutically important plant alkaloids.

PHARMACOLOGICAL ACTIVITY	ALKALOID
Central Nervous System:	Reserpine, Caffeine
Autonomic Nervous System:	
Cholinergic	Physostigmine, Pilocarpine
Cholinergic blocking	Atropine, Hyoscyamine
Adrenergic	Ephedrine
Ganglion blocking	Nicotine
Chemotherapeutic:	
Anticancer	Vinblastine, Vincristine, Harringtonine, Camptothecin
Antiamoebic	Emetine
Antimalarial	Quinine
Antibacterial	Berberine
Cardiovascular:	
Vasodilator	Papaverine, Theophylline, Vincamine
Hypotensive	Reserpine, Rescinnamine, Protoveratrines A and B
Antiarrythmic	Quinidine, Ajmaline
Analgesic:	Morphine, Codeine
Antitussive:	Glaucine, Noscapine
Anti-inflammatory:	Colchicine
Muscle Relaxant:	Tubocurarine
Local Anaesthetic:	Cocaine

World production of Cinchona bark is said to be in the order of
8-10,000 tonnes annually (Bisset 1982). Approximately half the
quinine isolated is converted to quinidine, which is used to treat
cardiac arrhythmias, while the remaining half is divided almost
equally between the pharmaceutical and food industries (Bisset 1982).

ALKALOID PRODUCTION BY PLANT CELL CULTURES
Over the last decade plant cell cultures have been
investigated as an alternative means of producing commercially
important secondary metabolites. The potential advantages over
traditional field methods of cultivation are clear, in particular,
independence from geographical, climatic and political problems
(Fowler 1983). In addition, with plant cell cultures, it may be
possible to optimise growing conditions, achieve more consistent
quality and to recover the product more easily.

It is clear from the scientific literature, that in the last decade a
wide range of alkaloids have been successfully detected in cell
cultures, covering almost all the main alkaloid types, e.g. indoles,
isoquinolines, quinolines, quinolizidines, tropanes, acridones and
purines (Anderson et al. 1985a).

Fig. 1   Examples of indole alkaloids from plant cell
cultures.

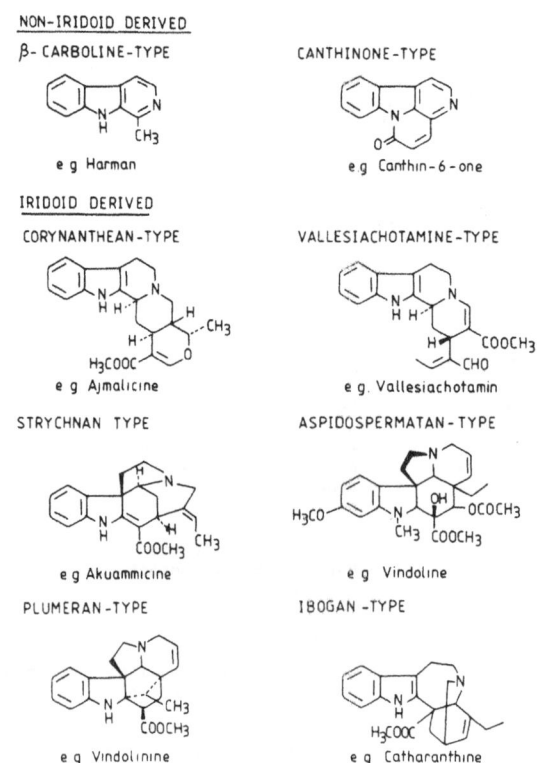

A great deal of interest has focussed on the indoles and isoquinolines
which represent the two major groups of plant alkaloids.

Indole Alkaloids
To date, some 47 or so indole alkaloids have been
identified in cell cultures, some being of the non-iridoid type, e.g.
harman (Sasse et al. 1982) and canthin-6-one (Anderson et al. 1983a)
but mainly they are those derived from iridoid precursors. Representatives of almost all the main types have been detected (Figure 1)
(Anderson et al 1985a). Studies with Catharanthus roseus have, not
surprisingly, been the main centre of scientific endeavour. Despite
this work, the actual number of indole producing species which have
been investigated in cell culture is relatively few. Although
enormous effort has gone into research on Catharanthus cultures, it
would appear that so far it has proved impossible to produce the important anti-leukaemic dimeric indoles, vinblastine and vincristine
(Figure 2). However, the report of two novel dimeric indoles,
voafrine A and B, being isolated from cell cultures of the related
Apocynaceous plant, Voacanga (Figure 2) is clear indication that dimeric alkaloids can be produced by cell cultures (Stockigt et al.1983)

Fig. 2   Examples of dimeric indole alkaloids

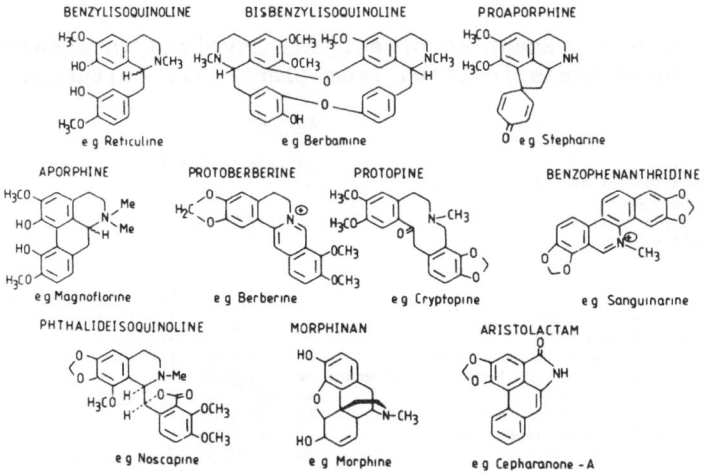

| Vinblastine | R = CH₃ |
| Vincristine | R = CHO |

From Catharanthus roseus

Voafrine A C-3′Hα configuration
Voafrine B C-3′Hβ configuration

From Voacanga species

Fig.3   Examples of isoquinoline alkaloids from plant cell
cultures.

BENZYLISOQUINOLINE   BISBENZYLISOQUINOLINE   PROAPORPHINE

e g Reticuline   e g Berbamine   e g Stepharine

APORPHINE   PROTOBERBERINE   PROTOPINE   BENZOPHENANTHRIDINE

e g Magnoflorine   e g Berberine   e g Cryptopine   e g Sanguinarine

PHTHALIDEISOQUINOLINE   MORPHINAN   ARISTOLACTAM

e g Noscapine   e g Morphine   e g Cepharanone - A

Isoquinoline alkaloids
Isoquinoline alkaloids have also been the subject of a
number of research investigations. About 45 isoquinolines have been
isolated from plant cell cultures (Rueffer 1985) and they represent
all of the major isoquinoline types (Figure 3). Some of the major
successes in the production of secondary metabolites from plant cell
cultures, both in terms of highest yielding cultures and biosynthetic
studies, have been obtained with isoquinoline alkaloids. It is ironic
therefore that the most important members of this group, from a
pharmaceutical viewpoint, the morphinans, have proved difficult to
produce.

ADVANTAGES OF ALKALOID PRODUCTION BY CELL CULTURE TECHNIQUES
Despite the difficulties encountered in producing
vinblastine, vincristine and the morphinans, such as morphine and
codeine, significant advances in our understanding of the biosynthesis
of both indole and isoquinoline alkaloids have been gained as a result
of studies with cell cultures (Anderson et al. 1985a).

Plant cell cultures offer a number of distinct advantages over whole
plants for biosynthetic studies, such as standardised conditions of
growth (Anderson et al. 1985a). They are not subject to seasonal
variations and cultures are often less complex in organisation and so
do not present the problems of incorporation that are frequently
encountered in entire plants. The aseptic environment needed for
culturing plant cells eliminates problems associated with
contamination by microorganisms and furthermore, purified enzymes and
active cell-free systems can be prepared more easily.

ENZYMES DERIVED FROM PLANT CELL CULTURES WHICH PRODUCE ALKALOIDS
The contribution of cell cultures to alkaloid biosynthetic
studies is reflected in the fact that no less than 15 key enzymes have
been isolated and identified in recent years (Table 2).

Table 2   Examples of enzymes involved in alkaloid
biosynthesis isolated from plant cell cultures.

ENZYME	SOURCE	
Strictosidine synthase	Catharanthus roseus	
Strictosidine glucosidase I and II	"	"
Geissoschizine dehydrogenase	"	"
Polyneuridine aldehyde esterase	Rauwolfia serpentina	
Vellosimine reductase	"	"
Vinorine synthase	"	"
(S)-Norlaudanosoline synthase	eg. Berberis, Eschscholtzia	
SAM:(R),(S)-norlaudanosoline-6-O-methyltransferase	eg. Berberis, Argemone	
SAM:(6-O-methylnorlaudanosoline)-4'-O-methyltransferase	eg. Berberis, Argemone	
SAM:(R),(S)-norreticuline-N-methyltransferase	eg. Berberis, Eschscholtzia, Fumaria	
SAM:(6-O-methylnorlaudanosoline)-5'-O-methyltransferase	eg. Argemone	
Berberine bridge-forming enzyme	eg. Macleaya, Berberis	
SAM:(S)-scoulerine-9-O-methyltransferase	eg. Berberis	
(S)-Tetrahydroprotoberberine oxidase (STOX)	eg. Berberis	
Methylenedioxy-ring-forming enzyme	eg. Berberis, Thalictrum	

One particular example of cell culture work has been the elucidation of the biosynthesis of berberine from the simple precursors, dopamine and 3,4-dihydroxyphenylacetaldehyde via intermediates requiring the co-operation of 8 enzymes which have been characterised (Zenk 1985). Berberine is the first alkaloid used medicinally, whose biosynthesis is now understood at the enzyme level and this achievement is one of the most important milestones in the applications of plant cell cultures to the elucidation of secondary metabolic pathways.

COMPARISON OF ALKALOID YIELD FROM WHOLE PLANTS WITH CELL CULTURES

Besides economic considerations, one of the main factors which has prevented cell cultures being considered as a serious rival to whole plants has been their relatively low yields of secondary metabolites when compared to the parent plant. Productivity of cell cultures is usually assessed by comparing the yield of the compound of interest from the culture with that from the whole or parts of the producing species and although this comparison is widely used it may be misleading. For example, good quality Cinchona bark, which is used as a source of quinine and quinidine, should contain more than 10% dry weight of total alkaloid. However, it must be borne in mind that the yield is expressed as a percentage of the bark which accumulates the alkaloids and not as a percentage of the whole plant, as such. Furthermore, in African plantations, it takes at least 7 years of cultivation before the bark can be harvested and in some instances optimum accumulation of the alkaloids can take up to 20 years.

Over the last 10 years, particularly with improved understanding of cultural requirements and the introduction of cell selection, there have been many examples where the level of secondary metabolites in cultures approaches and in some cases exceeds that in the whole plant.

Table 3   Examples of cell cultures which accumulate alkaloids at levels higher than the parent plant.

ALKALOID	SPECIES	YIELD (%d.wt.)	
		Cell Culture	Whole Plant
Caffeine	Coffea	1.60	1.60
Pseudoephedrine	Ephedra	2.25	0.60
Nicotine	Nicotiana	3.40	2.50
Protopine	Macleaya	0.40	0.32
Biscoclaurines	Stephania	2.29	0.92
Ajmalicine & Serpentine	Catharanthus	1.30	0.26
Canthin-6-ones	Ailanthus	1.27	0.01
Berberine	Coptis	11.40	10.00
Jatrorrhizine	Berberis	10.00	1-2

By 1982, it was estimated that at least 30 compounds were known to accumulate in cultures at levels higher than that of the plant (Staba 1982) and a number of these successes have been achieved with alkaloid producing species (Table 3). In some cases these yields are as a result of painstaking selection of high yielding cell lines, for example with ajmalicine and serpentine production in Catharanthus (Zenk et al. 1977) and jatrorrhizine in Berberis species (Hinz & Zenk 1981). The yield of jatrorrhizine from Berberis cultures, at levels of 10% dry weight of cells and which corresponds to 2.7 gl^{-1} is the highest ever reported yield of alkaloid from cell cultures (Hinz & Zenk 1981).

Comparison of the plant yield with that of the cell culture may well be misleading as in the case of Coptis cultures which produce slightly more alkaloid than the whole plant (Fukui et al. 1982). However, it must be pointed out that the 10% dry weight refers only to the root of the plant which accumulates this amount of alkaloid over a period of 5 years whereas the suspension cultures produce 11.4% dry weight in 3 weeks.

ALKALOID PRODUCTION BY SUSPENSION CULTURES
Ailanthus altissima
The first report of the production of canthin-6-one alkaloids by cell cultures was from Ailanthus altissima (Anderson et al. 1983a). The yield from the cultures is over 100 times that found in the whole plant (Table 3) and this is one of the most dramatic improvements in alkaloid yield ever reported for cultures when compared with entire plants. Ailanthus altissima is a member of the Simaroubaceae family which is well known in traditional medicine for a variety of uses. Species of Ailanthus, in particular have been used to treat amoebic dysentery, as an anthelmintic and to treat tumours. Studies have mainly been directed towards the quassinoids, which are a group of bitter principles derived biosynthetically from the degradation of triterpenes, e.g. ailanthinone (Figure 4). The quassinoids have a wide spectrum of biological activities including cytotoxic, amoebicidal, insecticidal effects and of considerable interest is the marked anti-leukaemic properties of some quassinoids. In recent year, the quassinoids have also been shown to possess potent antimalarial activity and a series of Simaroubaceous plants which are used in traditional medicine for the treatment of malaria have been the subject of several investigations (Bray et al. 1985, O'Neill et al. 1985).

Cultures of Ailanthus were originally established in our laboratories in an attempt to produce the quassinoid constituents which are present in the plant in very low levels. Both callus and suspension cultures, however, produced only traces of quassinoids but coincidentally, high levels of canthin-6-one alkaloids were discovered (Anderson et al. 1983a).

The whole plant produces canthin-6-one alkaloids together with simple β-carbolines as only minor constituents (Figure 4). The cultures were shown to produce the 1-methoxy derivative of canthin-6-one as the major alkaloid and canthin-6-one itself as the minor. The canthinone alkaloids are known to have antibacterial and antifungal properties and it has been shown that they are cytotoxic to guinea-pig keratinocyte cultures in vitro (Anderson et al. 1983a). Although these alkaloids have no obvious value it has been possible, because of the high yields, to study these cultures as a model system for alkaloid production in cell cultures.

Analyses of samples of leaves, young wood, old wood and seed pods from the tree which served as the initial source of Ailanthus cultures have revealed that the alkaloid levels in the parent plant parts are very low (Table 4). It is interesting to note that canthin-6-one is the major alkaloid in the leaf and wood samples but that the 1-methoxy analogue is the major alkaloid in the seed pods and indeed, in the cell cultures (Anderson 1985b). The yield from the cell cultures far exceeds that of the whole plant (Table 4).

Despite the ability shown by some cultures such as Ailanthus, Coptis and Berberis to accumulate significant levels of alkaloids, many cultures fail to produce alkaloids while others do so, but only in trace amounts. Many factors are known to affect alkaloid accumulation in cell cultures, including genetic, morphological, cultural

Fig. 4   Some constituents of Ailanthus altissima whole plants.

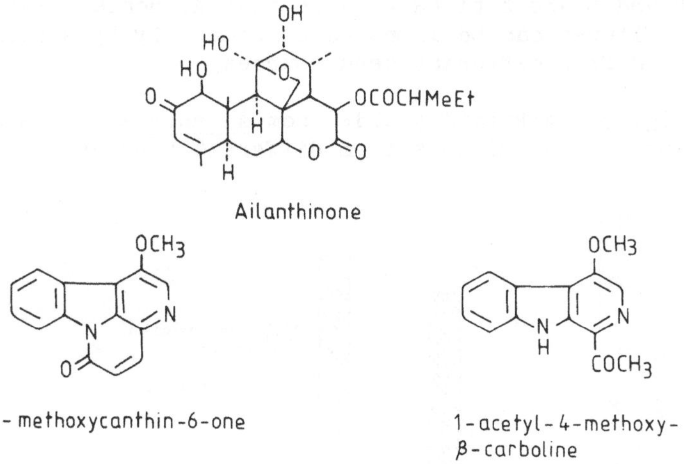

Ailanthinone

1-methoxycanthin-6-one

1-acetyl-4-methoxy-β-carboline

Table 4   Comparison of alkaloid yield from whole plant parts and cell cultures of Ailanthus altissima

% ALKALOID dry weight	LEAF	YOUNG WOOD	OLD WOOD	SEED PODS	CELL CULTURE
Canthin-6-one	0.0012	0.0037	0.0077	0.0003	0.05
1-Methoxycanthin-6-one	0.0001	0.0004	0.0006	0.0011	0.98

conditions and cell selection.

Considering genetic factors it is well known that individual plants of
a given species or cultivar may vary in their ability to produce and
to accumulate secondary metabolites. Ailanthus altissima was first
introduced to England from China in 1751, and from England was
introduced to Europe and to America where it has been widely used as
an ornamental tree and has become extensively naturalised. Studies
have shown that genetic differences now exist between seedlings of
American and Chinese Ailanthus and apparently American Ailanthus is
now not only genetically different from the Chinese Ailanthus but is
also genetically variable within its own population (Feret & Bryant
1974).

This information alerted us to possible variation between Ailanthus
seed from different sources and when seeds were examined from a number
of European trees, our attention was drawn to noticeable differences in
the seed sizes and weights. Cultures were initiated from different
seeds using identical conditions. Analysis of the alkaloids showed
significant differences between seeds from different sources (Figure
5). Compared to our original seed source A, two others B and C,
showed much improved alkaloid yields (Anderson 1985b). All three
lines produce 1-methoxycanthin-6-one as the major alkaloid and
canthin-6-one as the minor alkaloid. The ratio of 1-methoxycanthin-
6-one to canthin-6-one is clearly different in C and what is more the
yields from B and C are 6 times greater than A. Hence, alkaloid yield
in Ailanthus cultures can be improved considerably by screening
cultures derived from different seed sources.

Fig. 5   Alkaloid yields from Ailanthus altissima cell
suspension cultures from different sources

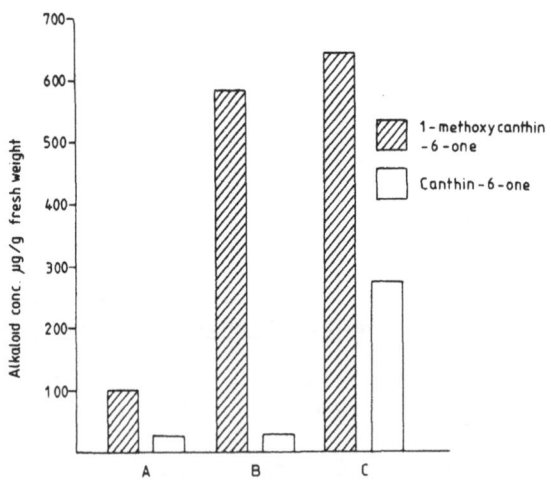

Papaver somniferum

Despite considerable investigation, many attempts to produce substantial amounts of morphinans in undifferentiated cultures of Papaver have failed. Some groups have reported morphine, codeine or thebaine but generally in low amounts (Constabel 1985) and morphinan production was often lost on repeated subculturing. Although morphinan production has been difficult, other isoquinoline alkaloids, such as benzophenanthridines, protopines and aporphine types, which are not typical of the plants have been found in Papaver cultures (Ikuta et al. 1974).

A similar approach to that used for Ailanthus was adopted in our studies with Papaver somniferum and cultures were established from 10 different seed sources. In this case, however, there was no significant difference, quantitatively or qualitatively, in alkaloid production. Furthermore, the Papaver cultures produced protopine-type alkaloids with cryptopine as the major component (Figure 6, Anderson et al. 1983b). Cryptopine was found to accumulate at higher levels in the medium than in the cells. In addition the presence of comparatively high levels of dopamine, one of the isoquinoline biosynthetic precursors, was found in the cells but not in the culture medium (Figure 6). Dopamine accumulates at higher levels than cryptopine and this build-up of precursor may reflect low activity of norlaudanosoline synthase, the first key enzyme in the benzylisoquinoline biosynthetic pathway.

Fig. 6   Cryptopine and dopamine content in Papaver
somniferum cell suspension cultures

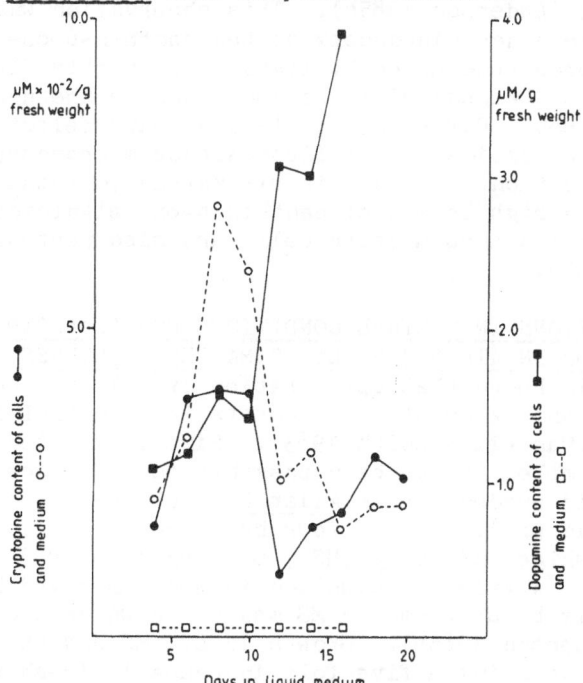

In the whole plant, morphinan alkaloids are primarily accumulated in
latex which is contained in specialised cells known as laticifers.
Whether these cells are the site of morphinan biosynthesis is disputed
and it has been suggested that synthesis may occur in surrounding
cells with the laticifers functioning as storage sites. Whilst the
production may not be wholly dependent on differentiation of
laticifers it would seem that their formation could substantially
increase the capacity of Papaver cultures to accumulate morphinan
alkaloids.

It has been demonstrated that isolated organelles obtained from
P. somniferum latex, rapidly and specifically accumulate large
quantities of morphinan alkaloids (Homeyer & Roberts 1984, 1985).
Similar organelles are being sought in Papaver cultures. Presumably,
for successful production of morphinan alkaloids in cell cultures it
is necessary to produce cells which are not only capable of producing
but also of storing morphinans. There is evidence that cell cultures
of other plants possess alkaloid accumulating cells notably in
Macleaya (Bohm 1978) and Catharanthus (Neumann 1983).

SELECTION OF CELL LINES OF AILANTHUS ALTISSIMA FOR ALKALOID PRODUCTION

Ailanthus altissima cultures grown in our laboratories
show a mixture of cell types when examined microscopically. Staining
with Dragendorff's reagent shows that the majority of the cells
contain alkaloids within their vacuole but close inspection reveals
the presence of non-alkaloidal cells which do not stain with
Dragendorff's reagent (Anderson 1985b). This observation was further
confirmed using fluorescence microscopy as the canthin-6-one alkaloids
show bright blue fluorescence under UV light. The highly fluorescent
cells stain orange with Dragendorff's reagent whereas the non-
fluorescent cells do not. Currently, cells are being selected from
these cultures using a combination of fluorescence microscopy and HPLC
with fluorescence detection. The aim is ultimately to establish cell
lines which accumulate high levels of canthin-6-one alkaloids and to
ascertain if the alkaloid accumulating cells are also responsible for
canthinone biosynthesis.

MODIFICATIONS IN CULTURE CONDITIONS FOR ALKALOID PRODUCTION IN AILANTHUS ALTISSIMA CELL CULTURES

It is well known that optimisation of cultural conditions,
both in the internal and external environments is essential for good
alkaloid production (Mantell & Smith 1983). Simply by altering the
basal medium, for example, it has been possible to influence
significantly alkaloid production in Ailanthus cultures. Comparison
of growth and alkaloid production in four basal media, viz. Linsmaer
and Skoog (LS), Murashige and Skoog (MS), Gamborg's B5 (B5) and Schenk
and Hildebrandt (SH). Alkaloid production is much improved by
transferring the cells to LS from the MS medium in which the cultures
were established (Anderson 1985b). Growth in LS, MS and B5 is
similar, the cells undergoing a five-fold increase in fresh weight in
one month (Anderson et al. 1985c). On the other hand growth in SH
medium is poor and there is a substantial drop in alkaloid production.

PRECURSOR FEEDING AND ALKALOID PRODUCTION IN CELL CULTURES
OF AILANTHUS AND CINCHONA

In some cases, supplementing the culture medium with one
of the biosynthetic precursors has been found to increase alkaloid
yield (Mantell & Smith 1983). Alkaloid yields from our Ailanthus
cell cultures and Cinchona root organ cultures have been improved by
feeding with L-tryptophan (Hay et al. 1986). Initial experiments
using L-[methylene-^{14}C] -tryptophan demonstrated that both Ailanthus
and Cinchona cultures could incorporate the radio-label into canthinone
alkaloids and quinine/quinidine, respectively. Uptake of
L-[methylene-^{14}C] -tryptophan into Ailanthus cells proceeded rapidly
with almost half the precursor taken up inside 15 minutes.
Radio-labelled alkaloids, however, were not detected until 48 hours
later (Anderson 1985b). In an attempt to increase alkaloid levels
unlabelled L-tryptophan was fed to Ailanthus cultures and at levels of
500 mgl^{-1} a 76% improvement in alkaloid yield was obtained.

Root organ cultures of Cinchona ledgeriana which produce low levels of
quinoline alkaloids, with quinine and quinidine as the major alkaloids
have been established in our laboratories (Anderson et al. 1982).
Although the alkaloid levels are very low in comparison to Cinchona
bark, alkaloid production has remained stable over the last three
years. The root organs grow well in culture forming small organs
composed of root-like projections, which grow to a length of
approximately 5 mm and then break off and bud to form new clusters
(Hay et al. 1986). Despite their high degree of organisation and
because they grow to a finite size, they are quite suitable for
growing on a large-scale unlike many other organised cultures and they
have been grown in 3L batches as well as immobilised on nylon
supports.

L-[methylene-^{14}C] -tryptophan is taken up by the Cinchona root organs
about 32 hours after feeding and the radio-label is almost immediately
incorporated into quinine and quinidine (Hay et al. 1986).
Supplementing the medium with unlabelled L-tryptophan at levels of 500
mgl^{-1} results in a surprising 5-fold increase in the yields of both
quinoline alkaloids.

CONCLUSIONS

Studies in our laboratories utilising suspension cultures
derived from species of Ailanthus, Cinchona and Papaver have
demonstrated that a number of factors significantly influence alkaloid
production. The original source of plant material and the cultural
conditions used can affect markedly the alkaloidal composition and
yield. Not only is it necessary to develop cell lines which are
capable of synthesising alkaloids but it is also essential that the
cells or tissues are capable of storing the alkaloids. It is obvious
that several fundamental questions remain still to be answered in
respect of alkaloid production (Heinstein 1985) and some of the major
areas of interest may be summarised as follows:

1. An understanding of the mechanisms by which stable high-yielding cell lines may be developed
2. Growth and production media need to be defined
3. Some alkaloids and other secondary metabolites are readily produced by cell cultures whereas others require the use of stress or elicitors in order to induce their production
4. The mechanisms by which alkaloids are produced and stored together with the factors which influence alkaloid accumulation and excretion require further studies
5. There is a need to understand the biosynthetic pathways which are involved in alkaloid production and this includes the identification of the enzymes which are responsible and their control mechanisms
6. Recognition and isolation of the plant genes which are responsible for the production of these enzymes may lead to their possible transfer to prokaryots.

In order to realise the true potential of plant cell cultures as alternatives to whole plants for the production of alkaloids and other secondary metabolites, it is essential that these basic areas of research receive further sustained investigation.

REFERENCES

Anderson, L.A., Keene, A.T. & Phillipson, J.D. (1982). Alkaloid production by leaf organ, root organ and cell suspension cultures of Cinchona ledgeriana. Planta Medica. 46, 25-27

Anderson, L.A., Harris, A. & Phillipson, J.D. (1983a). Production of cytotoxic canthin-6-one alkaloids by Ailanthus altissima plant cell cultures. Journal of Natural Products. 46(3), 374-78.

Anderson, L.A., Homeyer, B.C., Phillipson, J.D. & Roberts, M.F. (1983b). Dopamine and cryptopine production by cell suspension cultures of Papaver somniferum. J. Pharm. Pharmac. 35, 21P.

Anderson, L.A., Phillipson, J.D. & Roberts, M.F. (1985a) Biosynthesis of secondary products by cell cultures of higher plants. In Advances in Biochemical Engineering/Biotechnology, ed. A. Fiechter 31, pp. 1-36, Berlin, Heidelberg: Springer-Verlag.

Anderson, L.A. (1985b), unpublished observations.

Anderson, L.A., Phillipson, J.D. & Roberts, M.F. (1985c). Growth and alkaloid accumulation in cell suspension cultures of Ailanthus altissima. J. Pharm. Pharmac. 37, 45P.

Bisset, N.G. (1982). Economic aspects of medicinal and aromatic plants. In Proceedings of the fourth meeting of Turkish pharmacognosists. pp. 53-65, Eskisehir, Turkey.

Bohm, H. (1978). Regulation of alkaloid production in plant cell cultures. In Frontiers of Plant Tissue Culture 1978, ed. T.A. Thorpe, pp. 201-11, Calgary, Canada: IAPTC

Bray, D.H., O'Neill, M.J., Boardman, P., Phillipson, J.D. & Warhurst, D.C. (1985). Structure related in vitro antimalarial activities of some quassinoids. J. Pharm. Pharmac., 37, 142P.

Constabel, F. (1985). Morphinan alkaloids from plant cell cultures.
    In The Chemistry and Biology of Isoquinoline Alkaloids,
    ed. J.D. Phillipson, M.F. Roberts & M.H. Zenk, pp.
    257-64, Berlin, Heidelberg: Springer-Verlag.

Farnsworth, N.R. & Morris, R.W. (1976). Higher plants; the sleeping
    giant of drug development. Amer. J. Pharm., 147, 46-52.

Farnsworth, N.R. (1984). How can the well be dry when it is filled
    with water? Economic Botany, 38(1), 4-13.

Feret, P.P. & Bryant, R.L. (1974) Genetic differences between
    American and Chinese Ailanthus seedlings. Silvae Genetica
    23(5), 144-48.

Fowler, M.W. (1983). Commercial applications and economic aspects of
    mass plant cell culture. In Plant Biotechnology, ed. S.H.
    Mantell and H. Smith, pp. 3-37. Cambridge: Cambridge
    University Press.

Fukui, H., Nakagawa, K., Tsuda, S. & Tabata, M. (1982). Production of
    isoquinoline alkaloids by cell suspension cultures of
    Coptis japonica. In Plant Tissue Culture 1982, ed. A.
    Fujiwara, pp. 313-14, Tokyo: Japanese Association for
    Plant Tissue Culture.

Hay, C.A., Anderson, L.A., Roberts, M.F. & Phillipson, J.D. (1986). In
    vitro cultures of Cinchona species. Part I. Precursor
    feeding of C. ledgeriana root organ suspension cultures
    with L-tryptophan. Plant Cell Reports, 5, 1-4.

Heinstein, P.F. (1985). Future approaches to the formation of
    secondary natural products in plant cell suspension
    cultures. J. Nat. Prod., 48(1), 1-9.

Hinz, H. & Zenk, M.H. (1981). Production of protoberberine alkaloids
    by cell suspension cultures of Berberis species.
    Naturwissenschaften 68, 620-21.

Homeyer, B.C. & Roberts, M.F. (1984). Alkaloid sequestration by
    Papaver somniferum latex. Z. Naturforsch. 39c, 876-81.

Homeyer, B.C. & Roberts, M.F. (1985). Effect of pH on temperature
    dependent morphine uptake by Papaver somniferum and
    Papaver bracteatum latex organelles. J. Pharm. Pharmac.
    37, 140P.

Ikuta, A., Syono, K. & Furuya, T. (1974). Alkaloids of callus tissues
    and redifferentiated plantlets in the Papaveraceae.
    Phytochemistry 13, 2175-79.

Mantell, S.H. & Smith, H. (1983). Cultural factors that influence
    secondary metabolite accumulations in plant cell and
    tissue cultures. In Plant Biotechnology, ed. S.H. Mantell
    and H. Smith, pp. 75-110, Cambridge: Cambridge University
    Press.

Neumann, D. (1983). Indole alkaloid formation and storage in cell
    suspension cultures of Catharanthus roseus. Planta
    Medica, 48, 20-23.

O'Neill, M.J., Bray, D.H., Boardman, P., Phillipson, J.D. &
    Warhurst, D.C. (1985). Plants as sources of antimalarial
    drugs, Part I. In vitro test method for the evaluation of
    crude extracts from plants. Planta Medica 51, 394-97.

Rueffer, M. (1985). The production of isoquinoline alkaloids by plant
        cell cultures. In The Chemistry and Biology of
        Isoquinoline Alkaloids, ed. J.D. Phillipson, M.F. Roberts
        and M.H. Zenk, pp. 265-80. Berlin, Heidelberg: Springer
        Verlag.

Sasse, F., Heckenberg, U. & Berlin, J. (1982). Accumulation of
        β-carboline alkaloids and serotonin by cell cultures of
        Peganum harmala, Z. Pflanzenphysiol. 105(4), 315-22.

Staba, E.J. (1982). Production of useful compounds from plant tissue
        cultures. In Plant Tissue Culture 1982, ed. A. Fujiwara,
        pp. 25-26, Tokyo: Japanese Association for Plant Tissue
        Culture.

Stockigt, J., Pawelka, K-H, Tanahashi, T., Danieli, B. & Hull, W.E.
        (1983). Voafrine A and voafrine B, new dimeric indole
        alkaloids from cell suspension cultures of Voacanga
        africana Stapf., Helv. Chim. Acta., 66, 2525-33.

Zenk, M.H., El-Shagi, H., Arens, H., Stockigt, J., Weiler, E.W. &
        Deus B. (1977). Formation of the indole alkaloids,
        serpentine and ajmalicine in cell suspension cultures of
        Catharanthus roseus. In Plant Tissue Culture and its
        Biotechnological Application, ed. W. Barz, E. Reinhard and
        M.H. Zenk, pp. 27-43, Berlin, Heidelberg:
        Springer-Verlag.

Zenk, M.H. (1985). Enzymology of benzylisoquinoline alkaloid
        formation. In The Chemistry and Biology of Isoquinoline
        Alkaloids, ed. J.D. Phillipson, M.F. Roberts and M.H.
        Zenk, pp. 240-56. Berlin, Heidelberg: Springer-Verlag.

2. THE SYNTHESIS AND BIOTRANSFORMATION OF MONOTERPENES BY PLANT CELLS IN CULTURE

B. V. Charlwood & P. K. Hegarty
Department of Biology, King's College London,
London. WC2R 2LS. U.K.

K. A. Charlwood
Department of Life Sciences, University of London,
Goldsmiths College. London SE14 6NW. U.K.

There are over one thousand monoterpenes distributed widely throughout the plant kingdom and many of these natural products are used in the flavour, the pharmaceutical and the perfumery industries. The worldwide market for fragrance components alone is estimated at more that £3,000 m per annum, although most of these compounds are presently obtained through chemical synthesis. It is of commercial interest, therefore, to investigate the possibilities of producing monoterpenes by plant cells in culture (Klausner 1985). Furthermore, although the biosynthesis of the presumed precursors of the monoterpenes is well understood from studies on intact plants (Banthorpe & Charlwood 1980; Cori 1983), the detail of the later stages of the pathway is still largely unclear. It is envisaged that the relative simplicity of a tissue culture system, compared with the intact plant, may allow more detailed probing of the structural elaboration of monoterpene precursors.

In general, monoterpenes are not accumulated in finely-dispersed suspension cultures, and the possible reasons for this have been stated (Bohm 1982). It is clear that cultured cells still retain the genetic potential to synthesise monoterpenes since plants regenerated from callus tissue produce essential oils of similar composition to those of the parent plants (eg. Webb et al. 1984; Brown and Charlwood, 1986). Furthermore, the presence of all of the enzymes required to bring about the synthesis of geranyl pyrophosphate [GPP: 1] from acetyl coenzyme A [MeCOSCoA: scheme I] has been demonstrated in plant cell culture systems. For example, the reductase that converts 3-hydroxy-3-methylglutaryl coenzyme A [2] into mevalonic acid [MVA: 3] has been characterised in cultures of Anise (Huber & Rudiger 1978) and of Nepeta cataria (Arebalo & Mitchell, 1984); cultures of Tanacetum vulgare converted radio-labelled MVA into

Fig. 1.

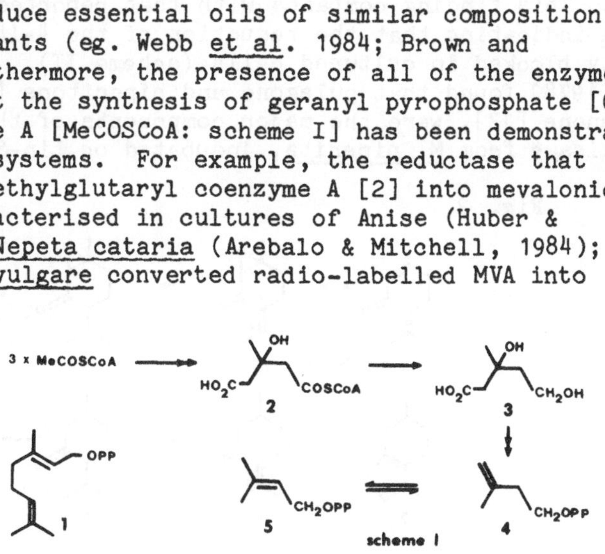

scheme I

isopentenyl pyrophosphate [IPP: 4] and dimethylallyl pyrophosphate
[DMAPP: 5] (Banthorpe & Wirz-Justice   1972); and extracts from
cultures of Rosa damascena synthesised geraniol and nerol from
exogenous IPP (Banthorpe & Barrow   1983).   This paper reviews the
literature (late 1985) concerning the subsequent expression of
synthesis, accumulation and transformation of monoterpenes by plant
cells in culture.

Monoterpene accumulation has been widely investigated in tissue
cultures derived from many species of Mentha.   Reports have indicated
that such cultures do not accumulate monoterpenes (eg. Wang and Staba
1962; Becker   1970; Suga et al. 1980b).   However, Bricout and
Paupardin (1975) found that cultures of M. piperita, grown on
Murashige and Skoog medium supplemented with glucose (3%) and
benzylaminopurine, synthesised pulegone [6] (30.4%) and menthofuran
[7] (12.3%) as the major components, whereas the intact plant
accumulates predominantly menthone [8] (23.8%) and menthol [9]
(34.6%).   The callus formed many rudimentary buds which were endowed
with secretory cells characteristic of the species.   Illumination of
the cultures was essential for oil production; increasing light
intensity increased the activity, but not the number, of secretory
cells.   Higher levels of glucose depressed monoterpene synthesis
(Paupardin et al.   1980).

A more comprehensive survey (Bricout et al. 1978b) of the genus
revealed that callus cultures of M. piperita, M. rotundifolia, M.
pulegium, M. viridis, M. aquatica and M. citrata all produced
monoterpenes.   Apart from the first two species listed, the cultures
accumulated oils with a similar composition to those found in the
intact plants but at much reduced yields.   M. rotundifolia grown in
vitro produced epoxypiperitenone [10] rather than the less oxidised
epoxypiperitone [11] which is characteristic of the species cultivated
in vivo.   This finding compares with that reported above for M.
piperita indicating that the reduction of the 4-(8) C=C bond is
partially blocked in cultured cells (scheme II).   In contrast, Kireeva
et al. (1978) found that pulegone and piperitone [12], but not
piperitenone [13], were the major components of the oil produced by
callus tissue from M. piperita, incubated on Lin-Staba medium

Fig. 2.

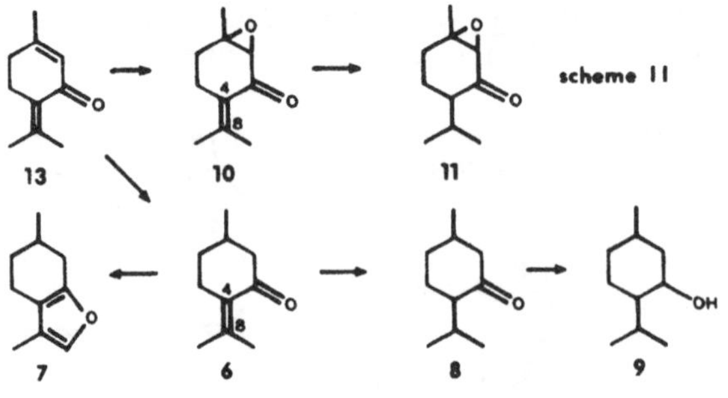

scheme II

supplemented with sucrose (3%) and 2,4-dichlorophenoxyacetic acid.
Menthol, isomenthol, menthone, isomenthone and menthyl acetate, but
not menthofuran, were also present in significant amounts.
Furthermore, the callus cultures produced, over a period of 60 days,
1.88% (dry wt.) total oil content which compares with 2.18% (dry wt.)
accumulated by the intact plant. Monoterpenes were detected both in
the callus cells and in the nutrient medium in approximately equal
proportions. Histochemical investigation showed that, within the
callus, oil was synthesised, not in specialised glandular structures,
but in giant callus cells of which the callus tissue was mostly
composed. Such giant cells may have resulted from the
dedifferentiation of the gland cells.

Accumulation of monoterpenes in cultures from a large number of Mentha
species and hybrids was very variable both between taxa and between
cell lines from the same species (Charlwood & Charlwood 1983). Many of
the cultures accumulated only early monoterpene precursors, but some
hybrid lines (eg. M. spicata x suaveolens) accumulated the major
terpene component of the intact plant. There was a distinct
correlation between the level of differentiation of the cultured cells
and the quantitative and qualitative accumulation of monoterpenes. A
similar situation exists for cultures derived from Pelargonium
variants (Brown & Charlwood 1986). Further discussion concerning the
level of differentiation and the accumulation of monoterpenes will be
found later in this volume.

In contrast, callus cultures from Citrus species do not generally
produce oil characteristic of the whole plant. Callus derived from C.
limonia initially produced only monoterpene hydrocarbons, but even
this ability was lost on subculture (Paupardin 1976). However, newly
formed pericarp cultures synthesised monoterpenes characteristic of
pericarp tissue including α-pinene [14], β-pinene [15], limonene [16],
linalol [17], citrals [18, 19] and geranyl acetate (Bricout &
Paupardin 1974). The oil accumulated in non-organised secretion cells
within the cultures; the complete formation of the secretion pockets
typical of lemon pericarp tissue was not essential (de Billy &
Paupardin 1971). The synthetic ability of these cultures was lost
during repeated subculture (over a 4 year period), but could be

Fig. 3.

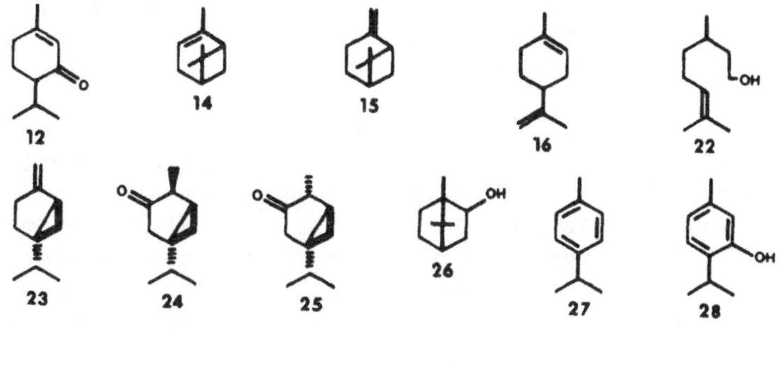

stimulated by the addition of geranyl acetate and ATP to the culture
medium (Paupardin 1974). Monoterpene synthesis in pericarp and
endocarp cultures was not enhanced by high temperatures (greater than
26oC), but both growth and synthesis were stimulated in the dark.

Similarly, monoterpenes were not produced in callus cultures from a
number of varieties of rose (Jones 1974). Kireeva et al. (1977),
however, claimed that callus tissues derived from petals, leaves and
sepals of Rosa damascena produced alcohols and glucosides similar to
those accumulated in the analagous organs of the intact plant (viz
linalol, geraniol [20], nerol [21], citronellol [22] and β-phenylethyl
alcohol). After 60 days in culture, callus from sepals contained half
as much essential oil as intact sepals, whilst callus from leaves
contained twice that of leaf tissue. The content of glycoside
derivatives was always lower than that of the intact plant. The
overall oil content of the cultures decreased during subsequent
subculture. More recently Banthorpe & Barrow (1983) have shown
that, whilst callus and suspension cultures of R. damascena did not
accumulate monoterpenes, extracts from the culture efficiently
interconverted the isoprenoid progenitors IPP and DMAPP. Newly formed
cultures were also able to synthesise geraniol from exogenous
[4-14C]-IPP and the enzyme activity appeared to be some 300-fold
greater than that of the intact plant. This clearly demonstrates that
the accumulation of a monoterpene by a cell line depends not only on
its synthetic capacity but also on its ability to regulate any further
processes of anabolism, catabolism and secretion or storage.

In contrast to the situation with rose, Banthorpe & Njar (1984)
reported monoterpene accumulation in cultures from Pinus radiata of up
to 40% of the levels found in needles and stem of the parent plant.
Whereas the plant produces 60-90% of β-pinene, with the residue almost
entirely the α-isomer, the illuminated cultures accumulated
essentially α-pinene (87-100%) within the cells. Trace components in
the cultures corresponded to those occurring in the parent plant. The
variation in the α- and β-pinene ratio in the cultures and the intact
plant is consistent with the hypothesis that two separate pathways
lead to these isomers. When cultures were grown in total-darkness,
toluene and acetone were also accumulated but the ratio of toluene to
α-pinene depended on the culture regime. It was suggested that the
toluene and acetone arose from the degradation of α-pinene under
non-illuminated conditions. Soluble extracts from the callus
efficiently converted [1-14C]-IPP into geraniol and nerol rather than
into the pinenes. Addition of the phosphatase inhibitor, sodium
fluoride, to the cell-free system enhanced pinene yields.

Similar, direct incorporation of [2-14C]-MVA into monoterpenes,
carotenoids and chlorophyll by green and colourless, partially
differentiated cultures of Tanacetum vulgare demonstrated the de novo
synthesis of isoprenoids in vitro (Banthorpe & Wirz-Justice 1972).
Colourless culture tissue accumulated monoterpenes in similar
proportions to those found in the parent plant but at only half of the
concentration. The main component of the culture oil was sabinene

[23] rather than the oxygenated derivative isothujone [24] found in the leaf. In contrast, the levels of thujone [25] were similar in culture and in the plant. When chlorophyll synthesis was inhibited, cultures still accumulated essentially the same monoterpenes in similar yields to the greening cells. It was claimed, therefore, that monoterpene synthesis is independent of the presence of functional chloroplasts.

Although undifferentiated and finely dispersed cell suspensions of celery did not accumulate any of the flavour components found in the intact plant, when the cultures were grown under conditions which limited their growth and induced greening, the synthesis of limonene and the phthalides was stimulated. (Watts et al. 1984, 1985). Limonene production was also increased by low-temperature stress and by the formation of aggregates, but none of the high yielding cultures exhibited morphological differentiation. Further, undifferentiated callus and, to a lesser extent, cell suspension cultures of Ocimum basilicum produced largely the same monoterpenes (eg. linalol, borneol [26], p-cymene [27] and thymol [28]) and phenylpropanoids as the parent plant (Lang & Horster 1977). Interestingly, however, the cultures accumulated nearly twice the amounts of glycosylated derivatives compared with the free monoterpenes.

The volatile oils of callus and suspension cultures from Perilla frutescens were essentially the same as those from the original plant leaves (Sugisawa & Ohnishi 1976; Nabeta at al. 1983) namely limonene, linalol, isoegomaketone [29] and the potent lung toxin, perillaketone [30]. Likewise callus derived from leaves of Eucalyptus citriodora initially produced monoterpenes which were identical to those of the source tissue but in lower concentrations (Gupta & Mascarenhas 1983). This synthetic ability was, however, lost over the first 10 subcultures. Callus raised from stem segments produced no oil and organogenesis was found to be an essential prerequisite of monoterpene production. Low levels of citronellal [31] were detected upon shoot

Fig. 4.

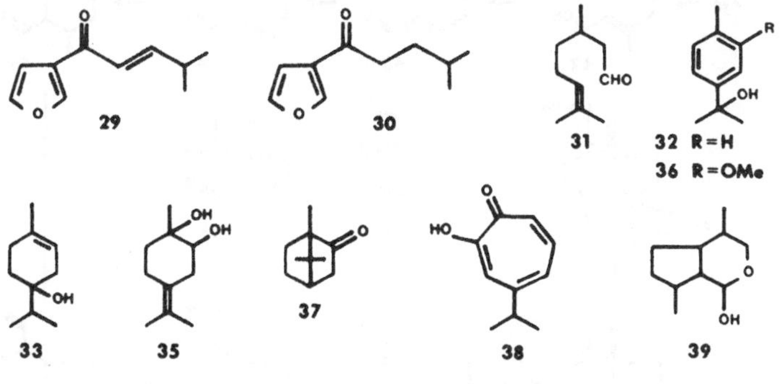

29   30   31   32 R=H
36 R=OMe

33   35   37   38   39

elongation, but after the regeneration of roots both citronellal and citronellol were formed.

Major differences were reported between the terpene content of cell cultures of Thuja occidentalis and the leaves of the whole plant (Witte et al. 1983). Whilst the cells mainly accumulated diterpenes, the monoterpenes p-cymen-8-ol [32], terpinen-4-ol [33], α-terpineol [34], menth-4(8)-en-1,2-diol [35], 2-methoxy-p-cymen-8-ol [36] and camphor [37] were excreted into the medium. Of these, only camphor was common to both the plant and the cell culture. Irregular monoterpenes, eg. thujaplicin [38], were detected in both the cells and the medium. No hydrocarbons were detected in the medium although, of course, they may have been lost by evaporation during culture. Nevertheless, these cultures were unable to synthesise the thujane group of monoterpenes which is characteristic of the plant.

The iridoids are a class of regular monoterpenes based on the iridane skeleton [39]. Compounds of this class are widely distributed in higher plants, and are usually transported to, and stored in, specialised oil glands as β-glucoside derivatives. Suspension cultures of Gardenia jasminoides retained their ability, over a 3 year period, to accumulate within the cells the iridoids gardenoside [40], geniposide [41: R=COOMe] and geniposidic acid [41: R=COOH] which are typical constituents of the parent plant (Ueda et al. 1981). In addition, the cultures synthesised tarennoside [41: R=CHO], an iridoid which has only been detected in the related plant Tarenna gracilipses. When [9-^{13}C]-10-hydroxygeraniol [42], a presumed precursor of the iridoids, was fed to these cultures (Uesato et al. 1983), label was incorporated into [40] and [41] with almost complete radomisation of carbons 9 and 10 of [42] between positions 3 and 11 of both glucosides (scheme III). No incorporation was obtained from the ^{13}C-labelled putative intermediates 10-hydroxycitronellol and 9,10-dihydroxy citronellol. It was thus concluded that the biosynthesis of the

Fig. 5.

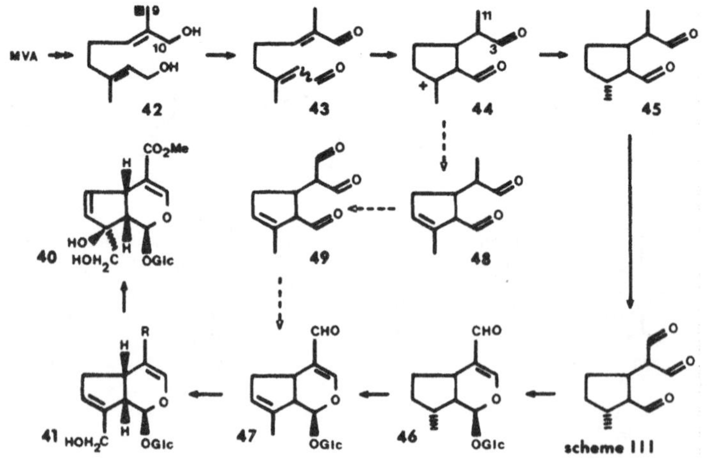

scheme III

iridane skeleton involves cyclization of 2E- or 2Z-10-oxocitral [43] to yield the iridodial cation [44]. Further steps in the biosynthesis of [40] and [41] were deduced from feeding ^{2}H- or ^{13}C-labelled iridodial, 8-epiiridodial [45], boschnaloside [46] and dehydroiridotrial glucoside [47] to the cultures (Uesato et al. 1984). The efficient incorporation of label from a deuterated dehydroiridodial [48] into [40] and [41] led to the hypothesis of an alternative pathway (dotted lines, scheme III) via dehydroiridotrial [49].

The valepotriates [50] are a group of iridoids which are partly responsible for the marked sedative and spasmolitic effects of many plants within the family Valerianaceae. Since there is a shortage of wild grown plant material in Western Europe to provide crude drug and isolated valepotriates, a widespread search has been made for valepotriate-producing cultures (Becker et al. 1977). Valtrate, acevaltrate and didrovaltrate, constituents of the intact plant, were identified in undifferentiated cultures of Valeriana wallichii as well as in cultures showing differentiation of roots and shoots. Of callus cell lines derived from a further eight plants of the Valerianceae, (viz V. officinalis, V. alliariifolia, V. sambucifolia, Fedia cornucopiae, Centranthus ruber, Valerianella dentata, V. coronata and V. locusta) all except those of Valeriana officinalis contained detectable amounts of the valepotriates typical of the parent plants (Becker & Schrall 1980; Schrall & Becker 1980). Interestingly, V. officinalis accumulates valepotriates exclusively in the roots, whilst the other plants contain the iridoids in the foliage as well. In some cases (for example F. cornucopiae) the level of valepotriates in culture was higher than that found in the intact plant.

It has recently been reported (Violon et al. 1984b) that both root-differentiated and callus cultures of C. macrosiphon and V. officinalis accumulate valepotriates. In these cultures, and those from C. ruber, a positive correlation between the levels of differentiation and of valepotriate production was demonstrated (Violon et al. 1984a). The iridoids were localised in intracellular lipid vesicles within the cultures (Violon et al. 1983), and these vesicles were closely related to the vacuolar membrane of the cell: the number of these vesicles was dependent upon the level of differentiation of the culture. These findings were confirmed by Becker et al. (1984a) who showed that fine cell suspensions, derived from iridoid-producing aggregates of V. wallichii, had lost their ability to produce.

The biosynthetic pathway to the valepotriates is largely unknown but is thought, by analogy with the route to loganin [51] and the gardenolides [40, 41], to involve the intermediacy of MVA, geraniol, 10-hydroxygeraniol and iridodial. In loganin biosynthesis, the formation of 10-hydroxygeraniol from geraniol is mediated by a cytochrome P-450 dependent mono-oxygenase. That a similar route to the valepotriates might be extant in cultures of C. macrosiphon was indicated by a report (Violon & Vercruysse 1985) that cytochrome P-450 levels increased in direct proportion with increasing amounts of valepotriates.

A non-mevalonoid route (involving the intermediacy of leucine) for the biosynthesis of monoterpenes has long been the subject of conjecture (see Charlwood & Banthorpe 1978), and recent evidence casts further doubt on the existence of such a pathway (Anastasis et al. 1983, 1985). Becker & Baumer (1983) have isolated cell lines of V. wallichii which are resistant to trifluoroleucine by virtue of their enhanced accumulation of leucine. However, even though the content of free leucine in the resistant cells was some 37-fold greater than that of the non-resistant lines, no increase in accumulation of valepotriates, volatile terpenes or sterols was observed in these cultures (Becker et al. 1984a).

Plant cells which have been selected for resistance to the polyene antibiotic nystatin show an increase in sterol synthesis, and it has been suggested that such augmentation of the pathway might also lead to an increase in monoterpene accumulation (Becker et al. 1984a). Two cell lines of V. wallichii with a high resistance to nystatin have been investigated: whilst both lines showed a 4-fold increase in sterol synthesis, valepotriate accumulation was enhanced in one line but reduced in the other.

The insecticidal esters, the pyrethrins [52], contain an irregular monoterpene moiety of the chrysanthemic acid-type [53] (Charlwood & Banthorpe 1978). Whilst pyrethrins have been identified in callus cultures of Tagetes minuta and T. erecta (Jain 1977), they could not be detected in root differentiated cultures and undifferentiated callus from Chrysanthemum cinerariaefolium (syn. Tanacetum cinerariifolium, Pyrethrum; Cashyap et al. 1978). However, shoots regenerated from the callus cultures were found to contain pyrethrins in similar ratios to those found in the four week old seedlings. In contrast, Zieg et al. (1983) report that the majority of cultures derived from a variety of genetically diverse pyrethrum plants and tissues are capable of producing low concentrations of pyrethrin. Positive correlations were observed between in vitro pyrethrin production and both the genotype of the plant and the level of tissue organisation. Callus cultures of pyrethrum have also been shown to produce chrysanthemic acid in very small amounts (Kueh et al. 1985). When [^{14}C]-labelled chrysanthemic acid was fed to these cultures it was efficiently converted into the β-glucoside suggesting that the acid is stored largely as a water soluble ester. In contrast, the cultures rapidly metabolised exogenous pyrethrins to unidentified products.

The use of two-phase culture systems to enhance the accumulation or recovery of monoterpenes has received much recent attention. Beiderbeck (1982) demonstrated that crown gall and habituated Matricaria chamomilla cell lines could be cultivated, with little effect on growth, in the presence of a lipophilic phase of a triacylglycerol with C8 and C10 fatty acid chains (Miglyol 812, Dynamit Nobel). The isoprenoids, characteristic of the plant, were accumulated in the two-phase culture system in high yield within 6 days, whereas no accumulation was detectable in single phase culture (Bisson et al. 1983). Similarly, when small, morphologically undifferentiated aggregates of Thuja occidentalis were grown in the presence of Miglyol or hexadecane (Berlin et al. 1984), the lipophilic phase accumulated the volatile hydrocarbons α- and β-pinene, myrcene [54], limonene and terpinolene [55] in addition to the menthane-type and tropolonate derivatives normally accumulated in single phase cultures (Witte et al. 1983). The total of monoterpenes excreted per day rose form 0.8 mg/g dry wt. in single phase culture to 3 mg/g dry wt. in the two-phase system, whilst accumulation of oxygenated derivatives doubled.

It was suggested that the observed increase in products was not due to enhanced synthesis but rather to the better solubility of the volatile compounds in the organic phase. This may not always be the case, however, since monoterpenes are generally cytotoxic and the removal of these products from the medium by the use of an artificial sink could well enhance synthesis and maintain cell viability. This is particularly important when a product is normally stored in a specialised gland in vivo. In undifferentiated cultures, a sink could provide the aspect of differentiation required for product sequestration. Becker et al. (1984c) have considered the utilisation of a second phase for the purpose of stabilising excreted products that would otherwise breakdown in the culture medium. Thus cell cultures of M. chamomilla and Pimpinella anisum, which did not accumulate essential oil in single phase culture, did produce in the presence of Miglyol or of a modified silicagel with lipophilic C8 side chains (Li-Chroprep RP-8, Merck). The production of valepotriates by a normally non-producing cell line of V. wallichii was also initiated by RP-8 (Becker & Herold 1983), whilst the total yield of valepotriates from a producing line was raised by 20% in a similar culture system.

Fig. 6.

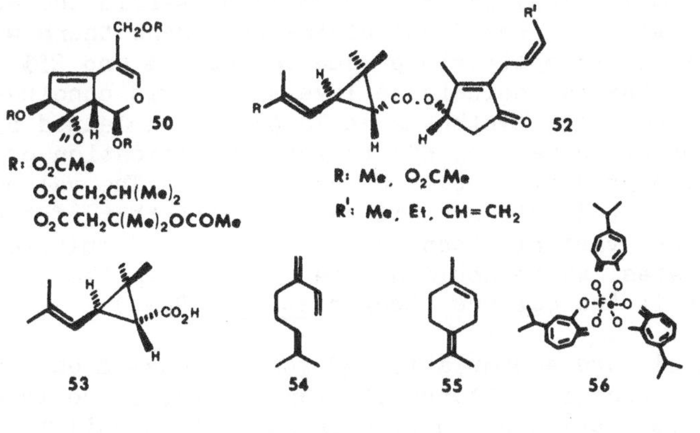

50   R: O₂CMe
      O₂C CH₂CH(Me)₂
      O₂C CH₂C(Me)₂OCOMe

52   R: Me, O₂CMe
      R¹: Me, Et, CH=CH₂

53   54   55   56

Two-phase culture does, however, present difficulties with respect to aeration and product separation particularly in a large scale system. Forche et al. (1984) have described a continuous extraction process to remove excreted monoterpenes from T. occidentalis cultures. Medium, containing the released product, was passed from the reactor vessel into a second column containing an adsorbing resin (Amberlite XAD4) where the monoterpenes were removed. The medium was recycled, although growth regulator and vitamins (that had been removed by the resin) had to be replaced on a weekly basis. The resin column itself was exchanged every week and, under these conditions, the culture retained its initial, low viability for up to 100 days. The production of monoterpenes, mainly terpinen-4-ol [33], α-terpineol [34] and menth-4(8)-en-1,2-diol [35], and the unusual tris-(thujaplicinato)-iron III complex [56], gradually declined during the culture period. Suprisingly, no terpinolene [55], the major oil component of the two-phase cultures, was detected in this system.

It has often been observed that polyploid plants accumulate higher yields of secondary compounds than do the corresponding diploids, although this is usually accompanied by a qualitative alteration in product profile. Mentha piperita plant-tissues grown for 15 days on medium supplemented with colchicine showed a 3-fold stimulation of monoterpene production (Bricout et al. 1978a) although the relative amounts of the major components (menthofuran and pulegone) varied considerably. The effects of colchicine on oil production were detectable for up to 2 months after treatment.

Chavadej & Becker (1984) have investigated ploidy levels and valepotriate production in cultures of V. wallichii treated with colchicine. The chromosome numbers shifted to polyploidy after the first colchicine treatment but returned almost to the initial pattern after 6 passages through alkaloid-free medium. After a second treatment with colchicine, however, the proportion of diploid cells never attained the level of the controls or of the cells treated only once. A 70-fold increase in valepotriate accumulation was sustained for over a year in cells which had received one treatment with 0.05% colchicine (Becker & Chavadej 1985; Becker et al. 1984a). A single treatment with 0.2% colchicine gave rise to a 54-fold increase in valepotriates. After a second colchicine treatment there was a further increase in valepotriate production of 45% and 25% respectively for the two colchicine levels. It was concluded that the increase in valepotriate synthesis could not be explained by higher ploidy level but may have been due to gene amplification or to a selection of high-producing cells by colchicine. The valepotriate content of the treated cultures did not resemble that of the original plant and was more variable than in the non-treated culture. Six new diene-valepotriates, which could not be detected in the parent plant, were identified in the culture (Becker et al. 1984b).

Although synthesis and accumulation of monoterpenes does not normally occur in fine suspension cultures of plant cells, there is evidence that such cultures retain a biotransformational capability.

Thus, whilst early reports (Staba et al. 1965 and references therein)
suggested that mint cultures could not generally transform exogenous
precursors, four cell lines derived from various Mentha chemotypes
were able to convert (+)-pulegone [57] (but not its enantiomer) to
(+)-isomenthone [58] (Aviv & Galun 1978). In each case the intact
plant of the strain tested accumulated derivatives of either carvone
[59] or menthone [8], and it was proposed that the enzyme system
responsible for saturation of the 4-(8) C=C bond was able to operate
on both C-2 and C-3 oxygenated intermediates. It was further
demonstrated (Aviv et al. 1983) that the transformation system from
Mentha cultures was effective on analogues of pulegone with an altered
complement of methyl substituents on the ring (eg. 2-isopropylidine
cyclohexanone [60] and both diastereomers of trans-6-methyl pulegone
[61, 62]). However, other 2,3-unsaturated ketones (viz mesityl oxide
[63], trans-6-t-butyl pulegone [64] and 3-isopropylidene-9-methyl-
dicyclodecalone-2 [65] were not so transformed. Cell suspension
cultures from a further two Mentha chemotypes (the intact plants of
which accumulated only those products synthesised early in the
biosynthetic pathway) were not able to metabolize exogenous pulegone
or its analogues.

None of six Mentha cell lines could effect the conversion of
(+)-neomenthone to the corresponding alcohol. In contrast, all these
lines were able to convert (-)-menthone [66] to (+)-neomenthol [67]
with strict overall stereospecificity (Aviv et al. 1981) and there was
no correlation between biotransformational ability and monoterpene
content of the parent plant. Immobilisation of cell lines in either
calcium alginate or polyacrylamide-hydrazide, cross-linked with
glyoxal (PAAH-G), did not affect the ability of the cells to transform
either (+)-pulegone or (-)-menthone (Galun et al. 1983). Furthermore,
monoterpenes were more readily released from PAAH-G entrapped cells
than from cells in a freely-suspended culture, and the immobilised
cells retained their transformational capabilities over 3 consecutive
batch-type additions of substrate. One interesting observation that
emerges from this work is that certain monoterpene substrates
disappear from both the medium and the cells within 24 hours. The
rate of such disappearance is specific for both cell line and

Fig. 7.

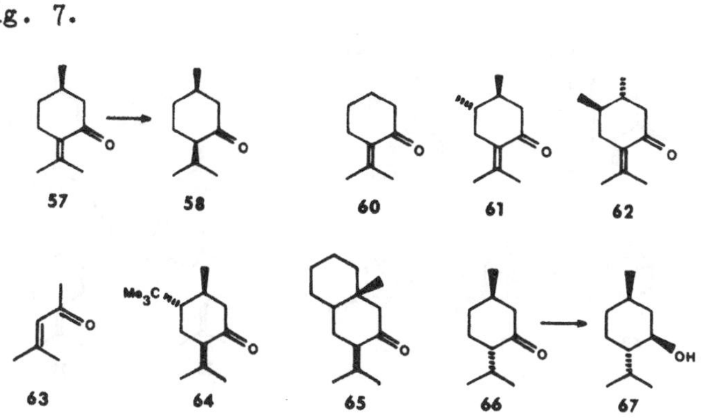

57   58   60   61   62

63   64   65   66   67

substrate and it has been suggested (Aviv et al. 1981) that enzymatic glycosylation of monoterpene is responsible. Indeed, from work carried out in our laboratories (unpublished), this phenomenon seems to be general for mint species, but we have found no evidence supporting the involvement of glycosylated derivatives.

Cell cultures of four genera of the Caprifoliaceae showed no detectable formation of the iridoids loganin [51] and secologanin [68] although the original plants contained substantial amounts of the latter (Tanahashi et al. 1984). However, some of the cell lines had the ability to convert, almost quantitatively, exogenous loganin to secologanin, the transformed product being retained within the cells. No stable intermediate could be detected between [51] and [68] although the transformation involves the oxidative opening of the methylcyclopentane ring.

It has been postulated that either geraniol or nerol provides the isoprenoid moeity of the psychoactive cannabinoids. Suspension cultures induced from Cannabis sativa were unable to biotransform these alcohols, in the presence of olivetol and olivetolic acid, into cannabinoids (Itokawa et al. 1977). The cell lines had also lost their ability to bring about de novo synthesis of the cannabinoids. More recently it has been demonstrated that cannabinoid acids are synthesised in callus cultures of C. sativa (Heitrich & Binder 1982). Notwithstanding, the former cultures were capable of interconverting geraniol [20] and nerol [21], and of oxidising both these primary allylic alcohols to citral a and b [18, 19], and the secondary cyclic, allylic alcohols trans- and cis-verbenol [69, 70] to verbenone [71]. The stereochemistry of these oxidations has been reported in detail (Takeya & Itokawa 1977). Non-allylic terpene alcohols were not biotransformed in this system. Cell suspension cultures of Muscat grapes were similarly able to interconvert and oxidise exogenous geraniol and nerol. In addition, these cultures reduced citral isomers to the corresponding alcohols, and in all cases a proportion of the geraniol formed was esterified to geranyl acetate (Ambid et al. 1982).

Fig. 8.

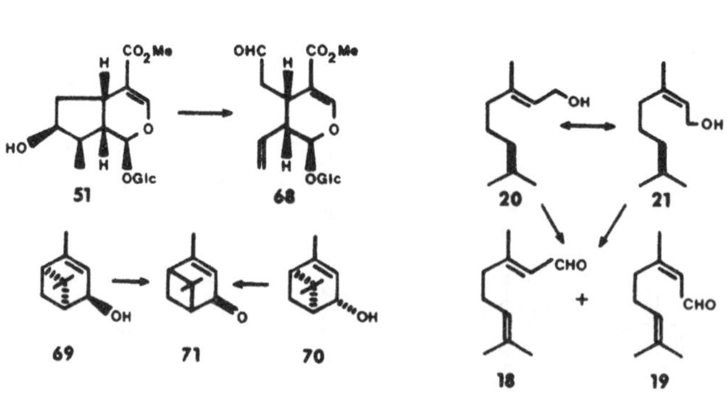

The transformation of monoterpenes by cell lines derived from plants that do not accumulate such secondary compounds has been studied in detail. Thus callus and suspension cultures of Nicotiana tabacum have the ability to hydroxylate selectively the trans-methyl group in the isopropylidene moiety of (-)-linalol [17], (-)-dihydrolinalol [72] and their acetates [73, 74] to give the corresponding 8-hydroxy derivatives [75 to 78; scheme IV]. Additionally, the acetates suffered extensive ester hydrolysis (Hirata et al. 1981). α-Terpineol [34], trans-β-terpineol [79] and the corresponding acetates [80, 81] were also hydroxylated at carbon atoms allylic to the C=C bond by the cell system (Suga et al. 1980a, 1982). However, the major transformation products of the acetates were the glycols [82, 83] indicating that the cells possessed the ability to hydroxylate, stereospecifically, the endocyclic and terminal C=C bond respectively (Hirata et al. 1982b; Suga et al. 1983). Similar stereo- and regioselective hydroxylations were observed (Lee et al. 1983) when the exogenous substrate, 1-acetoxy-p-menth-4(8)-ene [84], contained an exocyclic C=C bond.

In contrast to these findings, when carvone [59] and dihydrocarvone [85] were the foreign substrates the terminal C=C bond was not attacked (Hirata, et al. 1982a). The regio-and stereoselective reduction of the C=C bond adjacent to the carbonyl group of [59] was followed by the reduction of the carbonyl group of [85; scheme V].

Fig. 9.

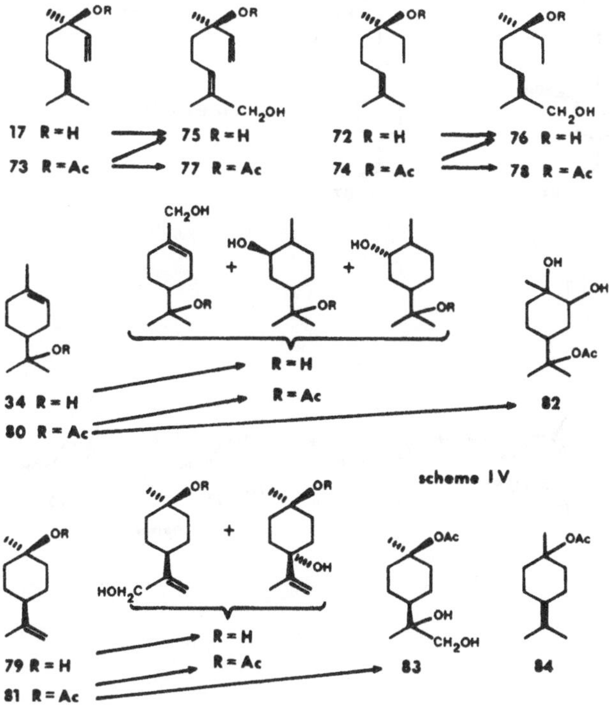

scheme IV

The conjugation of a carbonyl group with the C=C bond might be essential for the reduction of the C=C bond since no reduction of the endocyclic C=C bond occurred when carvoxime [86] or dihydrocarvoxime [87] were the foreign substrates (Suga et al. 1984). These oximes were hydrolysed by the Nicotiana cell system to the ketones [59, 85] which were subsequently reduced to the corresponding stereoisomers of the alcohol [88] in low yields.

Epoxidase and epoxide hydratase activities have been demonstrated in cell-free extracts from callus of Jasminum officinale (Banthorpe & Osborne 1984). Thus IPP [4] and isopentenol [89] were converted in low yields to the epoxide [90], the diols [91, 92] and the triol [93; scheme VI]. The cell-free system was also able to transform the isopropylidene C=C bonds of both geraniol and nerol to give analagous epoxides, diols and triols. It is unlikely that such oxidative metabolism is important in vivo since neither callus nor suspension cultures of J. officinale were able to bring about these biotransformations.

Fig. 10.

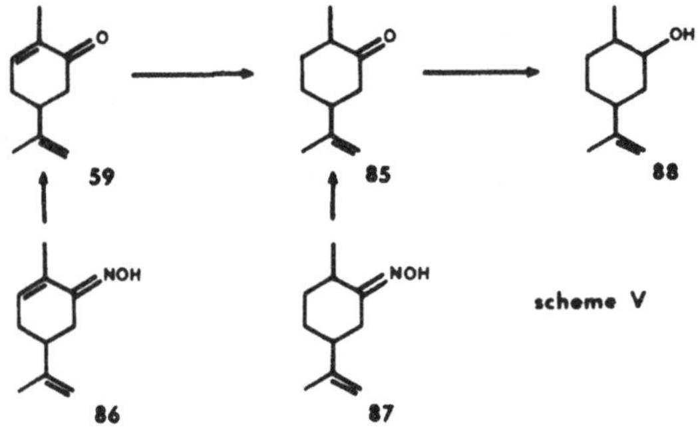

scheme V

Fig. 11.

scheme VI

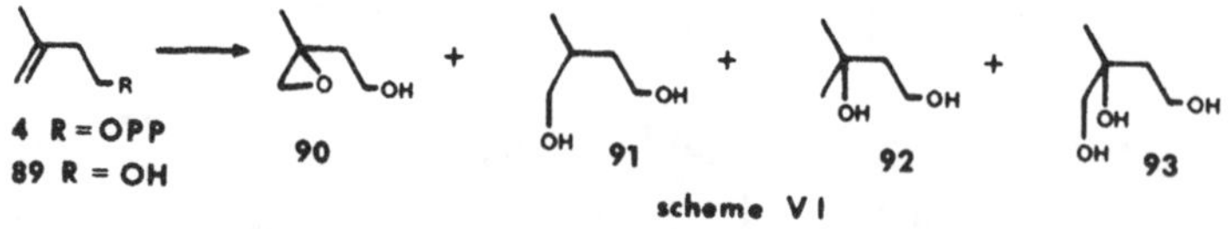

REFERENCES

Ambid, C., Moisseeff, M. & Fallot, J. (1982). Bioconversion of citral by a cell suspension culture of Muscat grapes. Plant Cell Reports, 1, 91-93.

Anastasis, P., Freer, I., Overton, K., Rycroft, D. & Singh, S.B. (1985). The role of leucine in isoprenoid metabolism. Incorporation of [3-^{13}C]-leucine and of [2-^{3}H, 4-^{14}C]-β, β-dimethylacrylic acid into phytosterols by tissue cultures of Andrographis paniculata. J. Chem. Soc. Chem. Commun., 148-49.

Anastasis, P., Freer, I., Picken, D., Overton, K., Sadler, I. & Singh, S.B. (1983). The role of leucine in terpenoid metabolism: Incorporation of [2-^{13}C]- and [3-^{13}C]-leucines into sesquiterpenoids by tissue cultures of Andrographis paniculata. J. Chem. Soc. Chem. Commun., 1189-91.

Arebalo, R.E. & Mitchell, E.D. (1984). Cellular distribution of 3-hydroxy-3-methylglutaryl coenzyme A reductase and mevalonate kinase in leaves of Nepeta cataria. Phytochem., 23, 13-18.

Aviv, D., Dantes, A., Krochmal, E. & Galun, E. (1983). Biotransformation of monoterpenes by Mentha cell lines: Conversion of pulegone-substituents and related unsaturated α-β ketones. Planta Medica, 47, 7-10.

Aviv, D. & Galun, E. (1978). Biotransformation of monoterpenes by Mentha cell lines: Conversion of pulegone to isomenthone. Planta Medica, 33, 70-77.

Aviv, D., Krochmal, E., Dantes, A. & Galun, E. (1981). Biotransformation of monoterpenes by Mentha cell lines: Conversion of menthone to neomenthol. Planta Medica, 42, 236-43.

Banthorpe, D.V. & Barrow, S.E. (1983). Monoterpene biosynthesis in extracts from cultures of Rosa damascena. Phytochem., 22, 2727-28.

Banthorpe, D.V. & Charlwood, B.V. (1980). The isoprenoids: the terpenoids. In Encyclopedia Plant Physiol., ed. E.A. Bell and B.V.Charlwood, vol. 8, pp. 185-220. Berlin: Springer-Verlag.

Banthorpe, D.V. & Osborne, M.J. (1984). Terpene epoxidases and epoxide hydratases from cultures of Jasminum officinale. Phytochem. 23, 905-7.

Banthorpe, D.V. & Njar, V.C.O. (1984). Light-dependent monoterpene synthesis in Pinus radiata cultures. Phytochem., 23, 295-99.

Banthorpe, D.V. & Wirz-Justice, A. (1972). Terpene biosynthesis VI: Monoterpenes and carotenoids from tissue cultures of Tanacetum vulgare L. J. Chem. Soc. Perkin Trans. I., 1769-72.

Becker, H. (1970). Studies on the formation of volatile substances in plant tissue cultures. Biochem. Physiol. Pflanzen, 161, 425-41.

Becker, H. & Baumer, J.J. (1983). Isolation and characterization of cell lines of Valeriana wallichii resistant to trifluoroleucine. Z. Pflanzenphysiol., 112, 43-51.

Becker, H. & Chavadej, S. (1985). Valepotriate production of normal and colchicine-treated cell suspension cultures of _Valeriana wallichii_. J. Nat. Prod., 48, 17-21.

Becker, J., Chavadej, S., Baumer, J. & Stoeck, M. (1984a). Isolation and characterisation of different cell-lines of _Valeriana wallichii_. Proc. 3rd Eur. Cong. Biotechnol., Munich, 1, 203-7.

Becker, H., Chavadej, S., Thies, P.W. & Finner, E. (1984b). The structure of new valepotriates from tissue cultures of _Valeriana wallichii_. Planta Medica, 50, 245-48.

Becker, H. & Herold, S. (1983). RP-8 Auxiliary phase for the accumulation of valepotriates from cell-suspension-culture of _Valeriana wallichii_. Planta Medica, 49, 191-92.

Becker, H., Reichling, J., Bisson, W. & Herold, S. (1984c). Two phase culture: A new method to yield lipophilic secondary products from plant suspension cultures. Proc. 3rd Eur. Cong. Biotechnol., Munich, 1, 209-13.

Becker, H. & Schrall, R. (1980). Valepotriates in tissue cultures of nine different Valerianaceae species in comparison to literature data of the intact plants. J. Nat. Prod., 43, 721-23.

Becker, H., Schrall, R. & Hartmann, W. (1977). Cultivation of tissue cultures of _Valeriana wallichii_ DC and first analytical determination. Arch. Pharm., 310, 481-84.

Beiderbeck, R. (1982). Two-phase culture: A method for the isolation of lipophilic substances from plant suspension cultures. Z. Pflanzenphysiol., 108, 27-30.

Berlin, J., Witte, L., Schubert, W. & Wray, V. (1984). Determination and quantification of monoterpenoids secreted into the medium of cell cultures of _Thuja occidentalis_. Phytochem., 23, 1277-79.

de Billy, F. & Paupardin, C. (1971). Sur l'evolution des huiles essentielles dans les tissue de pericarpe de Citron (_Citrus limonia_ Obseck) cultives _in vitro_. C. R. Acad. Sc. Paris. Ser. D., 273, 1690-93.

Bisson, W., Beiderbeck, R. & Reichling, J. (1983). Production of essential oils by cell-suspensions of _Matricaria chamomilla_ in a two phase system. Planta Medica, 47, 164-68.

Bohm, H. (1982). The inability of plant cell cultures to produce secondary substances. Proc. 5th Int. Cong. Plant Tissue Cell Culture, 325-28.

Bricout, J., Garcia-Rodriguez, M-J. & Paupardin, C. (1978a). Action de la colchicine sur la synthese d'huile essentielle par des tissue de _Mentha piperita_ cultives _in vitro_. C. R. Acad. Sc. Paris. Ser. D., 286, 1585-88.

Bricout, J., Garcia-Rodriguez, M-J., Paupardin, C. & Saussay, R. (1978b). Biosynthese de composes monoterpeniques par les tissus de quelques especes de Menthes cultivees _in vitro_. C. R. Acad. Sc. Paris. Ser. D., 287, 611-13.

Bricout, J. & Paupardin, C. (1974). Sur la composition de l'huile
    essentielle de tissus de pericarp de Citron (Citrus
    limonia Osbeck) cultives in vitro. C. R. Acad. Sci.
    Paris. Ser. D., 278, 719-22.
Bricout, J. & Paupardin, C. (1975). Sur la composition de l'huile
    essentielle de Mentha piperita L. cultivee in vitro:
    Influence de quelques facteurs sur la synthese. C. R.
    Acad. Sc. Paris. Ser. D., 281, 383-86.
Brown, J.T. & Charlwood, B.V. (1986). The control of callus formation
    and differentiation in scented Pelargoniums. J. Plant
    Physiol., 123, 409-17.
Cashyap, M.M., Kueh, J.S.H., Mackenzie, I.A. & Pattenden, G. (1978).
    In vitro synthesis of pyrethrins from tissue cultures of
    Tanacetum cinerariifolium. Phytochem., 17, 544-45.
Charlwood, B.V. & Banthorpe, D.V. (1978). The biosynthesis of
    monoterpenes. In Progress in Phytochemisty, ed. L.
    Reinhold, J.B. Harborne and T. Swain, vol. 5., pp. 65-125.
    Oxford: Pergamon Press.
Charlwood, B.V. & Charlwood, K.A. (1983). The biosynthesis of mono-
    and sesquiterpenes in tissue culture. Biochem. Soc. Trans.,
    11, 592-93.
Chavadej, S. & Becker, H. (1984). Influence of colchicine treatment
    on chromosome number and growth rate of tissue cultures of
    Valeriana wallichii DC. Plant Cell Tissue Organ Culture,
    3, 265-72.
Cori, O. (1983). Enzymic aspects of the biosynthesis of monoterpenes
    in plants. Phytochem., 22, 331-41.
Forche, E., Schubert, W., Kohl, W. & Hofle, G. (1984). Cell culture
    of Thuja occidentalis with continuous extraction of
    excreted terpenoids. Proc. 3rd Eur. Cong. Biotechnol.,
    Munich, 1, 189-92.
Galun, E., Aviv, D., Dantes, A. & Freeman, A. (1983). Biotrans-
    formation by plant cells immobilised in cross-linked
    polyacrylamide-hydrazide. Planta Medica, 49, 9-13.
Gupta, P.K. & Mascarenhas, A.F. (1983). Essential oil production in
    relation to organogenesis in tissue cultures of Eucalyptus
    citriodora Hook. Basic Life Sci., 22, 299-308.
Heitrich, A. & Binder, M. (1982). Identification of (3R,4R)-Δ1(6)-
    tetrahydrocannabinol as an isolation artefact of
    cannabinoid acids formed by callus cultures of Cannabis
    sativa L. Experientia, 38, 898-99.
Hirata, T., Aoki, T., Hirano, Y., Ito, T. & Suga, T. (1981). The
    biotransformation of foreign substrates by tissue cultures
    I: The hydroxylation of linalool and its related
    compounds with the suspension cells of Nicotiana tabacum.
    Bull. Chem. Soc. Jap., 54, 3527-29.
Hirata, T., Hamada, H., Aoki, T. & Suga, T. (1982a). Stereo-
    selectivity of the reduction of carvone and dihydrocarvone
    by suspension cells of Nicotiana tabacum. Phytochem., 21,
    2209-12.

Hirata, T., Lee, Y.S. & Suga, T. (1982b). The stereospecific hydroxylation of endocyclic ethylenic linkage in the biotransformation of α-terpinyl acetate with cultured suspension cells of Nicotiana tabacum. Chem. Lett., 671-74.

Huber, J. & Rudiger, W. (1978). Subcellular localisation and properties of 3-hydroxy-3-methylglutaryl coenzyme A reductase in plant cell suspension cultures of Anise. Z. Physiol. Chemie. 359, 277.

Itokawa, H., Takeya, K. & Mihashi, S. (1977). Biotransformation of cannabinoid precursors and related alcohols by suspension cultures of callus induced from Cannabis sativa L. Chem. Pharm. Bull., 25, 1941-46.

Jain, S.C. (1977). Chemical investigation of Tagetes tissue cultures. Planta Medica, 31, 68-70.

Jones, L.H. (1974). Plant cell culture and biochemistry: studies for improved vegetable oil production. In Industrial Aspects of Biochemistry, ed. B. Spencer, pp.813-33. London: FEBS.

Kireeva, S.A., Bugorskii, P.S. & Reznikova, S.A. (1977). Cultivation of Damask rose tissues and accumulation of terpenoids in them. Fiziol. Rast., 24, 824-31.

Kireeva, S.A., Mel'nikov, V.N., Reznikova, S.A. & Meshcheryakova, N.I. (1978). Essential oil accumulation in a peppermint callus culture. Fiziol. Rast., 25, 564-70.

Klausner, A. (1985). Common scents for biotech? Bio/technol., 3, 534-38.

Kueh, J.S.H., MacKenzie, I.A. & Pattenden, G. (1985). Production of chrysanthemic acid and pyrethrins by tissue cultures of Chrysanthemum cinerariaefolium. Plant Cell Reports, 4, 118-19.

Lang, E. & Horster, H. (1977). Sugar bound regular monoterpenes II: Production and accumulation of essential oils in Ocimum basilicum callus and suspension cultures. Planta Medica, 31, 112-18.

Lee, Y.S., Hirata, T. & Suga, T. (1983). Biotransformation of 1-acetoxy-p-menth-4(8)-ene with a suspension of cultured cells of Nicotiana tabacum. J. Chem. Soc. Perkin Trans. I., 2475-78.

Nabeta, K., Ohnishi, Y., Hirose, T. & Sugisawa, H. (1983). Monoterpene biosynthesis by callus tissues and suspension cells from Perilla species. Phytochem. 22, 423-25.

Paupardin, C. (1974). Sur l'evolution de l'huile essentielle dans des tissus de fruits de Citron (Citrus limonia Osbeck) cultives in vitro dans diverses conditions. Rev. Gen. Bot., 81, 223-41.

Paupardin, C. (1976). On the differentiation of secreting tissue and the formation of essential oil by plant tissues cultivated in vitro. C. R. Congr. Natl. Soc. Savant Sci. Lille, 1, 619-28.

Paupardin, C., Garcia-Rodriguez, M.J. & Bricout, J. (1980). Multi-
    plication vegetative de quelques plantes aromatiques:
    problemes poses par la production d'essence. C.R. Acad.
    Agric. France. 66, 658-66.
Schrall, R. & Becker, H. (1980). Valepotriates in tissue and
    suspension cultures of different Valerianaceae. Acta
    Hortic., 96, 75-83.
Staba, E.J., Laursen, P. & Buchner, S.A. (1965). Medicinal plant
    tissue cultures. In Proc. Int. Conf. Tissue Culture, ed.
    P.R. White, pp.191-210. Berkeley, California: McCutchan
    Publishing.
Suga, T., Aoki, T., Hirata, T., Lee, Y.S., Nishimura, O. & Utsumi, M.
    (1980a). Biotransformation of foreign substrates with
    callus tissues: Transformation of terpineols with tobacco
    suspension cells. Chem. Lett. 229-30.
Suga, T., Hirata, T. & Futatsugi, M. (1984). The biotransformation
    of carvoxime and dihydrocarvoxime with cell suspension
    cultures of Nicotiana tabacum. Phytochem., 23, 1327-28.
Suga, T., Hirata, T. & Lee, Y.S. (1982). The enantioselective
    biotransformation of α-terpineol and its acetate with the
    cultured cells of Nicotiana tabacum. Chem. Lett., 1595-
    98.
Suga, T., Hirata, T. & Yamamoto, Y. (1980b). Lipid constituents of
    callus tissues of Mentha spicata. Agric. Biol. Chem., 44,
    1817-20.
Suga, T., Lee, Y.S. & Hirata, T. (1983). The hydroxylation of β-
    terpineol and its acetate with the cultured cells of
    Nicotiana tabacum. Bull. Chem. Soc. Japan, 56, 784-87.
Sugisawa, H. & Ohnishi, Y. (1976). Isolation and identification of
    monoterpenes from cultured cells of Perilla plant. Agric.
    Biol. Chem., 40, 231-32.
Takeya, K. & Itokawa, H. (1977). Stereochemistry in oxidation of
    allylic alcohols by cell-free system of callus induced
    from Cannabis sativa L. Chem. Pharm. Bull., 25, 1947-
    1951.
Tanahashi, T., Nagakura, N., Inouye, H. & Zenk, M.H. (1984). Radio-
    immunoassay for the determination of loganin and the
    biotransformation of loganin to secologanin by plant cell
    cultures. Phytochem., 23, 1917-22.
Ueda, S., Kobayashi, K., Muramatsu, T. & Inouye, H. (1981). Studies
    on monoterpene glucosides and related natural products XL:
    Iridoid glucosides of cultured cells of Gardenia
    jasminoides f. grandiflora. Planta Medica. 41, 186-91.
Uesato, S., Ueda, S., Kobayashi, K. & Inouye, H. (1983). Mechanism
    of iridane skeleton formation in the biosynthesis of
    iridoid glucosides in Gardenia jasminoides cell cultures.
    Chem. Pharm. Bull., 31, 4185-88.
Uesato, S., Ueda, S., Kobayashi, K., Miyauchi, M. & Inouye, H. (1984).
    Biosynthetic pathway of iridoid glucosides in Gardenia
    jasminoides f. grandiflora cell suspension cultures after
    iridodial cation formation. Tetrahedron Lett., 25, 573-
    76.

Violon, C., Dekegel, D. & Vercruysse, A. (1983). Microscopical study of valepotriates in liquid droplets of various tissues from Valerian plants. Plant Cell Reports, 2, 300-3.

Violon, C., Dekegel, D. & Vercruysse, A. (1984a). Relation between valepotriate content and differentiation level in various tissues from Valerianeae. J. Nat. Product., 47, 934-40.

Violon, C., Sonck, W. & Vercruysse, A., (1984b). Comparative study of the essential oils in vivo and in vitro grown Valeriana officinalis L and Centranthus macrosiphon Boiss. by coupled gas chromatography-mass spectrometry. J. Chromatog., 288, 474-78.

Violon, C.J.I. & Vercruysse, A.A. (1985). Haemcytochromes in valepotriate producing tissue cultures of Centranthus macrosiphon.. Phytochem., 24, 2205-09.

Wang, C-J. & Staba, E.J. (1963). Peppermint and spearmint tissue culture II: Dual-carboy culture of spearmint tissues. J. Pharm. Sci., 52, 1058-62.

Watts, M.J., Galpin, I.J. & Collin, H.A. (1984). The effect of growth regulators, light and temperature on flavour production in celery tissue cultures. New Phytol., 98, 583-91.

Watts, M.J., Galpin, I.J. & Collin, H.A. (1985). The effect of greening on flavour production in celery tissue cultures. New Phytol., 100, 45-56.

Webb, J.K., Banthorpe, D.V. & Watson, D.G. (1984). Monoterpene synthesis in shoots regenerated from callus cultures. Phytochem., 23, 903-4.

Witte, L., Berlin, J., Wray, V., Schubert, W., Kohl, W., Hofle, G. & Hammer, J. (1983). Mono- and diterpenes from cell cultures of Thuja occidentalis. Planta Medica., 49, 216-21.

Zieg, R.G., Zito, S.W. & Staba, E.J. (1983). Selection of high pyrethrin producing tissue cultures. Planta Medica, 48, 88-91.

3.   POLYMERIC ADSORBENTS STIMULATE ANTHRAQUINONE PRODUCTION BY
CELL SUSPENSION CULTURES OF CINCHONA LEDGERIANA

R.J. Robins & M.J.C. Rhodes

Plant Cell Culture Group,
AFRC Institute of Food Research (Norwich Laboratory),
Colney Lane,
Norwich   NR4 7UA.   U.K.

INTRODUCTION
    Cells of Cinchona ledgeriana synthesise a wide range of
anthraquinones in both callus (Wijnsma et al., 1984) and liquid
suspension cultures (Robins et al., 1986) of which the major
structures have now been identified.   They are typical of the anthraquinones of the Rubiaceae in having a highly substituted A-ring, which
increases their solubility in water.   In the liquid cultures a
substantial proportion of the anthraquinones accumulate in the medium.
This property renders them a valuable system for studying the biotechnological production of metabolites by immobilised cells, in which
only that product released to the medium is recovered.   In such
systems, the productive capacity of the process may be more dependent
on the secretory capacity of the cells than on their biosynthetic
potential.

Because recovery of product from dilute downstream is important in
such processes, we have investigated the effectiveness of polymeric
adsorbents in performing this role (Robins and Rhodes, 1986).   This
paper discusses the effect adsorbents have on the metabolism of
anthraquinones in C. ledgeriana suspension cultures and in particular
the considerable enhancement of production which occurs in their
presence.

ANTHRAQUINONE PRODUCTION
    Time course and manipulation of anthraquinone production.
Anthraquinones are synthesised and secreted by suspension cultures of
C. ledgeriana at all stages of the growth cycle but show a marked
accumulation in the later stages (Fig. 1).

Time courses for anthraquinone synthesis differ between species,
possibly reflecting the extent to which product is secreted from the
cells, the extracellular space acting as an extension of the vacuole
and allowing greater accumulation than otherwise is possible.
Typically, 70-80% of that in late-stage cultures of Cinchona is
extracellular.

Altering the constituents of the medium had little effect on either
anthraquinone synthesis or their distribution between cells and medium
(Robins et al., 1986).   Low concentrations (0.2%) of N,Z-amine were
stimulatory, though higher concentrations (0.4%) were inhibitory,

while the precursor L-tryptophan   caused a progressive inhibition of
production over the range 0.1-5 mM, up to about 60% of the control.
While it is possible that these results are due to direct effects of
tryptophan on the shikimic acid pathway, limiting the supply of
chorismic acid, indirect effects, due to tryptophan toxicity at high
levels, may also contribute.

The effect of fungal or Streptomyces culture homogenates.
Flasks of culture contaminated by fungal growth often appeared very
highly coloured and it was found that, when fungal homogenates were

Fig. 1   Time course showing cell growth (O), anthraquinone production (●) and alkaloid production (■) on a
medium containing 2.4-D (0.5 mg/l) and benzyladenine (0.1
mg/l).   Each point represents duplicate samples.

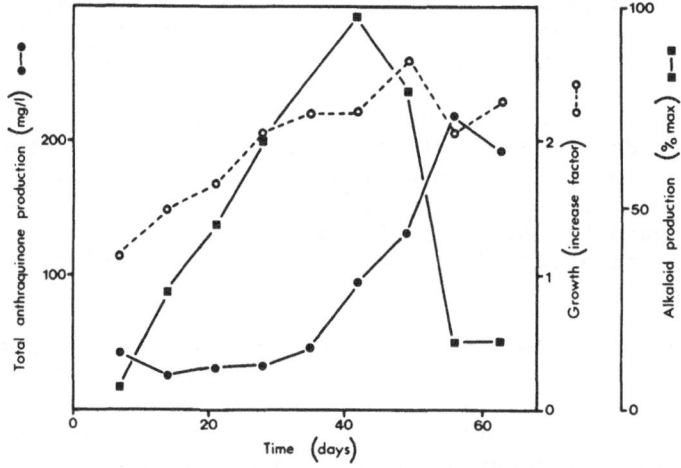

Fig. 2   The effect on anthraquinone production of adding
to cultures autoclaved homogenates of 1 = Cunninghamella
echinulata (NRRL 3655); 2 = Gongronella butleri (IMI 71628);
3 = Streptomyces platensis (NRRL 2364). C = Control. Total
stage I microbial cultures were homogenised and 0.1, 1.0
or 5.0 ml added to duplicate flasks, 3 days after
sub-culture and harvested at 18 days.

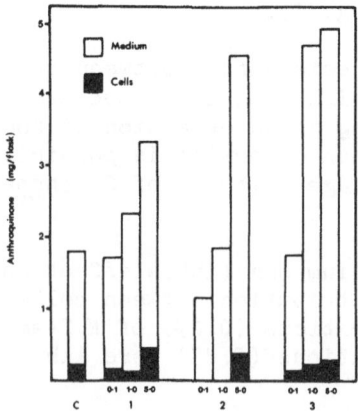

added to cultures, up to 3-fold more anthraquinone accumulated than in
the controls (Fig. 2). All the additional anthraquinones were
extracellular. It has now been shown that some of these
anthraquinones have anti-microbial properties and a role as a
phytoalexin has been proposed (Wijnsma et al., 1985).

ADSORPTION OF ANTHRAQUINONES
Cells of FRIN-CL1a produce so much extra-cellular
anthraquinone that it precipitates out of the medium onto the cells
and walls of the flask, forming a red-brown deposit. In order to
overcome this problem the effect of adding polymeric adsorbents to the
culture was investigated (Robins & Rhodes, 1986). Of a range of
adsorbents tested, the polyether foams were found to have a very high
binding capacity, but their affinity is low and the material is very
bulky. The Amberlite resins XAD-2 and XAD-7, however, have both a
high affinity and good adsorptive capacities. These two materials
were both tested further.

Effect of adsorbents on in vivo accumulation.
When tested in vivo (Table 1), not only did these adsorbents
effectively remove the anthraquinones from the medium, but in addition
a large stimulation of anthraquinone synthesis was observed. This was
most marked with XAD-4 and XAD-7, which caused a 10- to 15-fold
stimulation, much greater than that found with any chemical or
elicitor treatments. Furthermore, the majority of the anthraquinone
in the system was bound to the adsorbent (Table 2), from which it
could readily be recovered using alkaline methanol. Because of its
greater hydrophilicity, which improves contact with the medium, XAD-7
is the preferred material.

The amount of adsorbent to which the culture is exposed is critical,
the optimal yield of anthraquinones being obtained with 0.5-1 g/flask
of XAD-7 (10-20 g/l). Greater amounts are extremely detrimental to
both the production of anthraquinones and the growth of these
cultures.

Table 1   Production of anthraquinones in the presence of
various polymeric adsorbents. Adsorbent was added at 7
days and duplicate flasks harvested at 31 days following
sub-culture. a = Mean measured over days 0-31 following
subculture. b = 9.71 g fresh weight at 31 days. c =
Improvement relative to control.

Adsorbent	Growth	Mean rate of anthraquinone accumulation[a]	
(g/l)	(% of control)	(mg/l/day)	(g/g fr wt/day)
Control	100[b]	1.2(1)[c]	6.1(1)[c]
XAD-2 (10)	77	9.5(8.1)	63.8(10.5)
XAD-4 (10)	81	12.7(10.7)	123.8(20.3)
XAD-7 (10)	54	17.4(14.7)	166.5(27.3)
Foam FT60 (4)	80	7.7(6.5)	64.4(10.6)
Foam D (4)	83	8.6(7.3)	69.9(11.5)

Fig. 3   Time course showing the effect of various
adsorbents on (a) cell growth and (b) total anthraquinone
production.   Adsorbent illustrated are: none (▲), 10 g/l
XAD-2 (□), , 10 g/l XAD-7 (■), 4 g/l foam T60 (0) and 4
g/l foam D (●).   Cells were sub-cultured at day 0 and
adsorbent added at day 8.

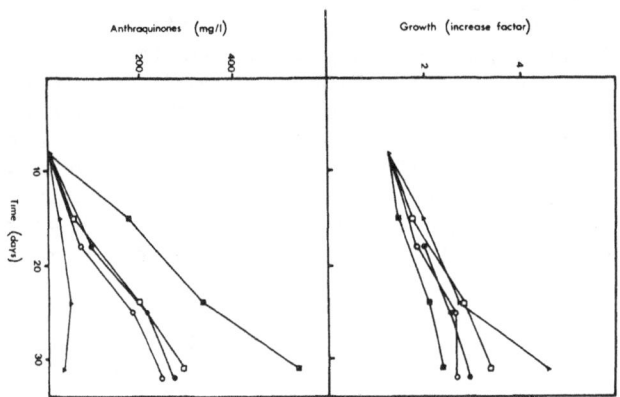

Fig. 4   The effect of the time of adding XAD-7 resin on
(a) cell growth and (b) anthraquinone production.   At the
number of days shown after sub-culture 50 g/l XAD-7 was
added.   Each bar represents duplicate flasks harvested at
28 days from sub-culture.

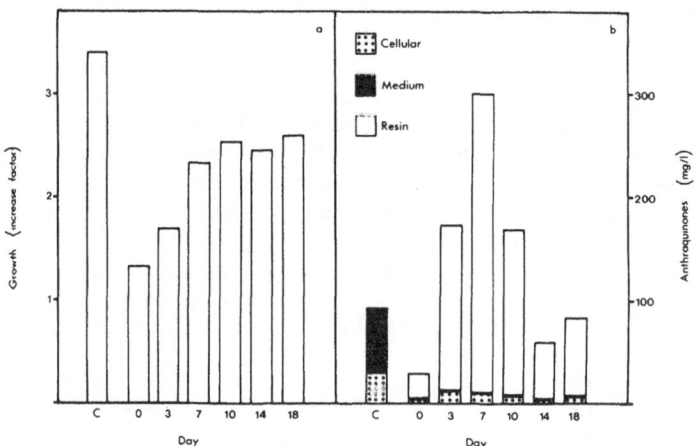

An immediate increase in the rate of accumulation of anthraquinones is found after adsorbent is added to cultures (Fig. 3). None of the adsorbents tested completely inhibited growth, though they all decreased the rate of growth relative to the control (Table 1; Fig. 3).

If XAD-7 is added too early in the growth cycle it is also highly detrimental to overall yield (Fig. 4), presumably because the resin absorbs essential factors from the medium. For a culture harvested at 28 days it is found that the best yield is obtained when adsorbent is added at day 7 (Fig. 4). In fact, if adsorbent is added at any time up to 35 days from sub-culture, the rate of anthraquinone synthesis is stimulated to a comparable extent, even though by about 21 days the controls are actually showing a net loss of anthraquinones from the system (Robins & Rhodes, 1986).

Table 2   The distribution of anthraquinones within the
system in the presence of various adsorbents.

Adsorbent (g/l)	Anthraquinone production (% distribution)			
	Cells	Medium	Adsorbent	Total (mg/l)
Control	14	86	–	36.6
XAD-2 (10)	17	2	81	294.6
XAD-4 (10)	7	1	92	354.2
XAD-7 (10)	5	1	94	539.0
Foam FT60 (4)	8	10	82	245.4
Foam D (4)	8	7	85	275.0

CONCLUSIONS
From the bio-technological view-point the adsorbents are extremely attractive. The high binding capacity ensures that the anthraquinone concentration in the medium is kept very low while the majority of the product in the system is concentrated onto the adsorbent, from which it may be readily recovered. The cells grow moderately well in the presence of adsorbent or, alternatively, adsorbent can be added after growth with equally good yields (Robins & Rhodes, 1986). Such a recovery process is only valuable in systems which naturally secrete a substantial proportion of their product, or which can be induced to do so by manipulation of their physico-chemical environment. The cultures of C. ledgeriana considered here fulfil this criterion, thus providing a valuable system for studying immobilisation technology. The adsorbents, however, are not ideal as they show no specificity towards the desired product. Thus, anthraquinones eluted with alkaline methanol from XAD-7 or polyether foams were only 15-20% and 50-60% pure respectively. Furthermore, the adsorbents are detrimental to growth, indicating that they affect other aspects of cellular metabolism than secondary product biosynthesis. A need for better, more specific materials is apparent.

The continual removal of anthraquinones from the extracellular environment stimulates the production of more of these compounds than would otherwise accumulate in the cell cultures. With the best adsorbents, the intracellular level is decreased and it is possible that feed-back controls of the biosynthetic pathway are being overidden. Alternatively, it may be that the anthraquinones are more resistant to degradation when bound to the adsorbent, allowing them to accumulate in the extracellular environment. It seems likely that the former mechanism is operative since, when adsorbent is added to cells from which a net loss of anthraquinone is occurring, the trend is reversed and accumulation occurs at a high rate. This switch presumably indicates an overall reversal of anthraquinone metabolism from degradation to synthesis.

REFERENCES
Robins, R.J. & Rhodes, M.J.C. (1986). The stimulation of anthraquinone production by Cinchona ledgeriana cultures with polymeric adsorbents. Appl. Microbiol. Biotechnol. 24, 35-41
Robins, R.J., Payne, J. & Rhodes, M.J.C. (1986). The production of anthraquinones by cell suspension cultures of Cinchona ledgeriana. Phytochemistry (in press).
Wijnsma, R., Verpoorte, R., Mulder-Krieger, Th. & Baerheim-Svendsen, A. (1984). Anthraquinones in callus cultures of Cinchona ledgeriana. Phytochemistry, 23, 2307-2311.
Wijnsma, R., Go J.T.K.A., Weerden, I.N. van, Haarkes, P.A.A., Verpoorte, R. & Baerheim-Svendsen, A. (1985). Anthraquinones as phytoalexins in cell and tissue cultures of Cinchona spec. Plant Cell Rep., 4, 241-244.

4.   BIOPOLYMER PRODUCTION BY ALGAL CELLS

D. Grey, G. Stepan-Sarkissian & M.W. Fowler

Wolfson Institute of Biotechnology,
University of Sheffield,
Sheffield   S10 2TN. U.K.

INTRODUCTION
Interest in algal cultures, their extracellular products and potentials as biomass sources, is relatively recent. Ready adaptation to environment and rapid growth rate of algal systems make them particularly suited for such uses.

Most of the early work in this field was carried out with macroalgae, i.e. seaweeds and related species. More recently, interest has focussed on microalgae, particularly as a source of natural products. Appropriate species of microalgae may be grown on a large scale to provide economically useful polysaccharides such as sulphated galactans. In addition, higher biomass yields can be obtained from microalgae than from macroalgae.

Patent literature (e.g. Golueke & Oswald, 1965; Schenck et al., 1976; Al'bitskaya et al., 1979) indicates that the green unicellular alga Chlamydomonas mexicana and the simple multicellular alga Porphyridium cruentum have potential as sources of biopolymers. For this reason it was decided to use these two species in our studies.

One of the earliest detailed reports concerning the production of extracellular polysaccharides by Chlamydomonas species was that of Lewin (1956). The biopolymer was reported to contain 25% of the total organic matter produced by the culture. No further major investigation on C. mexicana mucilages was undertaken until the patent of Schenck et al. (1976) described a nutrient medium and culture regime which yielded large amounts of high molecular weight biopolymer. In contrast, the extracellular polysaccharides synthesised by red algae have long been of interest because of their gelling properties. These substances are known commercially as agar, carrageenan, funoran, furcellaran and porphyran and appear to consist mainly of alternating units of 1,3-linked β-galactose and 1,4-linked β-galactose (Percival, 1978)

Culture Growth
Algal cultures were obtained from the Culture Collection of Algae and Protozoa (CCAP) in Cambridge, England. C. mexicana was grown almost exclusively on the medium of Schenck et al. (1976). P. cruentum cultures were grown on the medium used at the CCAP and code-named M11 (Asher & Spalding, 1982).

Small scale cultures (15 ml) of C. mexicana were set up to investigate
the effect on growth of exogenously supplied carbon sources. Cultures
were grown in the presence of 0.004% bicarbonate, 1.5% acetate and
1.5% glucose at 25°C under constant illumination (6000 lux). Glucose
appeared to support a higher growth rate and larger biomass
accumulation (data not shown).

More detailed studies of the effect of different carbon sources on the
growth and viscosity of P. cruentum were undertaken using trays
containing 25 wells each with a volume of 2.5ml. The carbon sources
used were acetate, whey, fructose, glucose, galactose, sucrose, brown
sugar and malt at concentrations of 0.5%, 1%, 2% and 3%. The highest
growth rates were achieved on 1% and 2% acetate (Fig. 1), followed by
0.5% acetate and 1% whey in which the medium viscosities reached were
2.3 and 2.5 centipoises respectively.

Fig. 1. The effect of different carbon sources on the
growth and viscosity of Porphyridium cruentum. Histograms
represent the cultures where growth was above that of the
controls. Significant changes in viscosity (figures in
brackets) are also shown. Viscosity values were measured
using an Ostwald viscometer (1mm diameter). The
reference solution was distilled water.

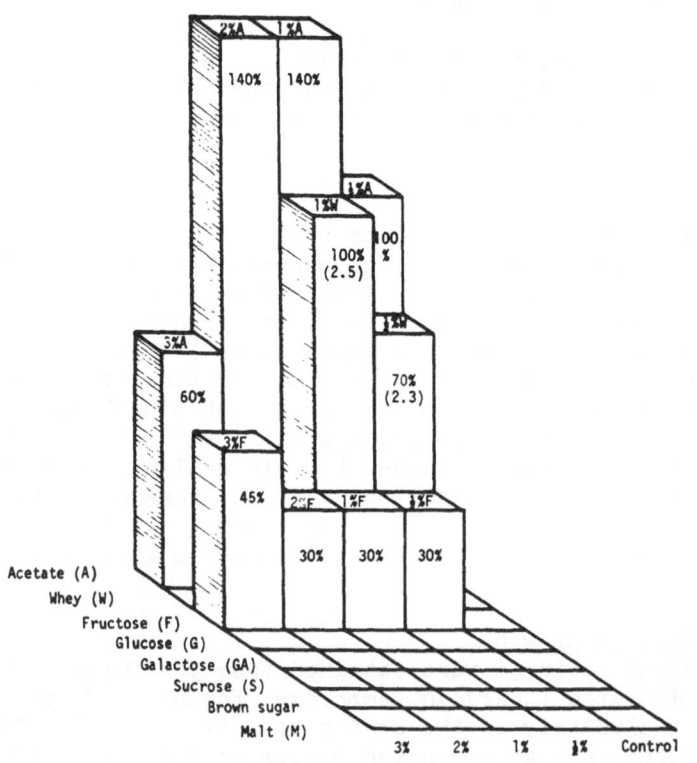

The effect of acetate on the growth of P. cruentum large-scale
cultures (1.7 L) was studied in the presence of 2% acetate and
compressed air under constant light. Control vessels were run in
parallel without compressed air (Fig. 2). The aerated cultures
achieved higher cell densities although in the early stages the growth
rates in the two experiments were similar.

The absence of viscosity in our C. mexicana cultures grown in the
medium of Schenck et al. (1976) prompted an experiment simulating in
every detail the conditions reported in their patent. Although the
final biomass accumulated during the stationary phases was similar in
both cases, the growth rate obtained by Schenck et al. (1976) was
significantly higher (Fig. 3). However, at the end of the 10 day
growth period the viscosity of our culture was 1.159 ± 0.036
centistokes compared to that of 22 centistokes reported in the patent.
Schenck et al. (1976) also noted that low nitrate-nitrogen levels (23
mg.l^{-1}) stimulated polymer production by C. mexicana cultures. In an
attempt to enhance biopolymer production in our cultures, medium
nitrate concentrations were altered in a series of experiments.
Nitrate levels below 100 mg.l^{-1} failed to support maximal growth rates
(Fig. 4). The viscosity of culture with the lowest nitrate-nitrogen
level (25 mg.l^{-1}) did not exceed 1.120 ± 0.026 centistokes.

Fig. 2.   Growth of Porphyridium cruentum in M11 medium
supplemented with 2% sodium acetate. Two sets of
experiments were set up in duplicate in the absence (open
and closed squares) or presence (open and closed circles)
of compressed air.

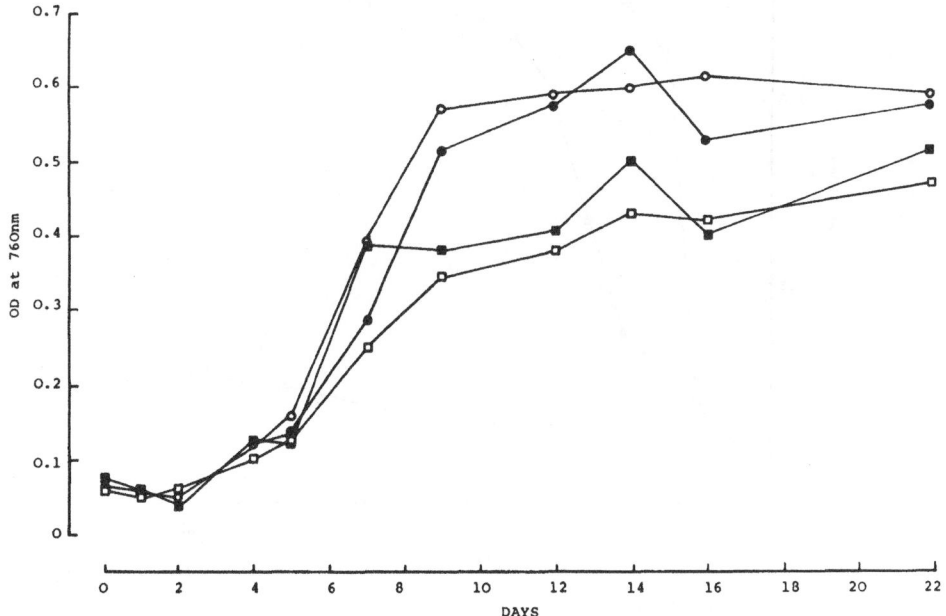

The effect of different sulphate concentrations on the growth and medium viscosity of C. mexicana cultures were also studied. The growth patterns appear to be unaffected by changes in initial sulphate concentrations (data not shown). However in each case the sulphate level in the medium rose above that originally supplied, perhaps as a result of sulphate excretion from the cells. This phenomenon may account for the absence of medium viscosity. At this stage it is neither possible to explain this phenomenon nor attribute it to a general characteristic of the cells.

Rates of CO_2 fixation by algal cultures were measured by supplying $H^{14}CO_3$ to 10ml of culture and incubating for 5h in an illuminated shaker. The results obtained from duplicate experiments indicated that C. mexicana has a higher rate of CO_2 absorption (1.132 nmoles.ml^{-1} h^{-1}) than P. cruentum (0.744 nmoles.ml^{-1} h^{-1}).

Fig. 3. Growth of Chlamydomonas mexicana in Schenck's medium supplemented with compressed air and 5% CO_2. Growth conditions were identical to those described by Schenck et al. (1976). The broken line represents the growth curve obtained in the above patent. The growth of cultures in the presence and absence of compressed air and 5% CO_2 is represented with closed squares and circles respectively.

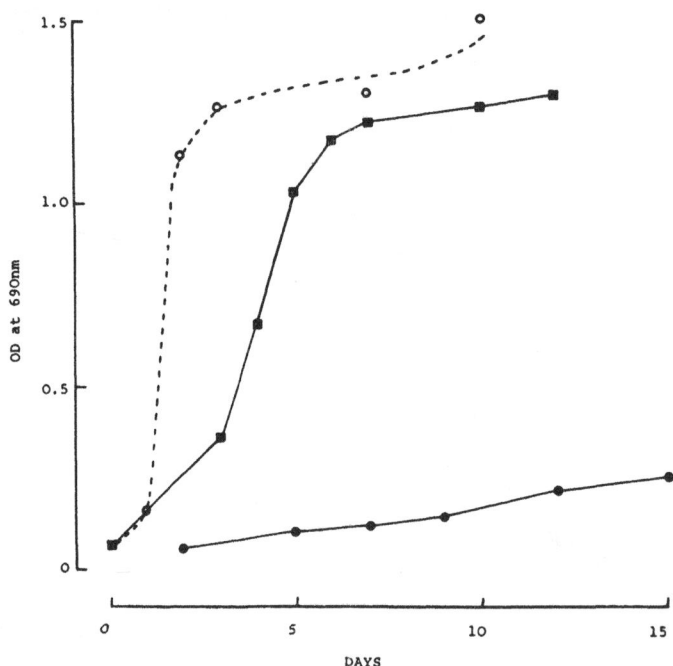

Biopolymer Production

The extracellular biopolymer from P. cruentum was
initially extracted according to the method described by Heaney-Kieras
et al. (1976).   Total biopolymer yields obtained from 39 and 43 day
old cultures of P. cruentum were 0.009 g.l⁻¹ and 0.21 g.l⁻¹
respectively using the detergent cetylpyridinium chloride to
precipitate the polysaccharide. A 7-fold increase in polymer
extraction was achieved when ethanol was used as a precipitating
agent. A third extraction method suggested by Dr. Elizabeth Percival
(London University) was also used. The procedure involved dialysis of
culture medium and evaporation under reduced pressure followed by
freeze-drying of the biopolymer. This method generally gave
biopolymer yields comparable to those from ethanol precipitation.

Chemical microanalysis of selected polymer samples from P. cruentum
revealed low levels of carbon and sulphur (<10%) in contrast to high
content of chlorine (>20%). Chlorine appeared to be present in a
covalently bound form since various steps of extraction procedures
would have removed the ionic species of this element. High levels of
chlorine would have inevitably masked the trace levels of carbon and
sulphur which are expected to be highly significant in a sulphated
polygalactan.

Fig. 4.   Growth of Chlamydomonas mexicana under different
nitrate-nitrogen regimes. The light regime was 15h
natural daylight.

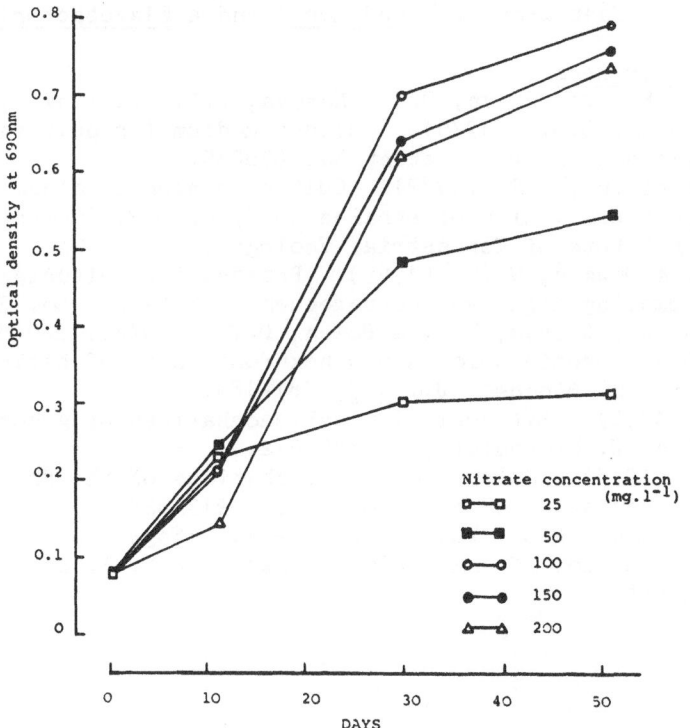

The presence of chlorine could also have been responsible for the relatively low viscosity values obtained from P. cruentum.

Comparison of the viscosity of the culture medium with that of the extracted and resuspended polymer showed the latter to have lower viscosity. It is possible that this loss of viscosity may be due to a process involved in the extraction procedure. Measurements taken before and after dialysis of the culture medium indicate a significant loss of sulphate from the medium during dialysis with the concentration falling from 64.5mM to 51.5mM, a drop of 20%. There was a similar fall (18%) in the viscosity of the medium over the dialysis period. It is difficult to determine if the loss of viscosity is due to the increase in the volume of dialysate since viscosity measurements are far from linear.

Isolation of Capsular Bacteria
Our attention was drawn by the CCAP at Cambridge to the presence of unidentified bacteria in the capsule of P. cruentum in cultures supplied by them. CCAP were not able to help with information of the identity of bacteria and a comprehensive literature search was inconclusive.

The bacteria were isolated from disrupted P. cruentum cells and were routinely grown on Yeast Extract Agar in an illuminated incubator at 20°C. Preliminary investigation showed the bacteria to be Gram-negative rods. Further studies indicated that two species of bacteria were involved; a fast growing Pseudomonas and a Flavobacterium.

References
Al'bitskaya, O.N., Goronkova, O.I., Nosova, L.P., Kirichkov, A.P. & Lunin, L.N. (1979). Culture medium for cultivation of microalgae. USSR Patent No. 678065.

Asher, A. & Spalding, D.F. (1982). Culture centre of algae and protozoa : List of strains 1982, pp 5.3. Cambridge : Institute of Terrestrial Ecology.

Golueke, C.G. & Oswald, W.J. (1965). Process for culturing and removing algae and carrageenan. US Patent No. 3,195,271.

Heaney-Kieras, J., Kieras, F.J. & Bowen, D.V. (1976). 2-0-Methyl-D-glucuronic acid, a new hexuronic acid of biological origin. Biochem. J., 155, 181-185.

Lewin, R.A. (1956). Extracellular polysaccharides of green algae. Can. J. Microbiol., 2, 665-672.

Percival, E. (1978). Sulphated polysaccharides of the Rhodophyceae -A review. ACS Symp. Ser., 77, 213-224.

Schenck, P., Foster, P.L., Walker, Jr., W.W. & Fogel, S. (1976). Production of algal bio-polymers. U.S. Patent No. 3,958,364.

5. ANTIMICROBIAL AGENTS FROM PLANT CELL CULTURES

J.S. Evans, E. Pattinson, & P. Morris

Wolfson Institute of Biotechnology,
University of Sheffield,
Sheffield. S10 2TN. U.K.

INTRODUCTION

There have been numerous broad based screening programmes initiated over the last twenty years, in which large numbers of plant species were evaluated for their antimicrobial activities (Farnsworth 1966; Mitscher et al. 1972, Misra & Dixit 1979, Leven et al. 1979). These screens showed that antibiosis is not uncommon amongst higher plants, so drawing attention to their potential as additional sources of antibiotics. The ability to culture plant cells in vitro has also led to investigations of their ability to accumulate new compounds with useful biological activities (Cambell et al. 1964; Khanna et al. 1971; Veliky & Lastta 1974; Misawa et al. 1985.). Antimicrobial agents from plant cell cultures have been detected in screens but relatively few studies have been published in which the initial results were followed by attempts to isolate and identify the active principles, or to characterize their production during cell culture.

ANTIMICROBIAL ACTIVITIES OF CELL CULTURES

Total methanol extracts from cell cultures of fifteen species of medicinal plants were screened for antimicrobial activity against six common micro-organisms, using a modification of the cup/plate diffusion method (Foster & Woodruff 1944). Initially 4-6 week old callus cultures and 14-20 day old suspension cultures were extracted with methanol, resuspended in 5% MeOH and the extracts tested against Bacillus subtilis (strain 7198 & 1604), Bacillus megaterium (strain 7581), E. coli (strain W31101), Saccharomyces cerevisiae (strains 745 & A3) Candida utilis (strain 927) and Agrobacterium tumefaciens (strain B6).

Of the limited number of species tested a high proportion (75%) showed antimicrobial activity against at least one of the test organisms (Table 1). The relative activities of extracts under standard conditions are shown in Table 2. These relative activities are justified from the linearity of the dose /response curves for crude extracts against S. cerevisiae and B. subtilis. Two of the highest activities against S. cerevisiae were found in extracts of Carthamus tinctoria and Papaver somniferum cultures.

CARTHAMUS TINCTORIUS (SAFFLOWER)

The growth of cell cultures of C. tinctorius on M&S medium containing 1.0 mg/l. 2.4-D, 0.1 mg/l. kinetin and 2% sucrose is shown in Fig. 1 (μ max 0.146 days^{-1}). Antimicrobial activity was found to correlate with total soluble cell phenolics. Maximum antimicrobial and phenolic production occurred in early exponential growth with activities four times greater than previously determined in 20 day old cells. The active component when fractionated according to the scheme shown in Fig. 3 was exclusively found in fraction I. A biogram of this fraction after two-dimensional TLC on cellulose in Butanol: Acetic acid: water and 6% acetic acid showed that the active component was a pale yellow compound, dark absorbing in UV light and showing several UV and visible adsorption peaks, a distinct bathochromic shift on addition of sodium methoxide and a blue reaction with Folin/Na$_2$CO$_3$ reagent. The compound was tentatively identified as a flavanoid, although it did not correspond to the common flavanoids by TLC. Insufficient material was available for further analysis.

Table 1   Antimicrobial activity against several test organisms, of MeOH extracts from cell cultures . (+)- inhibitory activity, (0)- no activity, (-)- not tested, (S)- suspension derived extracts, (c)- callus extracts.

SPECIES	SOURCE OF CELLS	AGROBACTERIUM TUMEFACIENS	BACILLUS SUBTILIS	BACILLUS MEGATERIUM	ESCHERICHIA COLI	SACCHROMYCES CEREVISIAE	CANDIDA UTILIS
ACER PSEUDOPLANTANUS	S	0	+	+	0	0	-
ARTEMISIA ANNUA	S	0	0	+	0	0	-
CARTHAMUS TINCTORIA	S	0	+	+	0	+	+
CATHARANTHUS ROSEUS	S	0	+	+	0	+	-
CINCHONA LEGERIANA	S	0	0	0	0	0	-
DATURA STRAMMONIUM	S	0	0	0	0	0	-
DAUCUS CAROTA	S	0	+	+	0	0	-
DIGITALIS LANATA	S	0	0	0	0	0	-
HUMULUS LUPULUS	S	0	0	+	0	+	-
NICOTIANA TABACUM	S	0	+	+	0	0	-
OENOTHERA BIENIS	S	0	0	+	0	0	-
PANAX GINSENG	C	0	+	+	0	0	-
PAPAVER SOMNIFERUM	C	0	+	+	0	+	+
PICRASMA QUASSOIDES	C	0	+	0	0	0	-
QUASSIA AMARA	S	0	0	0	0	0	-

Table 2   Relative activity of MeOH extracts from different plant cell cultures [cm^2/g fresh wt (mg carbenicillin equivalents /g fresh wt)] Media (M&S), 2N2K - 2 mg/l NAA + 2 mg/l kinetin, 2N2B - 2 mg/l NAA + 2 mg/l 6BAP, 1N1B-1 mg/l NAA + 1 mg/l 6BAP, M$_2$ - 1 mg/l 2,4-D + 0.1 mg/l kinetin.

SPECIES	SOURCE OF CELLS	BACILLUS SUBTILIS	BACILLUS MEGATERIUM	SACCHROMYCES CEREVISIAE
ACER PSEUDOPLANTANUS	S	2.6 (0.22)	3.0 (0.25)	0
ARTEMISIA ANNUA	S	0	3.1 (0.26)	0
CARTHAMUS TINCTORIA	S	1.8 (0.18)	1.7 (0.17)	1.6
CATHARANTHUS ROSEUS	S	1.4 (0.15)	1.1 (0.14)	0
DAUCUS CAROTA	S	0	0.1 (0.1)	0
HUMULUS LUPULUS	S	0	0.4 (0.11)	2.3
NICOTIANA TABACUM	S	0.3 (0.11)	0.5 (0.11)	0
OENOTHERA BIENIS	S	0	4.5 (0.41)	0
PANAX GINSENG	C	0.1 (0.1)	0.1 (0.10)	0
PICRASMA QUASSOIDES	C	0.8 (0.13)	-	-
PAPAVER SOMNIFERUM	C	0.1 (0.1)	0.1 (0.1)	0.4
PAPAVER SOMNIFERUM	CAPSULE	-	-	0.13
	S (2N2K)	-	-	0.10
	S (2N2B)	-	-	0.24
	S (1N1B)	-	-	1.73
	S (M$_2$)	-	-	0.14

PAPAVER SOMNIFERUM (OPIUM POPPY)
Cell growth
The growth of cell suspension cultures of P. somniferum in M₂ medium (see Table 2) at 25°C is shown in Fig. 2a. (μ max 0.15 d⁻¹). Antimicrobial activity against B.subtilis correlated with total soluble cell phenolics and peaked at 7-10 days. (Fig. 2b). Maximum activities were 75 times higher than in the 20 day old cultures screened previously. Acid hydrolysis of the cell wall fraction after exhaustive MeOH extraction revealed the presence of bound forms of the active compounds. The amount of bound activity remained fairly constant throughout the culture period (Fig. 2c). When expressed on the basis of culture volume, maximum levels of the antimicrobial agent reached 25 mg carbenicillin equivalents l.⁻¹ of culture (Fig. 2d).

Fractionation of crude methanol extracts
The crude MeOH extract from 12g dry wt. of cells was fractionated as in Fig. 3 to give fractions of varying polarities. Table 3a shows that the majority of the antimicrobial activity was found in the polar fractions III and IV. The low activities found in fractions I and II probably represent the presence of sanguinarine and chelerythrine alkaloids. These have known antimicrobial activities and have been isolated from our cultures. The activity in fraction III was non-dialysable and was not destroyed by acid or alkali hydrolysis. Reverse phase HPLC of fraction III from 0-100% MeOH showed all the activity to be eluted in the void volume (Table 3b). Fraction III was therefore separated by Sephadex G25 chromatography and the active fraction eluted as a single peak (Table 3).

Fig. 1   Growth and antimicrobial activity of cell suspension cultures of Carthamus tinctorius.

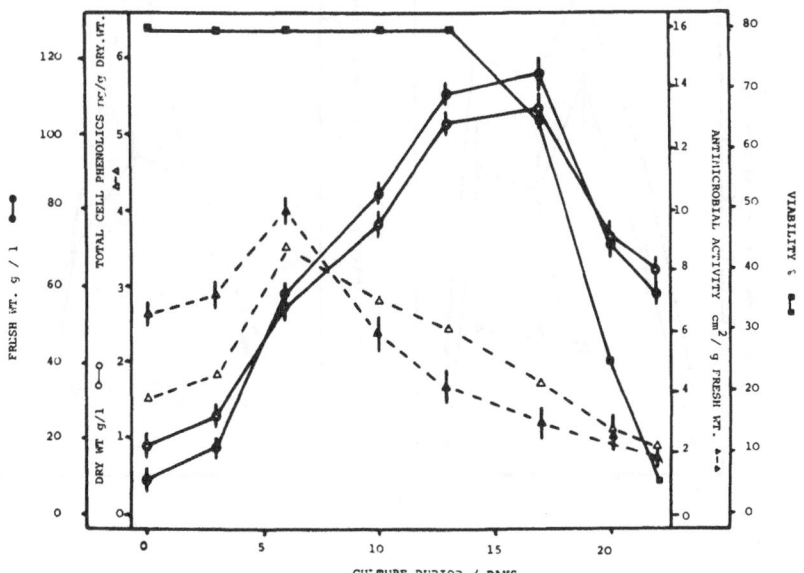

Fig. 2   Growth and antimicrobial activity of cell suspension cultures of <u>Papaver</u> <u>somniferum</u>

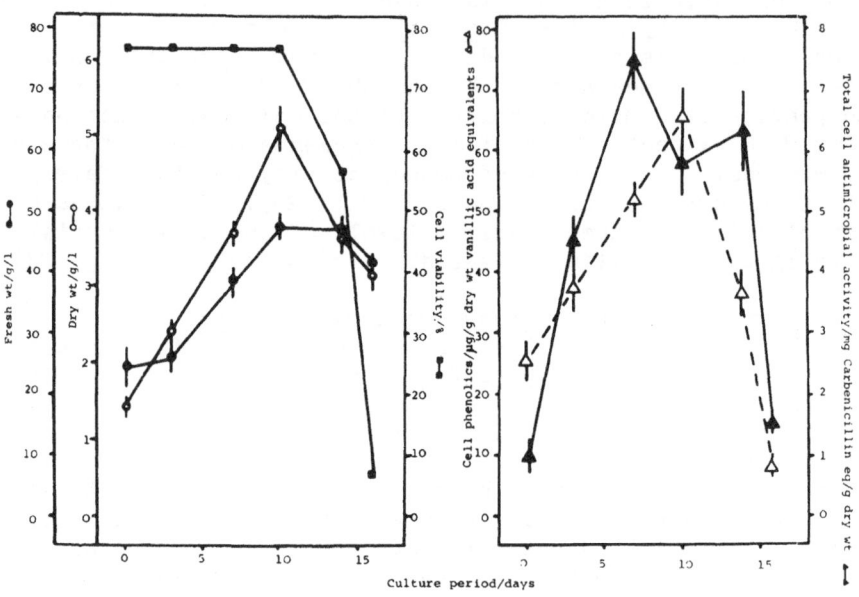

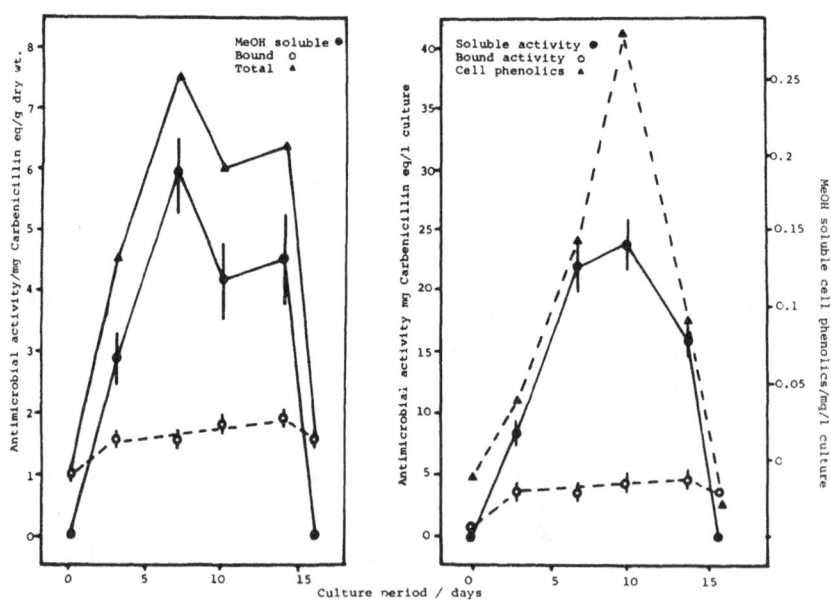

Fig. 3 Fractionation scheme

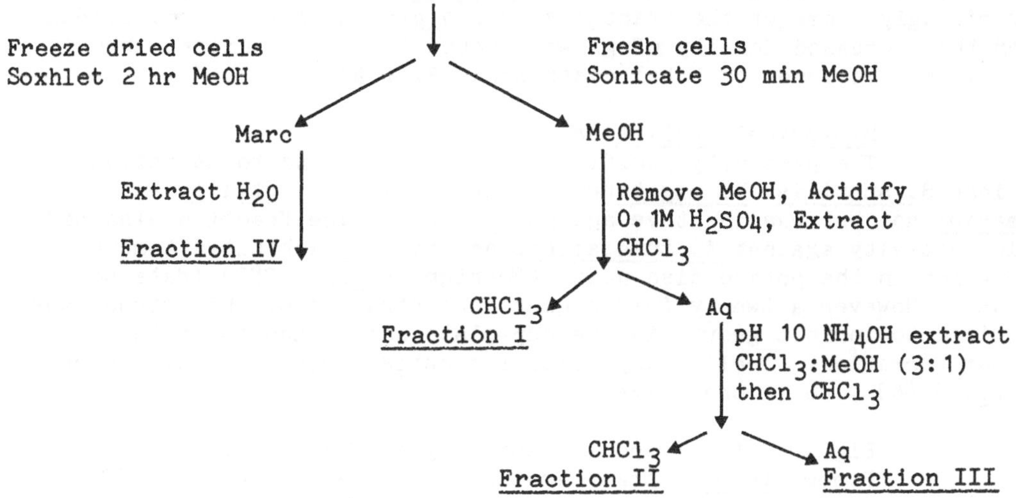

Table 3 Relative activity of fractions from P.somniferum cell cultures. Activity tested against B.subtilis. Inhibition zone in comparison to 100 μg carbenicillin (++++).

(a)

		% EXTRACT TESTED						
FRACTION	Vol EXTRACT	40	20	10	7	4	2	1
I	1 ml	++	0	0	0	0	0	0
II	1 ml	+	0	0	0	0	0	0
III	1.5 ml	/	/	++++	+++	++	+	+
IV	10.0 ml	/	/	++	+	0	0	0

FRACTION III		ACTIVITY	% EXTRACT TESTED
Dialysis	- Non Dialisable	++	5
	- Dialysable	0	100
Hydrolysis	- 1M HCl 1hr 100 °C	+	5
	- 1M NH_4Cl 1hr 100 °C	+	5
HPLC (MeOH: H_2O: 5% HOAC) (μBondapack C_{18})			
	- H_2O 5 mins	+++	4
	- 10-100 MeOH 20 mins	0	20
Sephadex chromatography (G25 2 x 50 cm)			
	- Fraction 14 (32.5-35ml)	0	25
	- Fraction 15 (35-37.5ml)	+++	25
	- Fraction 16 (37.5-40ml)	+++	25
	- Fraction 17 (40-42.5ml)	0	25

A preparative scale separation of fraction III on Sephadex and the
activity of each fraction against B.subtilis is shown in Fig. 4.
Surprisingly, some of the fractions caused growth stimulation, evident
from the increased density of growth around the well, whilst other
fractions gave zones of total clearance. (Fig. 4).

Biological activities
The partially purified fraction was found to be active
against B. subtilis, B. megaterum, S. cerevisiae, C. utilis, and
Fusarium sp. but was inactive against E. coli. The fraction also had
a low activity against A. tumefaciens and totally inhibited tumour
formation in the potato disc assay (Ferrigni et al. 1982) (data not
shown). However a twenty fold higher concentration of the extract was
required to inhibit growth in the cup-plate assay than to inhibit
tumour formation. Preliminary tests for antispindle activity (Somers
et al. 1986) were inconclusive.

Fig. 4   Antimicrobial activity of fractionated
P.somniferum extract against B.subtilis. Sephadex G25, 2
x 50 cm, eluted H_2O, 2.5 ml fractions.

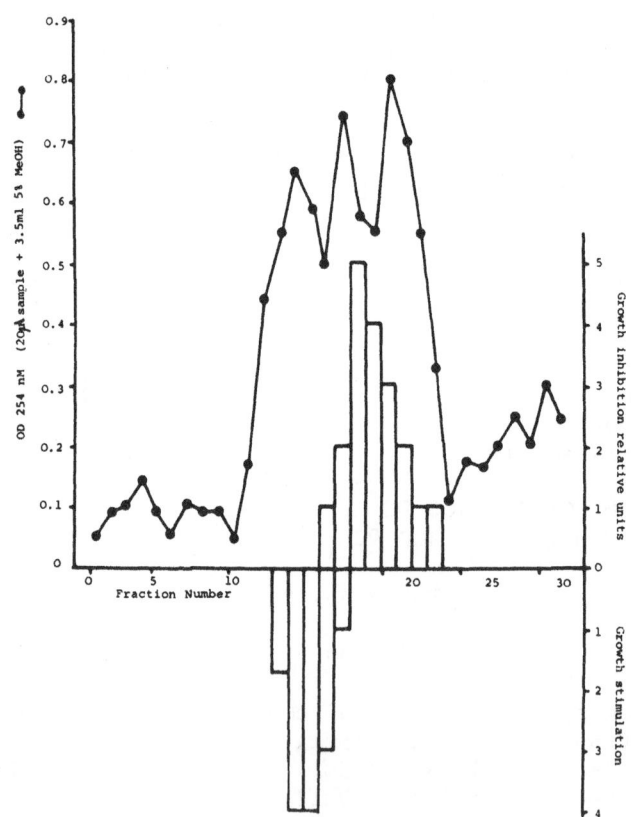

Chemical properties
Thin layer chromatography on cellulose in 6% HOAc showed the presence of four pale yellow phenolic compounds. No separation from the origin was found with other solvents on cellulose or on silica gel G. Preliminary MS analysis of the partially purified extract gave peaks at 198, 180, (100%), 168, 150 suggesting a polyhydroxylated monomer. Further mass spectral and NMR studies are currently being undertaken to determine the chemical structure of the active component.

REFERENCES
Cambell, G.C., Chan, E.C.S. & Banker, W.G. (1964). Growth of lettuce and cauliflower tissues in vitro and their production of antimicrobial metabolites. Can. J. Microbiology, 11, 785-89.

Farnsworth, N.R. (1966). Biological and phytochemical screening of plants. J. Pharm. Sci., 55, 225.

Ferrigni, N.R., Putnam, J.E., Anderson, B., Jacobson, L.B., Nichols, D.E., Moore, D.S. & McLaughlin, J.L. (1982). Modification and evaluation of the potato-disc assay and anti-tumour screening of Euphorbiaceae seeds. J. Nat. Prod. 45, 679-686.

Foster, J.W. & Woodruff, H.B. (1944) Procedure for the cup assay of Penicillin. J. Bact., 47, 43-58.

Khanna, P., Mohan, S. & Nag, T.N. (1971). Antimicrobials from plant tissue cultures. J. Nat. Prod. 30, 168-81.

Leven, M., Vanden-Berghe, D.A., Mertens, F., Vlietinck, A. & Lammers, E. (1979) Screening of higher plants for biological activities. 1. Antimicrobial activity. Planta Medica, 36, 311-21.

Mitscher, L.A., Lev, R.P., Bathala, M.S., Wu, W.N. & Beal, J.L. (1972). Antimicrobial agents from higher plants. I. Introduction, rationale and methodology. J. Nat. Products, 35, 157-65.

Misawa, M., Hayashi, M. & Takayama, S. (1985). Accumulation of antineoplastic agents by plant cell cultures. In Primary and Secondary metabolism of plant cell cultures. ed. K-H. Neumann, W. Barz & E. Reinhard. pp. 235-46 Berlin: Springer-Verlag.

Misra, S.B. & Dixit, S.N. (1979) Antifungal activity of leaf extracts of some higher plants. Acta Botanica. Indica, 7, 147-50.

Somers, A., Parry, J.M., Stafford, A. & Kelly, S.L. (1986). Detection of natural products that induce aberations of the mitotic spindle. Mutagenesis (in press).

Veliky, I.A. & Lastta, R.K. (1974). Antimicrobial activity of cultured plant cells and tissues. J. Nat. Prod., 37, 611-20.

SECTION 2

GROWTH AND DIFFERENTIATION IN RELATION TO
SECONDARY METABOLISM

6. FLAVOUR PRODUCTION IN ONION TISSUE CULTURE

H.A. Collin & D. Musker
Department of Botany, University of Liverpool,
Liverpool, L69 3BX. England

G. Britton
Department of Biochemistry, University of Liverpool,
Liverpool, L69 3BX. England

INTRODUCTION

Onion flavour is the most commonly used flavour additive
for pre-cooked and convenience foods. Although there is an easily
accessible natural source of supply, the industries concerned with
food flavours are showing an increasing interest in the use of tissue
culture technology to produce flavours, or to improve flavour
production in intact plants (Van Brunt, 1985). The composition of
onion flavour has had a great deal of attention (Whitaker, 1976), but
the biosynthesis of onion flavour compounds in the plant has had a
limited investigation. If tissue culture and gene manipulation
technology are going to be used in the improvement of onion flavour,
it is important to understand the factors controlling the accumulation
of flavour in the cells of the onion leaf and bulb. This reasoning
applies to the use of tissue cultures for the synthesis of many other
secondary plant products. Unfortunately many of the commercially
important secondary products have complex and little known pathways of
biosynthesis. Since the secondary pathway in onion is short and
involves relatively simple compounds, the onion also provides a good
model for investigating the control of secondary product synthesis in
plant tissue cultures.

PRODUCTION OF ONION FLAVOUR

Most flavours are made up of very many compounds and are
therefore difficult to reproduce synthetically or produce from plant
tissue culture. Onion flavour however is derived from four compounds,
the alkyl cysteine sulphoxides, S-methyl, S-propyl and S-ethyl-L-
cysteine sulphoxides and S-trans-prop-1-enyl-L-cysteine sulphoxide,
which is the major flavour precursor in onion.

When the onion cells are disrupted an enzyme, alliinase, is released
which hydrolyses the alkyl cysteine sulphoxides producing the volatile
lachrymatory compound, ammonia and pyruvate. Secondary reactions
produce disulphides which contribute to the milder onion odours. The
release of pyruvate in this reaction provides a method for estimating
the total amount of flavour precursors present (Schwimmer &
Guadagni, 1962) and also an assay method for measuring the activity of
the alliinase (Selby & Collin, 1976).

The biosynthesis of the flavour precursors have been examined by
Granroth (1970), who established that S-methyl, S-propyl and
S-ethyl-L-cysteine sulphoxides were derived from serine, whereas the

major flavour precursor in onion, S-trans-prop-1-enyl-L-cysteine sulphoxide was derived from valine and cysteine, following a totally different biosynthetic route (Fig. 1). When Selby & Collin (1976) examined the production of flavour in onion tissue cultures, the tissue cultures were found to contain alliinase, but no onion odour. A comparison of the alliinase from onion bulb and callus, using substrate specificity and Km values, showed that alliinase from the two sources were identical (Selby, et al., 1979). A detailed examination of the flavour precursors in the callus using electrophoresis - tlc methods for separating the compounds (Bieleski & Turner, 1966), showed the tissue cultures contained a very limited amount of S-methyl-L-cysteine sulphoxide only (Fig. 2a, b). The analysis of flavour precursors in the onion bulb and culture tissue has been confirmed by HPLC (D. Musker - personal communication).

The absence of secondary products in the onion tissue cultures is a situation which is characteristic of many other species when in culture.

Fig. 1.   Biosynthesis of the major flavour precursor in onion, S-trans-prop-1-enyl-L-cysteine sulphoxide.

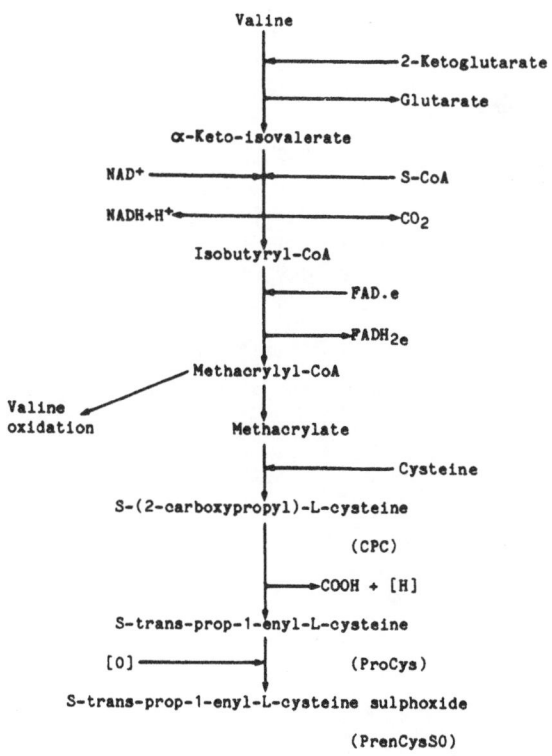

It has been suggested by Yeoman et al. (1981) that the marked change
in biochemical activity in tissue cultures is caused by changes in
the:-
1.   Relative activity of the primary and secondary pathways.
2.   Supply of intermediates to the primary and secondary pathways.
3.   Activity of the enzymes on the secondary pathways.

The effect of the supply of intermediates on the activity of the
secondary pathway leading to synthesis of s-trans-prop-1-enyl-L-
cysteine sulphoxide and the regulation of this secondary pathway in
onion have been examined.

> Fig. 2.   Thin layer separation of onion extract (a)
> showing presence of S-trans-prop-1-enyl-L-cysteine
> sulphoxide (PrenCysSO) and smaller amounts of S-methyl-
> L-cysteine sulphoxide (MeCysSO) and (b) callus extract
> showing limited amounts of S-methyl-L-cysteine
> sulphoxide.

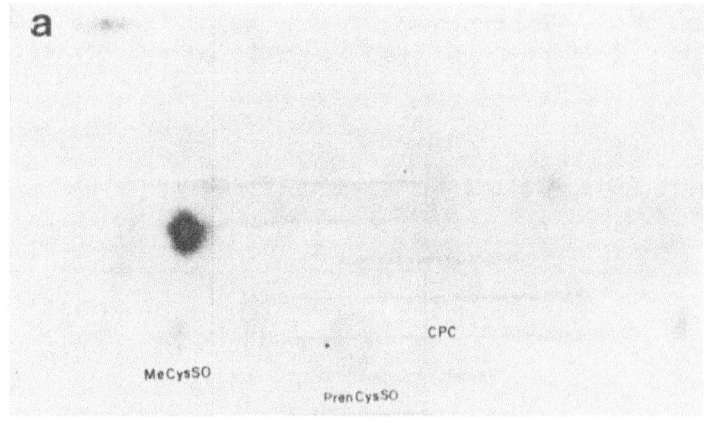

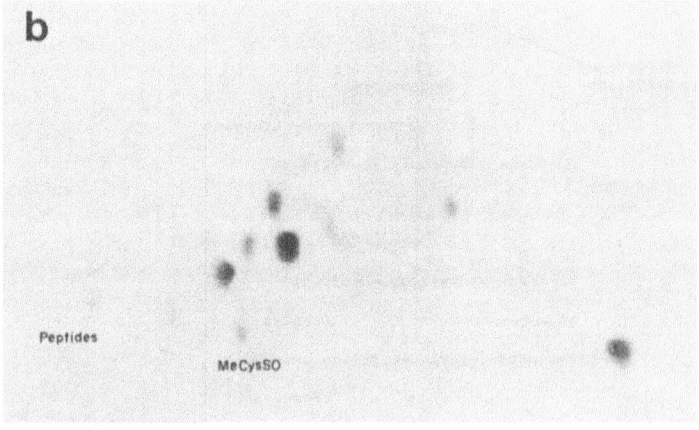

SUPPLY OF INTERMEDIATES TO FLAVOUR PRECURSOR BIOSYNTHESIS

In the intact onion the intermediates that contribute to the secondary pathway leading to synthesis of the major flavour precursor are sulphate, valine and the more direct intermediates methacrylate and cysteine. There are two approaches to examining the role of intermediates in the control of secondary product formation. The most direct approach is to supply the intermediates to the source tissue as a component of the medium then observe any increase in secondary product formation. The second approach is to examine the internal control of supply of specific intermediates. Initially the intermediates, sulphate, valine, methacrylate and cysteine were added to callus medium at concentration of up to 100mM. After 6 weeks incubation the callus had developed no detectable odour (Selby, et al., 1980). In a more detailed analysis the callus was exposed to low concentrations (1 mM and 3 mM) of cysteine and valine then after 6 weeks the callus extracted and separated by electrophoresis-tlc. No flavour compounds could be detected. Despite increasing the external supply of intermediates the production of flavour precursors was unaffected.

An alternative approach was to examine the control of synthesis of the two substrates for the secondary pathway i.e. methacrylate and cysteine. In bacterial and mammalian systems the rate limiting enzymes for the production of methacrylic acid is a-keto-isovalerate dehydrogenase (Aberhart & Tann, 1979; Ikeda & Tanaka, 1983). In animal cells this enzyme is bound to the inner mitochondrial membrane, and forms part of the route of valine oxidation. Granroth (1970) showed that the pathway from valine to methacrylate is active in onion since leaves exposed to C^{14} valine developed radioactivity in the intermediates of the flavour precursor pathway and in the precursor, trans-prop-1-enyl-L-cysteine sulphoxide. Further evidence for the existence of the pathway in plants is provided by Parry et al. (1985) who fed C^{14} isobutyrate to asparagus and found radioactivity in S-(2-carboxypropyl)-L-cysteine, which is a key intermediate in the synthesis of secondary products in asparagus. The key regulatory enzyme for the synthesis of methacrylate, a-keto-isovalerate dehydrogenase, has not been located in plants. It is important to establish whether this enzyme is present in plant mitochondria, or is located in the chloroplasts, or proplastids. Finally the site of location must be established in the callus.

The enzyme regulating cysteine production, cysteine synthase, has had much more attention in plants (Giovanelli et al., 1980). It will be more straightforward therefore to measure the activity of this enzyme in callus, and compare it with that in intact tissue. Since the bulb tissue is colourless, the enzyme may well be located in the proplastids in both intact plant tissue and in callus tissue.

BIOSYNTHESIS OF FLAVOUR PRECURSOR

The presence of the secondary pathway in onion leading to formation of trans-prop-1-enyl-L-cysteine sulphoxide was established by Granroth (1970) in a series of experiments in which C^{14} valine and

C^{14} cysteine were fed to excised leaf tips. Radioactivity appeared in the cysteine derivatives S-(2-carboxypropyl)-L-cysteine, S-trans-prop-1-enyl-L-cysteine and S-trans-prop-1-enyl-L-cysteine sulphoxide. When the precursors valine and cysteine were fed to callus no onion odour developed, or precursor was detected (Turnbull, et al. 1980). However, when intermediates on the secondary pathway, such as S-(2-carboxypropyl)-L-cysteine and S-trans-prop-1-enyl-L-cysteine and the final precursor compound S-trans-prop-1-enyl-L-cysteine sulphoxide, were fed to the callus, the flavour precursor accumulated as shown by the production of onion odour and the presence of S-trans-prop-1-enyl-L-cysteine sulphoxide (Selby, et al. 1980).

The feeding experiments of Granroth (1970) were repeated by Selby et al. (1980) using leaf tips and callus. C^{14} valine was relatively metabolically inert with little distribution of radioactivity in the flavour precursors in both shoot and callus. However, using C^{14} cysteine much more radioactivity was found in the flavour precursor in the shoot, while in the callus there was limited radioactivity in the intermediates of the secondary pathway and in the flavour precursor, S-trans-prop-1-enyl-L-cysteine sulphoxide. This distribution of radioactivity showed that the path of synthesis to S-trans-prop-1-enyl-L-cysteine sulphoxide was operating in the callus but at a much lower level than in the onion. The labelling of the major precursor was always much lower than the other precursor showing radioactivity, S-methyl-L-cysteine sulphoxide, indicating that even though the main flavour precursor was being formed, the situation in the callus was not the same as in the intact plant where the S-methyl-L-cysteine is only a minor metabolite.

Fig. 3.   Inhibition of alliinase enzyme by hydroxylamine.

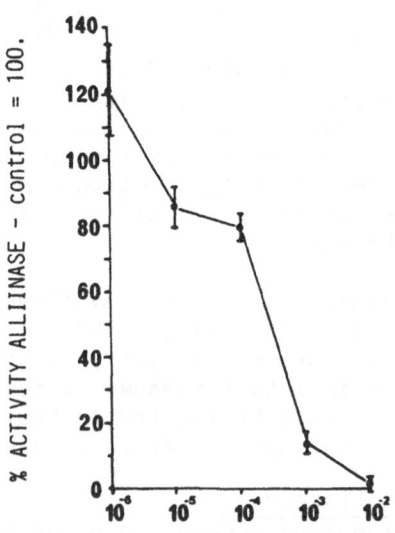

LOG$_{10}$ MOLARITY HYDROXYLAMINE

The presence of radioactivity in the S-trans-prop-1-enyl-L-cysteine
sulphoxide showed that the inhibition to the synthesis of this flavour
precursor in the callus was not complete. The fact that valine,
cysteine and methacrylic acid were unable to stimulate flavour
precursor synthesis whereas intermediates of this flavour precursor
pathway led to precursor synthesis suggested that the pathway was
inhibited at the stage of S-(2-carboxypropyl)-L-cysteine formation.
This stage of the pathway in callus needs to be examined in more
detail.

One of the major problems in attempting to measure the presence and
activity of the enzyme controlling formation of S-(2-carboxypropyl)-L-
cysteine, or any other stage on the secondary pathway, is the
degradative effect of alliinase on the final precursor. However the
reaction sequence can now be examined in vitro, since the addition of
an enzyme inhibitor, hydroxylamine, will inhibit the activity of the
alliinase (Fig. 3). The activity of all the enzymes controlling the
secondary pathway will be measured now using an extraction and assay
procedure incorporating hydroxylamine.

SITE OF SYNTHESIS AND ACCUMULATION OF FLAVOUR PRESENCE
The flavour producing system in onion requires spatial
separation of alliinase and flavour precursors to prevent a continuous
breakdown of the precursors. It is possible that there are specialized
structures in the cell which contain and therefore separate both
components. The lack of flavour precursor synthesis in the onion
callus may be due to the absence of these structures. When callus
tissue is differentiated into roots and shoots, the capacity to
produce an onion odour reappears (Turnbull et al., 1981). An
examination of the cells of the redifferentiated tissue showed the
presence of a range of vesicles in the cytoplasm (Fig. 4) which were

Fig. 4.   Onion bulb cell with vesicle (v) (x 3,000)

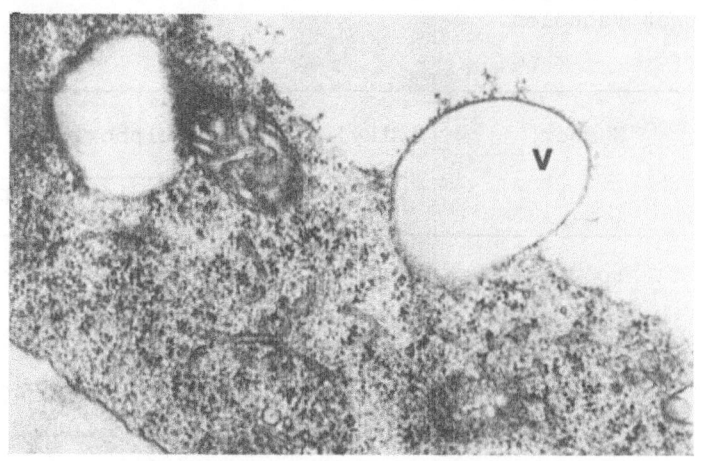

Table I -   Alliinase activity in the vacuoles, cytoplasm and protoplasts of onion. [a] Comprising contents of 12×10^7 protoplasts and 7×10^7 burst vacuoles.

Organelle	Number	Protein/ organelles (mg)	Pyruvate organelles		Alliinase specific activity (µmol/ mg protein)
			µmol	mg	
Protoplast	12×10^7	6.2	30.9	3.4	4.9
Vacuole	5×10^7	2.6	10.9	1.2	4.2
Cytoplasm	_[a]	3.5	16.3	1.8	4.6

Table II -   Amount of S-alkyl substituted cysteine sulphoxides in the vacuoles, cytoplasm and protoplasts of onion.
[a] Comprising contents of 40×10^7 protoplasts and 32×10^7 burst vacuoles.

Organelle	Number	Alkyl cysteine sulphoxide	
		µg/organelles	in protoplast %
Protoplast	40×10^7	880	100
Vacuole	8×10^7	trace (<2)	0.5-1.0
Cytoplasm	_[a]	80	10

largely absent from the undifferentiated callus tissue (Turnbull et al., 1981).   In order to locate the site of accumulation of the precursor, isolated protoplast and vacuoles were prepared from bulb tissue then analysed (Lancaster & Collin, 1981).

The vacuoles contained the same amount of alliinase as the protoplasts, whereas the flavour precursors were present in only negligible amounts in the vacuoles (Table I and II).   The conclusion made by Lancaster & Collin (1981) was that the alliinase is compartmented in the vacuole, whereas the flavour precursors are in the cytoplasm.   It is possible therefore, that the vesicles noted in the differentiated tissue are sites of synthesis, or accumulation of the flavour precursor.   This suggestion concerning the role of the vesicles can only be confirmed by micro-autoradiography using labelled precursor, or by isolating the vesicles and examining the contents.

It is important to recognize that specialised structures may be required for a secondary pathway.   These structures might not be as specialised as those found in the onion cell, but may be proplastids or chloroplasts.   In an examination of secondary product synthesis it is important to establish the site of location of the secondary pathway and site accumulation of the final product.   The reduced synthesis of the major flavour precursor in onion callus is due partly to an inhibition of synthesis of S-(2-carboxypropyl)-L-cysteine.   The enzyme responsible for this stage is currently being isolated and its activity assayed, and the factors which control the supply of methacrylate and cysteine to this reaction are also being examined further.

REFERENCES

Aberhart, D.J. & Tann, C.H. (1979).   Stereochemistry of the conversion of methacrylate to β-hydroxyisobutyrate in Pseudomonas putida.   Journal of the Chemical Society Perkin Trans. 1, 939-942.

Bieleski, R.L. & Turner, N.A. (1966).   Separation and estimation of amino acids in crude plant extracts by thin layer electrophoresis and chromatography.   Analytical Biochemistry, 17, 278-293.

Giovanelli, J., Mudd, S.H. & Datko, A.H. (1980).   Sulfur amino acids in plants.   In Amino acids and derivatives.   ed. B.J. Miflin, pp. 453-505.   The Biochemistry of Plants, Vol. 5. eds. P.K. Stumpf & E.E. Conn, London, New York, Academic Press.

Granroth, B. (1970).   Biosynthesis and decomposition of cysteine derivatives in onion and other Allium species.   Annales academiae scientiarum fennica, Ser. AZ 154, 1-71.

Ikeda, Y. & Tanaka, K. (1983).   Purification and characterisation of isovaleryl CoA dehydrogenase from rat liver mitochondria. Journal Biological Chemistry 258, 1077-1085.

Lancaster, J.E. & Collin, H.A. (1981).   Presence of alliinase in isolated vacuoles and of alkyl cysteine sulphoxides in the cytoplasm of bulbs of onion (Allium cepa).   Plant Science Letters, 22, 169-176.

Parry, R.J., Mibusawa, A.E., Chiu, I.C., Naidu, M.V. & Ricciardone, M. (1985). Biosythesis of sulfur compounds. Investigations of the biosynthesis of asparagusic acid. Journal of American Chemical Society, 107, 2512-2521.

Selby, C. & Collin, H.A. (1976). Clonal variation in growth and flavour production in tissue cultures of Allium cepa L. Annals of Botany, 40, 911-918.

Selby, C., Galpin, I.J. & Collin, H.A. (1979). Comparison of the onion plant (Allium cepa) and onion tissue culture. I. Alliinase activity and flavour precursor compounds. New Phytologist, 83, 351-359.

Selby, C., Turnbull, A. & Collin, H.A. (1980). Comparison of the onion plant (Allium cepa) and onion tissue culture. II. Stimulation of flavour precursor synthesis in onion tissue cultures. New Phytologist, 84, 307-312.

Schwimmer, S. & Guadagni, D.G. (1962). Relation between olfactory threshold concentration and pyruvic acid in onion juice. Journal of Food Science, 27, 94-97.

Turnbull, A., Galpin, I.J. & Collin, H.A. (1980). Comparison of the onion plant (Allium cepa) and onion tissue culture. III. Feeding of C^{14} labelled precursors of the flavour precursor compounds. New Phytologist 85, 483-487.

Turnbull, A., Galpin, I.J., Smith, J.L. & Collin, H.A. (1981). Comparison of the onion plant (Allium cepa) and onion tissue culture. IV. Effect of shoot and root morphogenesis on flavour precursor synthesis in onion tissue culture. New Phytologist, 87, 257-268.

Van Brunt, J. (1985). Nibbling at the flavour market. Biotechnology, 3, 525-538.

Whitaker, J.R. (1976). Development of flavour odour and pungency in onion and garlic. Advances in Food Research, 22, 73-133.

Yeoman, M.M., Lindsey, K., Miedzybrodyka, M.B. & McLauchlan, W.R. (1981). Accumulation of secondary products as a facet of differentiation in plant cell and tissue cultures. In Differentiation in vitro 4th Symposium British Society Cell Biology, ed. M.M. Yeoman & D.E.G. Truman, pp. 65-81. Cambridge: Cambridge University Press.

7. <u>KINETICS OF GROWTH AND ALKALOID ACCUMULATION IN</u>
<u>CATHARANTHUS ROSEUS CELL SUSPENSION CULTURES</u>

P. Morris

Wolfson Institute of Biotechnology,
University of Sheffield,
SHEFFIELD   S10 2TN. U.K.

INTRODUCTION
        Despite the large amount of work which has been carried
out on alkaloid biosynthesis in cell cultures of <u>Catharanthus roseus</u>
(Zenk <u>et al</u>., 1977, Kurz <u>et al</u>., 1980, Knoblock & Berlin, 1980, Kohl
<u>et al</u>., 1981, Merillion <u>et al</u>., 1984) little information is available
on the growth and product accumulation kinetics of different culture
systems. In order to be able to manipulate the productivity of cell
cultures in multi-litre vessels, detailed comparisons of the growth
and product relationships of cells grown under different conditions
are required. In this paper the kinetics of growth and alkaloid
accumulation in cell cultures of <u>Catharanthus roseus</u> grown on B5
maintainance medium, on Zenks production medium and on M_3 production
medium in small scale (100ml) cultures are reported.

Growth and Product Kinetics on Maintenance Medium
        Cell suspension cultures (cell line C11C) routinely
maintained on Gamborg's B5 medium with 1 mg l^{-1} 2,4-D, 0.1 mg l^{-1}
kinetin and 2% sucrose for 4 years as described previously (Morris &
Fowler, 1980) were used in this study. Growth and alkaloid production
in these cells is shown in Fig. 1.

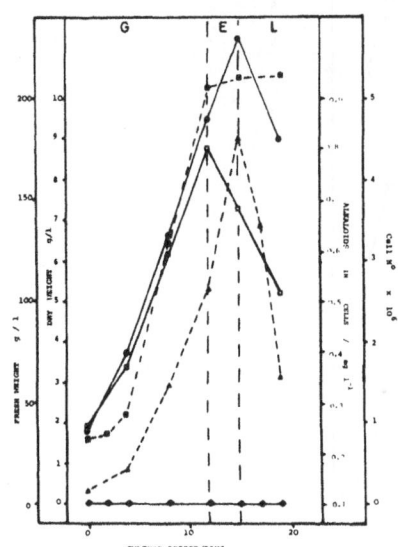

Fig. 1.   Kinetics of growth and
serpentine accumulation in the cell
line C11C maintained on B5 medium for
48 subcultures. Cultures maintained at
25°C in 250ml flasks. Fresh wt (●),
dry wt (□), cell number (■),
serpentine (▲), ajmalicine (◆).
G - growth phase, E - expansion phase,
L - lysis phase.

Under these conditions growth (fresh wt., dry wt. and cell number) was balanced with a specific growth rate of 0.21 days^{-1}. Alkaloid levels were low (< 1 mg l^{-1}), restricted to serpentine only, and were uncoupled from growth. This cell line was used for investigating growth and alkaloid accumulation on different production media.

Growth and Product Kinetics on Production Medium
Zenks production media
The above cell line (CIIC) maintained on 2,4-D containing B5 medium when transferred to Zenk's production medium containing 5% sucrose exhibited unbalanced growth in relation to fresh and dry weight increases (μ dry wt = 0.18, μ fresh wt = 0.083 d^{-1}) but high levels of alkaloid accumulation (18 mg l^{-1} ajmalicine at 20 d and 35 mg l^{-1} serpentine at 40 d). Ajmalicine was accumulated during the growth phase. These cells also had an increased capacity to accumulate starch and this gave rise to a long stationary phase during which serpentine was the major alkaloid accumulated (Fig. 2).

Addition of 2,4-D to this production media caused an inhibition of serpentine yield, a decline in the rate of serpentine accumulation and an increase in the lag phase before serpentine accumulation began

Fig. 2.   Kinetics of growth and alkaloid accumulation in cell line C11C transferred to Zenks production medium. Fresh wt. (○), dry wt. (□), serpentine (△), ajmalicine (▲).

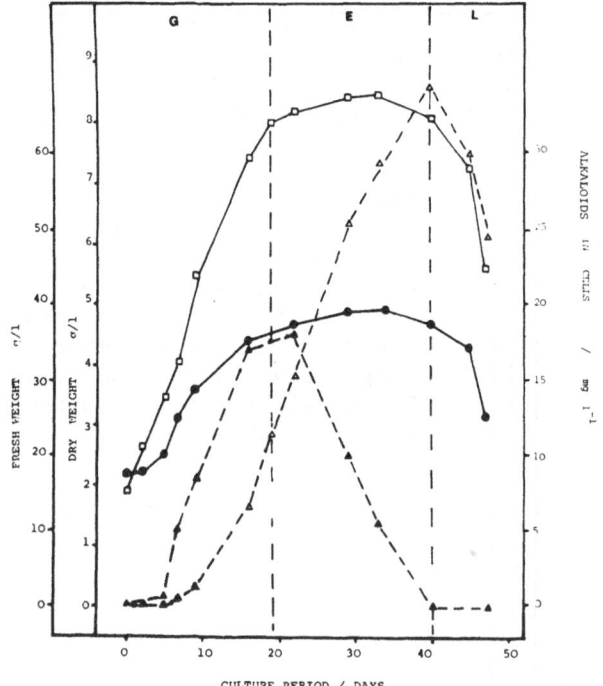

CULTURE PERIOD / DAYS

(Fig. 3). However, 2,4-D addition had little effect on growth rates. The presence of 2,4-D therefore drastically affected the growth/product relationship on this medium. As the cells were grown on 2,4-D containing medium prior to transfer to production medium (Fig. 2), carry-over of 2,4-D may have caused some inhibition of serpentine accumulation. The effect of 2,4-D carry over is shown in the control (Fig. 3) where cells were washed at low pH for several hours prior to inoculation into production medium. Under these conditions no lag phase was observed, maximum alkaloid yields increased from 2.0 to 6.0 mg g dry wt^{-1} serpentine, and serpentine accumulation became coupled to dry weight increase.

Production medium M3

There are several disadvantages to the two-stage system described above. Firstly, cells do not survive on production medium when subcultured (Morris, 1986), secondly, 2,4-D removal is difficult to achieve when working with large culture volumes and thirdly long culture periods are required to reach maximum alkaloid yields. As the important parameter for productivity of a cell culture is the amount of product accumulated / 1 culture / day, high yielding cultures (in terms of amount of product / g dry wt) can be less productive than lower yielding cultures, if the rate of accumulation of product or the final biomass yields are lower. Thus a cell culture which produced perhaps lower maximum alkaloid levels but which has higher rates of product accumulation, a shorter culture cycle and which has growth associated alkaloid accumulation kinetics may be more productive and could be useful for mass cultivation.

Such a system is shown in Fig. 4. Cell line CIIC grown on B5 medium when transferred to M&S medium containing 1 mg l^{-1} NAA, 0.1 mg l^{-1} kinetin and 2% sucrose was found to give a sustainable cell culture which has maintained its alkaloid productivity for over 2 years (see paper 34). In these cells growth was balanced (μ max 0.28 d^{-1} at 25°C) with respect to fresh wt, dry wt and cell number increases and

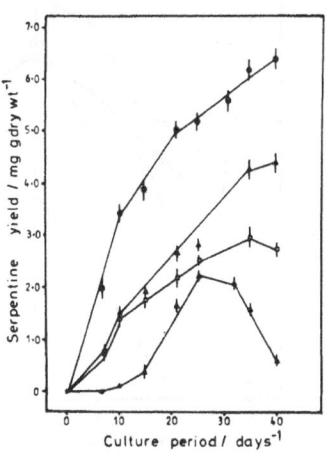

Fig. 3.   Effect of added 2,4-D on serpentine accumulation in cell line C11C grown on Zenks production medium. Control (●), 0.01 mg l^{-1} 2,4-D (△), 0.1 mg l^{-1} 2,4-D (□), 1.0 mg l^{-1} 2,4-D (▲). Cells washed for 2 hours at pH 4 prior to inoculation into production medium.

Fig. 4.   Kinetics of growth and alkaloid accumulation at 25°C in the cell line C87 maintained on M$_3$ production medium for 28 subcultures. Fresh wt. (▲), dry wt. (■) cell number (□), serpentine (●), ajmalicine (○), cell viability (■).

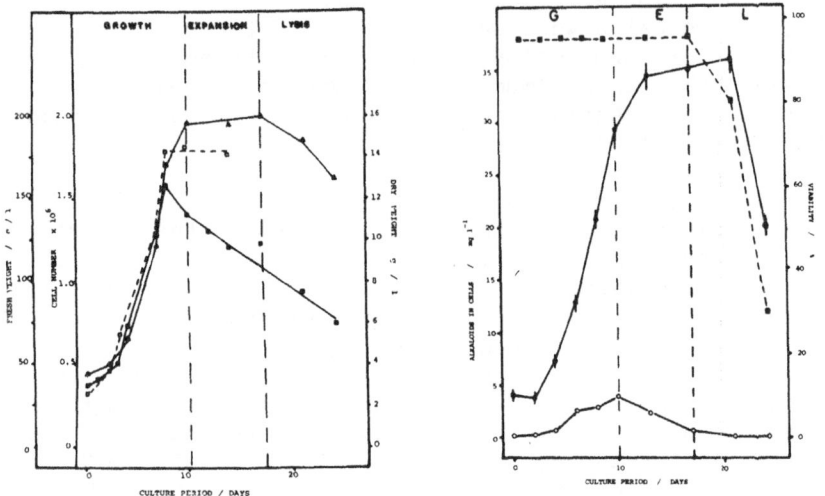

Fig. 5.   Kinetics of growth and alkaloid accumulation in the C87 cell line on M$_3$ medium at (a) 35°C, (b) 15°C. Fresh wt. (●), dry wt. (○), serpentine (△), ajmalicine (▲).

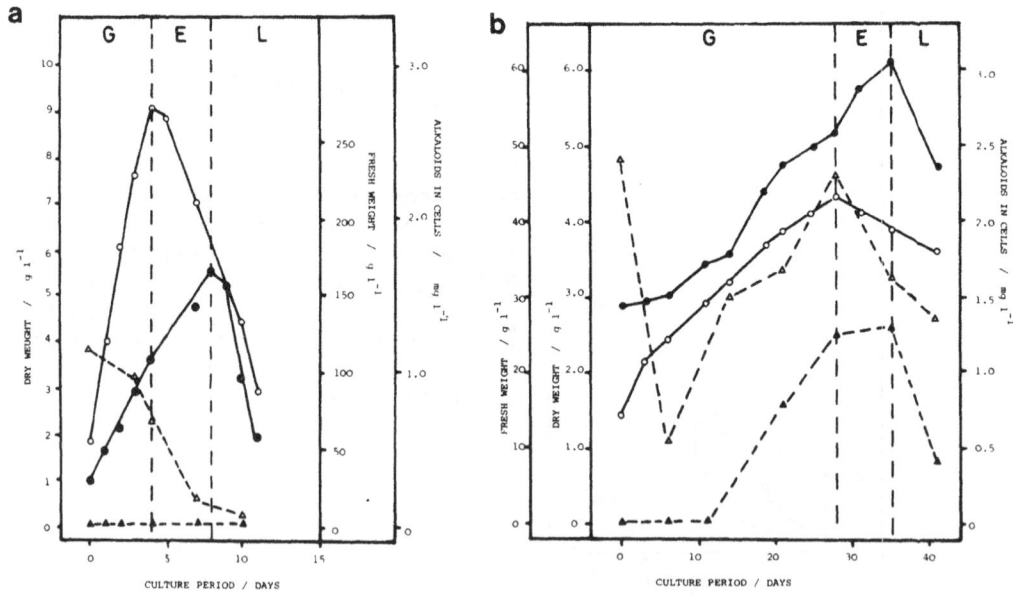

high serpentine yields were found (2 mg/g dry wt). Serpentine
accumulation was coupled to growth with rapid accumulation during the
growth phase and before cell expansion.

These growth and production kinetics were however very sensitive to
culture temperature (see paper 34). This is illustrated in Fig. 5
where the kinetics of growth and alkaloid accumulation at 15°C and
35°C are compared (cf Fig. 4 at 25°C). At low temperatures
growth and alkaloid production were suppressed, although they remained
coupled. However, the serpentine to ajmalicine ratio was greatly
reduced (Fig. 5a). At high temperatures fresh wt and dry wt increases
were unbalanced and no serpentine or ajmalicine accumulation was found
(Fig. 5b). Thus in this system the exact relationship between growth
and alkaloid accumulation was dependant upon culture temperature.

CONCLUSION
It can be concluded from these studies that the kinetics
of growth and alkaloid accumulation are not fixed at least in
Catharanthus cultures and can be manipulated by varying culture
conditions. It is important therefore when scaling-up cultures into
larger vessels to appreciate that changing cultural conditions may
well alter the growth / production kinetics. Furthermore higher
productivity in cultures may be achieved by choosing a system with the
optimal growth and product accumulation kinetics.

REFERENCES
Kohl, W., Witte, B. & Hofle, G. (1981). Alkaloids from Catharanthus
      roseus tissue culture. Z. Natforsch, 37, 1346-51.
Knoblock, K-H. & Berlin, J. (1980). Influence of medium composition
      on the formation of secondary compounds in cell suspension
      cultures of Catharanthus roseus (L) G. Don. Z.
      Naturforsch., 35, 551-56.
Kurz, W.G.W., Chatson, K.B., Constable, F., Kutney, J.P., Choi,
      L.S.L., Kolodziejczyk, P., Sleigh, S.K., Stuart, K.L. &
      Worth, B.R. (1980). Alkaloid production in Catharanthus
      roseus cell cultures: initial studies on cell lines and
      their alkaloid contents. Phytochem., 19, 2583-87.
Merillon, J.M., Rideau, M. & Chenieux, J.C. (1984). Influence of
      sucrose on levels of ajmalicine and tryptamine in
      Catharanthus cells in vitro. Planta Medica, 50, 497-501.
Morris, P. & Fowler, M.W. (1980). Sucrose utilization by cell
      suspension cultures of Catharanthus roseus G. Don.
      Biochem. Soc. Trans., 8, 638-9.
Morris, P. (1986). Regulation of product synthesis in cell cultures
      of Catharanthus roseus. II   Comparison of production
      media. Planta Medica (in press).
Zenk, M.H., El Shagi, H., Arens, H. & Stockigt, J. (1977). Formation
      of the indole alkaloids serpentine and ajmalicine in cell
      suspension cultures of Catharanthus roseus.In Plant Tissue
      Culture and its Biotechnological Applications (ed. W.
      Barz, E. Reinhard & M.H. Zenk) pp27-43. Berlin. Springer-
      Verlag.

8. DIFFERENTIATION AND MONOTERPENE BIOSYNTHESIS IN PLANT CELL CULTURES

J.T. Brown & B.V. Charlwood

Department of Biology
King's College London,
London.   WC2R 2LS   U.K.

The use of plant cell cultures for the production of monoterpenes has long been considered an attractive solution to the problems and uncertainties encountered with traditional cropping techniques (Schuler et al., 1984). It has been suggested (Zenk 1978) that, since plant cells are totipotent, all of the necessary genetic and physiological potential for natural product formation should be present in an isolated cell. According to this theory, cultured cells obtained from any part of a plant might be expected to yield secondary compounds similar to those of the plant grown in vivo. In practice, although a wide range of secondary metabolites have been successfully produced in plant cell cultures, in only a few cases has production of the desired components been comparable to that of the parent plant (Fowler 1983).

It has often been suggested that the failure to produce secondary compounds in culture is in some way linked to the level of differentiation exhibited by the cultured cells, and that a degree of differentiation is required for the expression of the genes associated with secondary metabolism (Becker 1970). This may be particularly relevant in the case of monoterpene production since, in the intact plant, these and other lipophilic compounds tend to be secreted into specialised cells or tissues. Such accumulation sites are found in Pelargonium, Mentha and Cannabis where oil is secreted into glandular hairs, in Pinus and Thuja, where monoterpenes are contained within resin ducts, and in lemon pericarp tissue in which essential oils are secreted into schizogenous glands.

The members of the genus Pelargonium are unique with respect to the wide spectrum of monoterpenes that they produce and hence provide an ideal model system in which to investigate the production of monoterpenes by plant cell cultures. Table 1 lists the main monoterpene components of a range of the Pelargonium variants which are currently being studied in our laboratory. We have developed techniques (Brown & Charlwood 1986a) which facilitate the initiation and maintenance of Pelargonium tissue cultures. No monoterpenes were detected when pale and friable callus, or fine suspension cultures were grown on maintenance medium (Murashige & Skoog medium supplemented with benzylaminopurine [BAP; 5 mg/l], naphthalene acetic acid [NAA; 1 mg/l] and sucrose [3%]). The lack of production of monoterpenes in these cultures may have been due to either a loss in

genetic ability or to a repression of the relevant genes under the
culture conditions. The first possibility could be investigated by
regenerating plants from non-productive tissue and assessing the
capacity of the regenerants for monoterpene accumulation. In order to
facilitate the precise control of both morphological and cellular
differentiation, initial experiments involving the manipulation of
plant growth regulators were carried out. Figure 1 represents the
mean response of callus tissue from 10 Pelargonium variants incubated
on media with varying supplements of BAP and NAA (Brown & Charlwood
1986a). Our results indicate that the most efficient regeneration
medium is MS supplemented with BAP [0.5 mg/l], NAA [0.05 mg/l] and
sucrose [3%]. In each case where plant regeneration was achieved, the
regenerated plants were capable of synthesising monoterpenes with
product profiles that were very similar, both qualitatively and
quantitatively, to those of the parent plants when cultivated under
identical environmental conditions. Figure 2 shows the monoterpene
profiles by glc of Pelargonium crispum (a) parent plant, (b) callus
maintained on 'maintenance medium' and (c) regenerated plant. This
confirms the views of Bohm (1972) that the lack of accumulation of
secondary products in plant cell cultures is not generally due to a
loss or mutation of the relevant genes.

Organogenesis has often been used as a means of inducing monoterpene
production in tissue cultures which fail to produce when they are
maintained in a morphologically undifferentiated conditions. Studies
on Centranthus macrosiphon, C.ruber and Valeriana officinalis (Violon

Table 1   The major components of some Pelargonium
           variants.

Variant	Monoterpene
P. australe	1,8-cineol, myrtenol
P. citriodorum	α-terpineol, citronellol
P. crispum	α-terpineol, borneol, citral
P. echinatum	1,8-cineol
P. filifolium	1,8-cineol, thujone
P. fragrans	α-terpinene, α-pinene, β-pinene, limonene, neomenthol, carvone, isomenthol
P. graveolens	citronellal, citronellal, fenchol, geraniol
P. quercifolium	myrcene, geranyl acetate
P. radula	borneol, geranyl acetate, citronellal
P. tomentosum	phellandrene, menthone, isomenthone, neomenthol
P. "Attar of Roses"	geranyl acetate, citronellal, borneol
P. "Chocolate tomentosum"	phellandrene, limonene, menthone, isomenthol, isomenthone, menthyl acetate, menthol
P. "Crispum variegatum"	1,8-cineol, borneol, citral
P. "Duke of York"	α-terpineol
P. "Joy Lucille"	citronellal, 1,8-cineol
P. "Lillian Pottinger"	geranyl acetate, carvone, pulegone, lavandulyl acetate
P. "Mabel Grey"	borneol, geranyl acetate

<u>et al</u>. 1984) showed that valepotriate production could be increased
to the levels found in the parent plant by the induction of root
morphogenesis in callus cultures.   The increased production could be
correlated to the formation of intracellular oil containing vesicles
in the newly formed roots. These droplets were shown to be closely
associated with the vacuolar membrane of the cell (Violon <u>et al</u>.
1983).   Cytochromes P-450 and b5 are thought to be involved in the
biosynthesis of valepotriates and their levels have been monitored
(Violon & Vercruysse 1985) in both non-producing callus cultures and
producing, root-differentiated cultures.   A biochemical link between
the differentiation of roots and valepotriate production was revealed
with very high levels of cytochrome P-450 present in the root
differentiated cultures but little or none in callus cultures of the
same age.   It was concluded that this probably reflects one of the
underlying biochemical differences between the two morphological
conditions.   The effects of morphogenesis on monoterpene production
have also been studied in <u>Mentha piperita</u> (Paupardin 1976).
'Primitive' tissue (ie. callus tissue containing developing shoot
buds) was found to contain higher amounts of oil than undifferentiated
callus tissue.   There was, however, a difference in the major
components produced in that the primitive tissue synthesised mainly
menthol and menthone whereas the callus produced mainly pulegone and
menthofuran.   It was noted that the buds of the primitive tissue gave
rise to rudimentary leaves endowed with secretory hairs which are the
characteristic monoterpene accumulation sites in the Labiates.   The
effect of increasing light from 400 to 4000 lux was to stimulate the
oil content by 5-fold but this was found to be due to an increase in
activity in the secretory hairs but not in their number.

Fig. 1   The control of differentiation in Pelargonium
callus cultures.

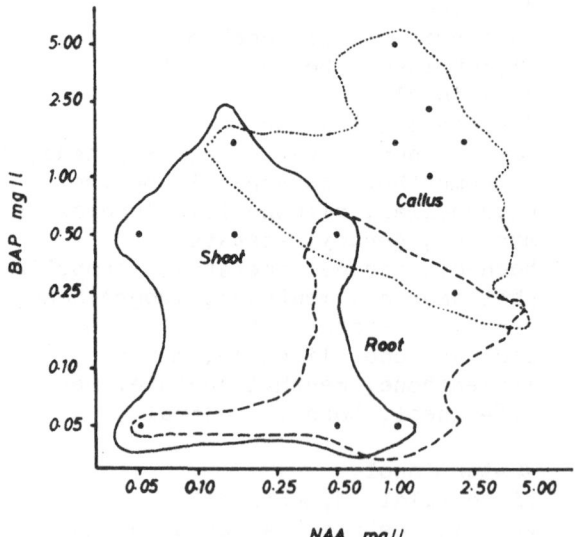

Monoterpene accumulation in tissue cultures has also been linked to
the presence of isolated specialised cells and tissues rather than
complete organogenesis. Thus, studies using lemon (<u>Citrus limonia</u>)
pericarp cultures (Paupardin 1974) showed that whilst oil pockets
remained in the cultured tissue monoterpene accumulation continued,
but as these schizogenous glands disappeared on extended subculture
monoterpene accumulation diminished accordingly. On the other hand,
callus cultures of lemon pericarp contained oil in concentrated
islands of parenchyma tissue in the heart of the callus. Schizogenous
glands are formed by cell senescence caused by a build-up of toxic
product. On lysis, such cells release their contents and affect their
neighbours thus continuing the process. True glands may not have been
observed in the callus system since monoterpene accumulation was low.

Organogenesis and differentiation of distinct tissue types are not,
however, pre-requisites of monoterpene accumulation. Callus cultures
of <u>Mentha piperita</u> (Kireeva <u>et al</u>., 1978) showed monoterpene
production in giant parenchyma cells within the callus. Although the
level of production was comparable to that of the plant, there was a
variation in the main components of the oil. A study of six <u>Mentha</u>
species (Bricout <u>et al</u>., 1978) confirmed that callus cultures have the
ability to produce monoterpenes without the presence of defined
structures. Further, callus and suspension cultures of <u>Perilla</u>

Fig. 2   Comparison of monoterpene content in plant,
callus and regenerated plant of <u>Pelangonium crispum</u>.

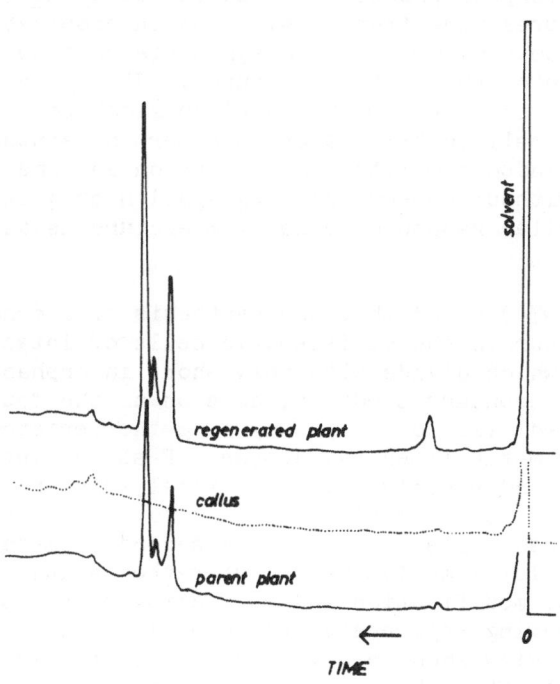

frutescens (Nabeta et al., 1983) produced oil with major components identical to those of the parent plant. Monoterpene synthesis has also been reported in parenchyma cells of Rosa damascena (Kireeva et al., 1977).

The monoterpene limonene is a flavour component of celery (Apium graveolens). The production of limonene and other flavour components can be induced by the formation of aggregates in celery suspension cultures (Watts et al., 1984). Similar induction has been observed in cultures of Pelargonium (Brown and Charlwood, 1986b) and in Thuja (Witte et al., 1983). In the latter example there was no correlation between aggregate size and productive capability. In all cases the aggregates exhibited no sign of either organogenesis or specialised tissue formation.

It is suggested that aggregation induces monoterpene production by effecting a change in the environmental conditions experienced by the cells. It may be that this arrangement is closer to that experienced by cells of the intact plant, a possibility that has been fully discussed by Yeoman et al (1982). Chloroplasts may play an important role in terpenoid biosynthesis (Wooding & Northcote, 1965) and it has been noted that greening of aggregates often accompanies monoterpene synthesis (eg. in celery cultures - Watts et al., 1985). However, this was not the case with Thuja cultures in which the aggregates were dark brown in appearance. Further, studies in our laboratory using P. fragrans have shown that greening is not a requirement for monoterpene production. Banthorpe et al (1972) showed that when tansy cultures were treated with the chloroplast inhibitors, actidione and streptomycin, monoterpene synthesis continued at levels comparable with those of the control cultures. It may be that greening is purely coincidental with the biochemical complexity necessary within the cell to bring about monoterpene synthesis. Indeed, cold stress induces monoterpene formation in fine suspension cultures of celery without concomitant aggregation or greening (Watts et al, 1984). A similar response to low temperature is also evident in the intact plant.

Constabel et al., (1974) noted that the synthesis of secondary products does not occur in the meristematic cells of intact plants. Whilst these cells, which divide with only short interphases, do not normally accumulate secondary products, as soon as the daughter cells become 'differentiated' (ie. develop away from the meristematic condition) secondary metabolites may appear. Fast growing, pale, friable callus cells and rapidly dividing, finely-dispersed suspension cultures are both held in the meristematic state. However, Constabel suggested that cultured cells could produce secondary metabolites when they ceased to be strictly meristematic but rather acquired a certain degree of biochemical modification and maturation of the type exhibited in slow-growing, aggregate and immobilised cultures. Stange (1984) restates this view when he suggests that cells can only exhibit specialised functions after there has been a diversion away from the cell proliferation cycle.

It can be concluded, therefore, that differentiation is necessary for
monoterpene accumulation in plant cells in culture, provided that
differentiation is defined in terms of a movement away from
meristematic activity towards cell maturation at the biochemical
level. Although fine, fast growing suspension cultures and friable
callus cultures facilitate a rapid increase in biomass, they offer
little in the way of opportunity for cells to become biochemically
mature.

REFERENCES

Banthorpe, D.V. & Wirz-Justice, A. (1972). Terpene biosynthesis VI:
    Monoterpenes and carotenoids from tissue cultures of
    Tanacetum vulgare L. J. Chem. Soc. Perkin Trans. I, 1769-
    1772.

Becker, H. (1970). Studies on the formation of volatile substances in
    plant tissue cultures. Biochem. Physiol. Pflanzen, 161,
    425-441.

Bohm, H. (1972). The inability of plant cell cultures to produce
    secondary substances. Proc. 5th Int. Cong. Plant Tissue
    Cell Culture, 325-328.

Bricout, J., Garcia-Rodriguez, M-J., Paupardin, C. & Saussay, R.
    (1978). Biosynthese de composes monoterpeniques par les
    tissus de quelques especes de Menthes cultivees in vitro.
    C. R. Acad. Sc. Paris. Ser. D., 287, 611-613.

Brown, J.T. & Charlwood, B.V. (1986a). The control of callus
    formation and differentiation in scented Pelargoniums. J.
    Plant Physiol., 123, 409-17.

Brown, J.T. & Charlwood, B.V. (1986b). Studies on the production of
    monoterpenes in tissue cultures of scented Pelargoniums.
    (in preparation).

Constabel, F., Gamborg, O.L., Kurz, W.G.W. & Steck, W. (1974).
    Production of secondary metabolites in plant cell
    cultures. Planta Medica, 25, 158-168.

Fowler, M.W. (1983). Commercial applications and economic aspects of
    mass plant cell culture. In Plant Biotechnology, ed. S.H.
    Mantell and H. Smith, pp.3-37. Cambridge: Cambridge
    University Press.

Kireeva, S.A., Bugorskii, P.S. & Reznikova, S.A. (1977). Cultivation
    of Damask rose tissues and accumulation of terpenoids in
    them. Fiziol. Rast., 24, 824-831.

Kireeva, S.A., Mel'nikov, V.N., Reznikova, S.A. & Meshcheryakova, N.I.
    (1978). Essential oil accumulation in a peppermint callus
    culture. Fiziol. Rast., 25, 564-570.

Nabeta, K., Ohnishi, Y., Hirose, T. & Sugisawa, H. (1983). Monoterpene
    biosynthesis by callus tissues and suspension cells from
    Perilla species. Phytochem., 22, 423-425.

Paupardin, C. (1974). Sur l'evolution de l'huile essentiele dans des
    tissus de fruits de Citron (Citrus limonia Osbeck)
    cultives in vitro dans diverses conditions. Rev. Gen.
    Bot., 81, 223-241.

Paupardin, C. (1976). On the differentiation of secreting tissue and
        the formation of essential oil by plant tissues cultivated
        in vitro. C. R. Congr. Natl. Soc. Savant Sci. Lille, 1,
        619-628.
Shuler, M.L., Pyne, J.W. & Hallsby, G.A. (1984). Prospects and
        problems in the large scale production of metabolites from
        plant cell tissue cultures. J. Amer. Oil Chem. Soc., 11,
        1724-1728.
Stange, L. (1984). Cellular interactions during early differentia-
        tion. In Encyclopedia Plant Physiol., ed. H.F. Linskens
        and J. Heslop-Harrison, vol. 17, pp.424-452. Berlin:
        Springer-Verlag.
Violon, C., Dekegel, D. & Vercruysse, A. (1983). Microscopical study
        of valepotriates in liquid droplets of various tissues
        from Valerian plants. Plant Cell Reports, 2, 300-303.
Violon, C., Dekegel, D. & Vercruysse, A. (1984). Relation between
        valepotriate content and differentiation level in various
        tissues from Valerianeae. J. Nat. Prod., 47, 934-940.
Violon, C.J.I. & Vercruysse, A.A. (1985). Haemcytochromes in
        valepotriate producing tissue cultures of Centranthus
        macrospiphon. Phytochem., 24, 2205-2209.
Watts, M.J., Galpin, I.J. & Collin, H.A. (1984). The effect of growth
        regulators, light and temperature on flavour production in
        celery tissue cultures. New Phytol., 98., 583-591.
Watts, M.J., Galpin, I.J. & Collin, H.A. (1985). The effect of
        greening on flavour production in celery tissue cultures.
        New Phytol., 100, 45-56.
Witte, L., Berlin, J., Wray, V., Schubert, W., Kohl, W., Hofle, G. &
        Hammer, H. (1983). Mono- and diterpenes from cell
        cultures of Thuja occidentalis. Planta Medica, 49, 216-
        221.
Wooding, F.B.P. & Northcote, D.H. (1965). The fine structure of the
        mature resin canal cells of Pinus pinea. J. Ultrastruct.
        Res., 13, 233-244.
Yeoman, M.M., Lindsey, K., Miedzybrodzka, M.B. & McLauchlan, W.R.
        (1982). Accumulation of secondary products as a facet of
        differentiation in plant cell and tissue cultures. In
        Differentiation in vitro: Brit. Soc. Cell Biol. Symp. 4,
        ed. M.M. Yeoman and D.E.S. Truman, pp.65-82. Cambridge:
        Cambridge University Press.
Zenk, M.H. (1978). The impact of plant cell culture on industry. In
        Frontiers of Plant Tissue Culture, ed. T.A. Thorpe, pp.1-
        13. Calgary: University of Calgary.

9. THE EFFECT OF OSMOTIC STRESS ON GROWTH AND ALKALOID ACCUMULATION IN CATHARANTHUS ROSEUS

K. Rudge & P. Morris

Wolfson Institute of Biotechnology,
University of Sheffield,
Sheffield   S10 2TN. U.K.

INTRODUCTION
The _in vitro_ culture of plant cells by definition provides artificial conditions for plant growth and it is possible that the imposition of such conditions may induce stresses which contribute to changes in the levels of secondary metabolite formation. A study of the effects of stress on plant cell growth may therefore provide information concerning the general nature of stress response in cultured plant cells and, in particular, highlight the sensitivity of secondary product metabolism to environmental changes.

Osmotic stress was chosen for this investigation as it is well documented, at least in whole plants, as having far reaching effects on primary metabolism (Hanson & Hitz, 1982). In cultured plant cells the majority of research has concentrated on the imposition of osmotic stress in order to induce drought tolerance in species of agricultural or horticultural importance (Bressan _et al_. 1981; Handa _et al_. 1983). However some work has been done on other aspects of osmotic stress such as the effects on callus differentiation (Brown _et al_. 1979; Kimball _et al_. 1975) and cell acidification (Marigo _et al_. 1983).

FIGURE 1.   Dry weight accumulation for cells grown in different concentrations of mannitol. Values represent the average of three samples ± one S.E.M.

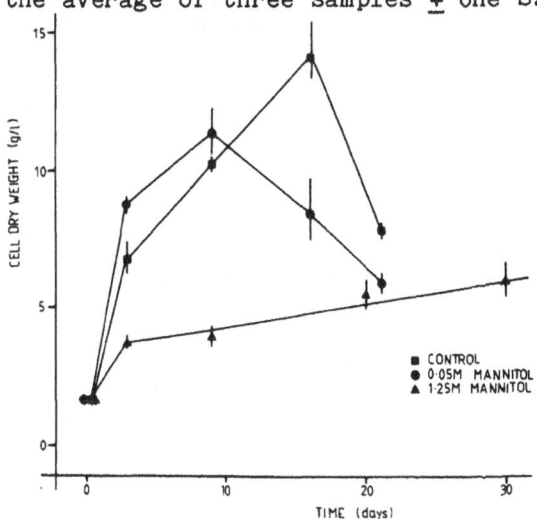

INITIAL GROWTH AND PRODUCT ACCUMULATION

In a preliminary study growth and alkaloid production were followed with cells grown in mannitol supplemented media over the range 0.05 to 1.5M mannitol.   A stable, four year old, alkaloid producing cell line of <u>Catharanthus roseus</u> was used, routinely maintained on a production medium (Morris, 1986).   Alkaloids were measured by HPLC after methanol extraction (Morris <u>et al</u>, 1985).

The results from this initial investigation indicated three categories of stress response. At low mannitol levels (0.05 to 0.2M), cells grew faster than controls, leading to an early dry weight peak (Fig. 1), and a faster decline in viability.   Serpentine reached 0.1% dry weight (Fig. 2), with rates of production similar to controls; however maximum accumulated totals were less, given the shorter growth period of stressed cells.   The second group of cells, grown at 0.3 to 0.6M mannitol, showed slower growth rates than controls giving a later although comparable dry weight peak (Fig. 5).   Cell viability was increased in stressed cells with 0.6M cells still fully viable at the final harvest of 41 days; compared with control cells which had fallen to 25% viability by 25 days.   Stressed cells also remained small and

FIGURE 2.   The effect of increasing medium osmolarity on maximum cell alkaloid levels.   Points represent the mean ± S.E.M. of three samples.

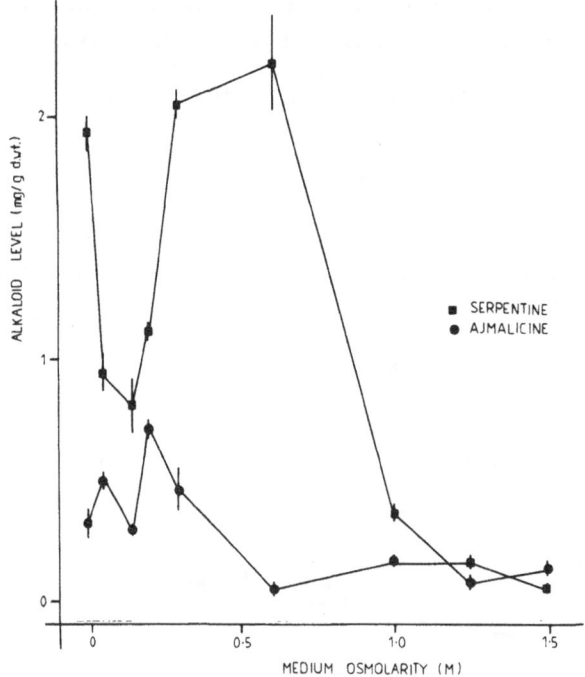

densely packed with starch granules long after starch utilisation and
cell expansion had occurred in unstressed cells (Fig. 3). Serpentine
levels were similar to controls, reaching 0.2% dry wt. at 25 days
(Fig. 2). In the last group of cells, grown at 1.0 to 1.5M mannitol,
both growth and product level were severely inhibited by the degree of
osmotic stress imposed (Figs. 1 & 2). From these results a stress
level of 0.6M mannitol was chosen to standardise on for further
investigation, as this degree of stress appeared to be inhibiting
carbohydrate utilisation whilst not affecting alkaloid levels.

CELL DIVISION

Determinations of cell numbers showed cell division to be
inhibited in 0.6M mannitol stressed cells; both in rate of division
and total cell population. Control cells had a doubling time of 1.42
days, reaching a final concentration of 4.2×10^6 cells/ml compared
with 5.15 days and 1.3×10^6 cells/ml for treated cells.

When considered in conjunction with the similar maximum dry weights
(Fig. 5), this threefold difference in final cell number indicates a
smaller, heavier, stressed cell population. This is in agreement with
the apparent morphological changes (Fig. 3).

As a consequence of the differing final cell numbers and weight per
cell, it is clear that a better reflection of cellular events is given
when results are expressed on a cell number basis rather than a dry
weight basis. An example which will be discussed in more detail
below, is that total amino acid levels are similar when compared in
mmoles/g dry weight but can be shown to have increased five-fold when
expressed per cell.

FIGURE 3.   Electron micrographs of untreated (left) and
0.6M mannitol treated cells (right), x 4400. Samples were
fixed and stained with uranyl acetate/lead citrate.

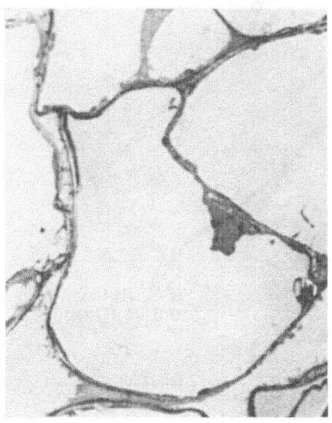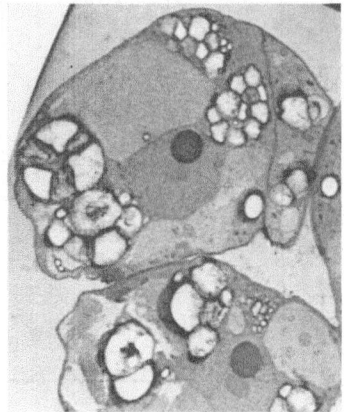

FREE AMINO ACIDS
Intracellular proline levels were measured using the method of Bates et al (1973), with total amino acid analysis on selected samples. Proline has frequently been reported as a preferentially accumulated compatible solute; being non-inhibitory to enzymes, even at high concentrations (Fitter & Hay, 1981). With tomato cell cultures, Handa et al (1983), found proline and other

Table 1:   Percentage composition of total free amino-acids for control and 0.6M mannitol stressed cells at 20 days.

AMINO ACID	CONTROL (%)	0.6M (%)
Serine	43.6	34.3
Arginine	13.6	16.6
Glutamine	3.4	13.8
Histidine	5.2	6.2
Aspartate	8.2	3.9
Phenylalanine	5.1	1.9
Proline	0.4	3.9
Others	20.3	19.4
Total Content		
(mmoles/g.dry weight)	0.5	0.4
(mmoles/10^6 cells)	0.6	2.9

FIGURE 4.   Cell starch levels in unstressed and 0.6M mannitol treated cells.   Points represent the mean $\pm$ S.E.M. three samples.

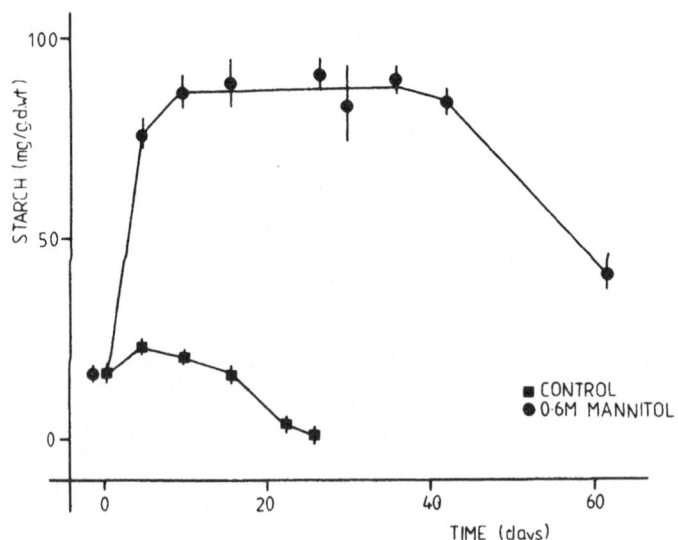

amino acids increased in response to an elevation in external NaCl
levels.

In C. roseus, using control and 0.6M mannitol cells, proline increased
with time, peaking at 23 and 29 µmoles/g dry weight for control and
treated cells respectively.

However, closer analysis of 20 day old samples showed proline not to
be the predominant amino acid in either cell type. The major amino
acid in both was serine but with osmotic stress causing a 25% fall in
the level of serine and small upward adjustments in the levels of
several other amino acids (Table 1). Overall, osmotic stress
increased free amino acid levels from 0.55 to 2.86 mmoles.10^6 cells, a
five-fold increase and it is possible that a greater osmoregulatory
role can be assigned to these molecules.

STARCH ACCUMULATION

Starch levels were measured using an aminoglucosidase
assay (Kerr et al. 1982). The results of this showed a maximum starch
content four times greater in stressed cells (Fig. 4). Taking into
account the smaller cell number for stressed cells, this gives starch
levels of 58 and 727 µg/10^6 cells for control and stressed cells
respectively. This measured increase in starch levels agrees with the

FIGURE 5.   Dry weight accumulation for unstressed and
0.6M mannitol treated cells. Points represent the mean ±
S.E.M. of three samples.

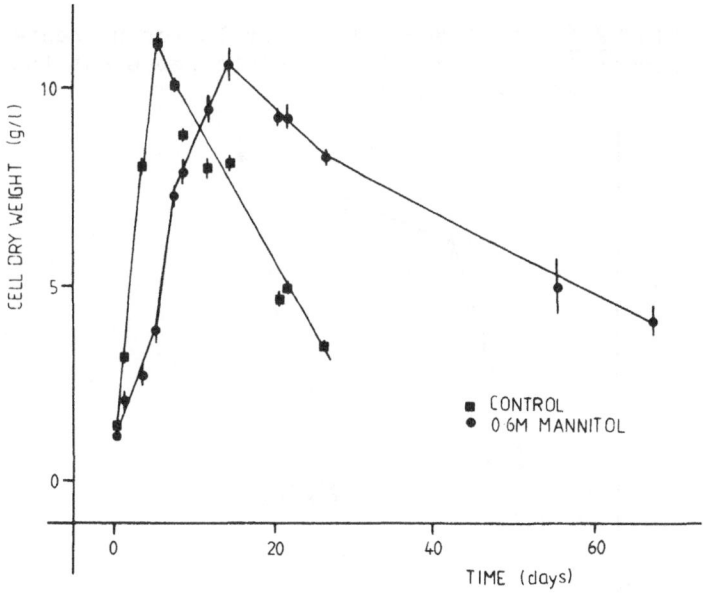

noted morphological changes (Fig. 3).

ALKALOID PRODUCTION
Further alkaloid analysis showed serpentine to be the major alkaloid in both cell types, with ajmalicine of lesser importance; a result consistent with previous data (Fig. 2). However when compared over a much longer time scale, taking into account the extended viability of stressed cells, the mannitol treated cells showed increased maximum serpentine levels (Fig. 6). When re-expressed on a per cell basis, this gives final levels of 1.4 and 21.5 mg/10^6 cells for control and stressed cells. Rates of production also differed, being 0.13 mg/10^6 cells/day for control cells and 1.43 mg/10^6 cells/day for stressed cells, a ten-fold increase. Ajmalicine levels were also affected in mannitol treated cells although less so than for serpentine.

DISCUSSION
Overall, the results show that osmotic stress is having two main effects on the cells. Firstly, inhibition of cell division gives a smaller cell population which therefore has a greater carbon resource per cell. Whether the increase in cell lifespan is due to this alone, or is compounded by a decrease in rates of carbon utilisation and thus primary metabolism, is currently being investigated further.   Secondly, osmotic stress has the effect of increasing cell alkaloid production, both in rate and in overall level of accumulation. When expressed as total production in mg per litre of culture, stressed cells gave levels three times those of untreated

FIGURE 6. Cell serpentine levels for untreated and 0.6M mannitol stressed cells. Points represent the mean ± S.E.M. of three samples.

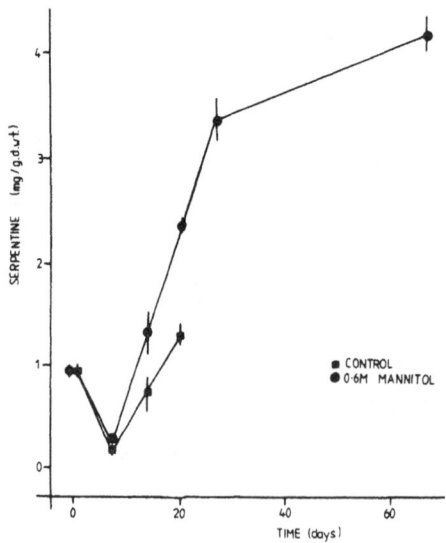

cells. Again this is aided by increased cell longevity. Thus stress
could be seen as an effector for increased alkaloid production, a
phenomenon also reported for other forms of stress (Pate, 1983;
Heinstein, 1985). Again, further study is necessary for a more
precise elucidation of the responses of cells to osmotic stress.

REFERENCES
Bates, L.S., Waldren, R.P. & Teare, I.D. (1973). Rapid determination
        of free proline for water stress studies. Plant and Soil.
        39 205-7.
Bressan, R.A., Hasegawa, P.M. & Handa, A.K. (1981). Resistance of
        cultured higher plant cells to PEG induced water stress.
        Plant Sci. Lett. 21 23-30.
Brown, D.C.W., Leung, D.W.M. & Thorpe, T.A. (1979). Osmotic
        requirement for shoot formation in tobacco callus.
        Physiol. Plant. 46 36-41.
Fitter, A.J. & Hay, R.K.M. (1981). Environmental Physiology of
        Plants. London Academic Press.
Handa, S., Bressan, R.A., Hasegawa, P.M. & Handa, A.K. (1982).
        Relationship between proline accumulation and osmotic
        stress in cultured plant cells. Plant Physiol. 69 Supple-
        ment Number 4. p.59.
Handa, S., Bressan, R.A., Handa, A.K., Carpita, C. & Hasegawa, P.M.
        (1983). Solutes contributing to osmotic adjustment in
        cultured plant cells adapted to water stress. Plant
        Physiol. 73 834-42.
Hanson, A.D. & Hitz, W.D. (1982). Metabolic responses of mesophytes
        to plant water deficits. Ann. Rev. Plant Physiol. 33
        163-207.
Heinstein, P.F. (1985). Future approaches to the formation of
        secondary natural products in plant cell suspension
        cultures. J. Nat. Prods. 48 1-9.
Kerr, P.S., Huber, S.C. & Israel, D.W. (1982). Effect of N-Source on
        soybean leaf sucrose phosphate synthase. Starch formation
        and whole plant growth. Plant Physiol. 75 483-8.
Kimball, S.L., Beversdorf, W.D. & Bingham, E.T. (1975). Influence of
        osmotic potential on the growth and development of soybean
        tissue cultures. Physiol. Plant. 46 36-41.
Marigo, G., Delorme, Y.M., Luttge, U. & Boudet, A.M. (1983). Role de
        L'acide malique dans la regulation de pH vacuolaire dans
        le cellules de Catharanthus roseus cultivees in vitro.
        Physiol. Veg. 21 1135-1144.
Morris, P. (1986). Regulation of product synthesis in cell cultures
        of Catharanthus roseus II. Comparison of production media.
        Planta Medica (in press).
Morris, P., Scragg, A.H., Smart, N.J. & Stafford, A. (1985).
        Secondary product formation by cell suspension cultures.
        In Plant Cell Culture. A Practical Approach. Ed.
        R.A.Dixon. I.R.L. Press. Oxford. pp.127-66.
Pate, D.W. (1983). Possible role of ultraviolet radiation in
        evolution of cannabis chemotypes. Econ. Bot. 37 396-405.

10. THE RELATIONSHIP BETWEEN TROPANE ALKALOID PRODUCTION AND
     STRUCTURAL DIFFERENTIATION IN PLANT CELL CULTURES OF
     ATROPA BELLADONNA AND HYOSCYAMUS MUTICUS

M.A. Collinge & M.M. Yeoman

Botany Department,
University of Edinburgh,
King's Buildings, Mayfield Road,
Edinburgh. EH9 3JH. U.K.

INTRODUCTION
     The tropane alkaloids atropine (L-hyoscyamine) and
scopolamine are widely used anticholinergic drugs which are obtained
from plants. Atropa belladonna (Deadly Nightshade) is the species
most commonly cultivated for production of atropine. The dried leaves
contain 0.3-0.5%, and the roots 0.4-0.6% of tropane alkaloids. The
major alkaloid is L-hyoscyamine. Other alkaloids including
scopolamine, and the N-oxide of hyoscyamine and scopolamine are
present in varying proportions in different tissues, and at different
stages of the growth cycle (James, 1950; Phillipson & Handa, 1975,
1976). A richer source of hyoscyamine is Hyoscyamus muticus (Egyptian
Henbane) which is gathered from the wild in Egypt, and is cultivated
in Nigeria and on a small scale in northern India. The dried leaves
and flowering tops contain 0.35-1.39% of total alkaloid, 90% of which
is hyoscyamine (Morton, 1977). The immediate precursors of
hyoscyamine are tropine and tropic acid, which are derived from
ornithine and phenylalanine respectively. Tropane alkaloids are
secondary metabolites and as such are associated with differentiation.

In this study cell cultures of A. belladonna and H. muticus are being
used to study the factors controlling tropane alkaloid formation and
accumulation and to develop a procedure to increase the amount of
alkaloid produced, which could feasibly be used for the production of
hyoscyamine in vitro, on an industrial scale.

Previous work has shown that generally the more highly aggregated slow
growing and structurally differentiated cultures produce higher levels
of secondary metabolites (Yeoman et al., 1980; Yeoman et al., 1982;
Lindsey & Yeoman, 1983a). Immobilisation of plant cells results in
slow growth, higher cell-cell contact, and the establishment of
chemical and physical gradients; and yet allows manipulation of the
culture environment (Lindsey & Yeoman, 1983b, 1984). Cells are
immobilised in polyurethane foam by a simple procedure (Lindsey et
al., 1983). Polyurethane foam blocks (1 cm^3) are added to a newly
subcultured suspension culture. During the following 14 to 21 days,
the cells are entrapped and grow in the pores until the blocks are
densely packed. The pattern of growth during the immobilisation
process is illustrated in Fig. 1.

METHODS OF ANALYSIS
The system of analysis used must be able to measure accurately small amounts of alkaloids, and to separate the different alkaloids and their precursors. High performance liquid chromatography (HPLC) best satisfies these requirements. The extraction method and HPLC system described here have been derived from that of Baumann (personal communication).

Extraction Procedure
The tissue was macerated with 5% ammonia in methanol and left to stand overnight. The filtered extract was concentrated under reduced pressure and taken up in 15ml 0.1N hydrochloric acid. The acid extract was filtered and made alkaline with a buffer composed of 10ml 0.2M ammonium chloride and 9.6ml 25% ammonium hydroxide; made up to 20ml with water, and added to the top of an "Extrelut" column (Merck). After 20 minutes, the column was eluted with 40ml chloroform. The chloroform was evaporated to dryness, and the residue stored in methanol at approximately 5°C. The extracts were resuspended in the HPLC mobile phase before analysis.

Chromatographic Analysis
HPLC analysis was performed with a Gilson 302 liquid chromatograph, fitted with a Gilson u.v. detector, and a Shimadzu Chromatopac C-R1B data processor. 20μl samples were separated at ambient temperature on a Spherisorb S5 Octyl 25cm x 46mm column (Phase Separations Ltd.). The column was eluted with an isocratic mobile phase consisting of 22.5% acetonitrile and a buffer containing 50mM potassium dihydrogen orthophosphate adjusted to pH 3.0 with orthophosphoric acid. Detection was a 254nm.

Typical HPLC traces are shown in Fig. 2. Standards of the three most abundant alkaloids; hyoscyamine, scopolamine and hyoscyamine-N-oxide are clearly resolved by this system, as shown in Fig. 2a. Figs. 2b and c show the separation of hyoscyamine from the other components of

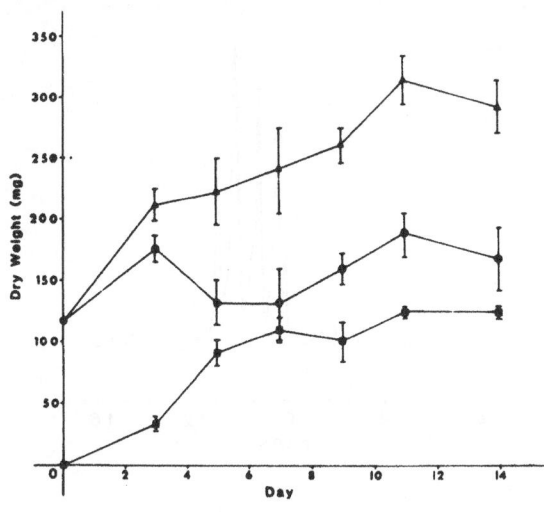

Figure 1. The changes in the dry weight of total (triangles), suspended (circles), and immobilised (squares) cells of H. muticus during immobilisation. Each value is the mean of three replicates + S.E.

a

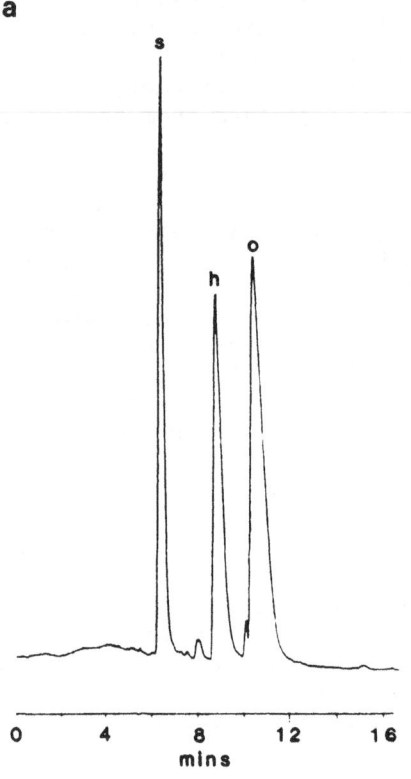

Figure 2a.   Typical HPLC separation of standard drugs: scopolamine(s), hyoscyamine (h), and hyoscyamine-N-oxide (o).

Figure 2b.   HPLC profile of an extract from _Atropa_ callus containing hyoscyamine (h).

Figure 2c.   HPLC profile of the same _Atropa_ callus extract after the addition of an internal standard of hyoscyamine.

b

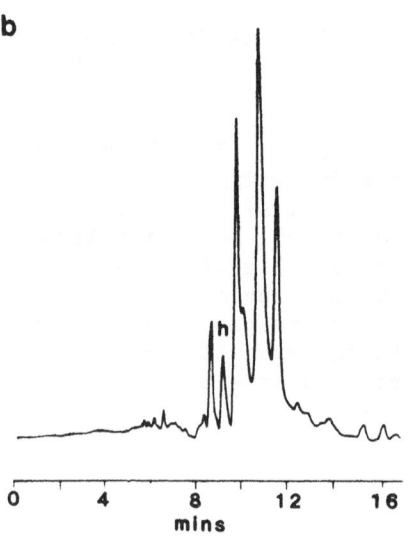

c

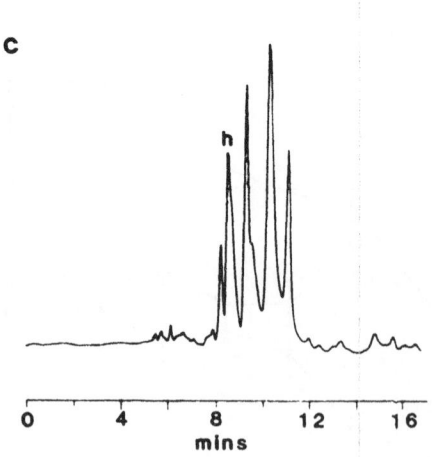

a tissue culture extract and its identification by the use of an
internal standard.   The callus extract alone was analysed first (Fig.
2b).   Then an internal standard of hyoscyamine was added to the
extract, resulting in the trace shown in Fig. 2c, where the area of
the hyoscyamine peak has increased.   The identity of the hyoscyamine
peak in extracts has also been confirmed by spectral analysis of the
peak using the diode array detector of a Hewlett Packard 1090 liquid
chromatograph, and by thin layer chromatography.

ALKALOID PRODUCTION IN DIFFERENTIATED CULTURES

Early work with A. belladonna showed that alkaloids were
only accumulated in excised root and root callus cultures (West &
Mika, 1957) or in tissue cultures in which roots had differentiated
(Thomas & Street, 1970).   However, alkaloids have been produced in
callus cultures (Khanna et al. 1976; Eapen et al. 1978a,b; Sharma &
Khanna, 1982), and in suspension cultures (Lindsey & Yeoman, 1983);
but higher levels were always found in slow growing or highly
differentiated cultures.   Abnormal alkaloid patterns in callus, but
not in differentiated shoot buds, were noted by Eapen et al. (1978a,
b).

By inducing morphogenesis in immobilised cultures, yields may be
improved, and the alkaloid profile may be closer to that in vivo.
Reports in the literature (Konar et al. 1972); Thomas & Street, 1972;
Gosch et al. 1975) have indicated that the induction of
differentiation in A. belladonna is readily achieved, but there are no
similar studies with H. muticus.   In this investigation experiments
have been performed to find conditions which induce morphogenesis, and
to characterise the types of structures formed.

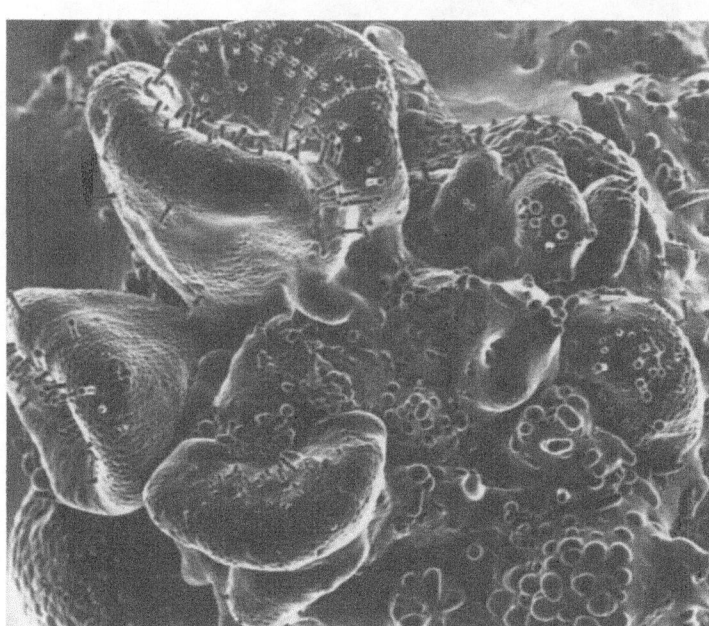

Figure 3a.   Scanning
electron micrograph
of cryopreserved H.
muticus callus
showing a mass of
callus in which a
number of shoots
have developed.

Organogenesis can be induced in Murashige & Skoog medium (Flow
Laboratories).   Cells are grown first in a primary induction medium
which contains 1.0 mg.ml^{-1} para-chlorophenoxyacetic acid, 0.5mg.ml^{-1}
2,4-dichlorophenoxyacetic acid and 0.1mg.l^{-1} kinetin.   The secondary
induction medium, in which organogenesis occurs contains only
0.1mg.ml^{-1} kinetin.   H. muticus usually exhibits more differentiation
than A. belladonna under these conditions.   Both suspension and
immobilised cell cultures form roots and undeveloped primorida.   In
callus both roots and shoots were formed.   Scanning electron micrographs of cryopreserved developing shoots in H. muticus callus are
shown in Fig. 3.

At present, the effect of this structural differentiation on the
pattern and extent of alkaloid production is being determined, and
compared with that of cultures which do not display structural
differentiation.   In this way, some understanding of the effect of
differentiation, and the form it must take to achieve, or approach the
alkaloid metabolism of the intact plant will be gained.   If a positive
correlation is obtained, this induction of structural differentiation
could be incorporated into the final in vitro alkaloid production
system together with other treatments such as nutrient, limitation or
precursor feeding in order to increase the yield of hyoscyamine from
the cultured cells.

Figure 3b.   Scanning electron micrograph of cryopreserved H. muticus callus showing a single developing
shoot at high magnification.

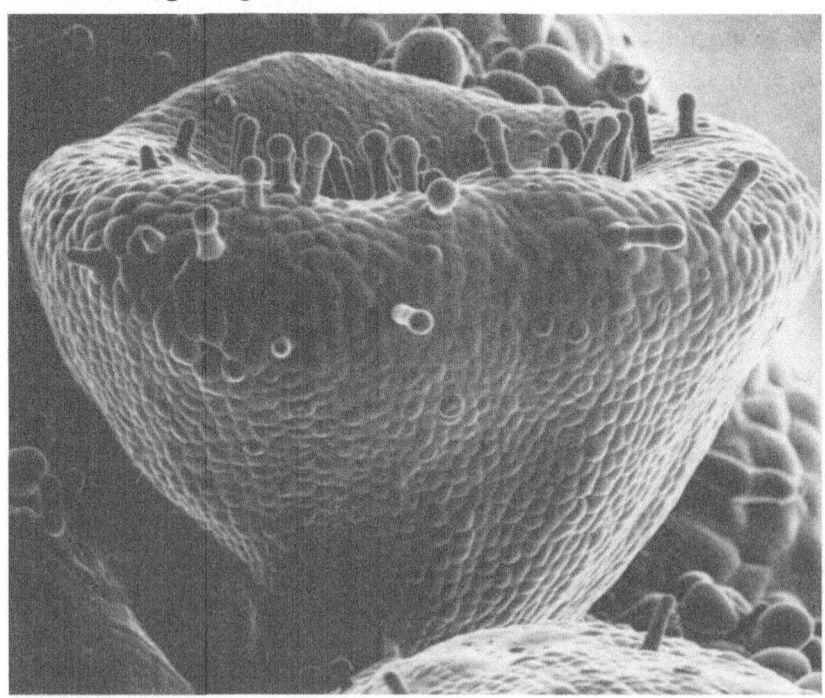

Acknowledgements
We gratefully acknowledge financial support from the
S.E.R.C. Biotechnology Directorate, and would like to thank Phase
Separations Ltd. for their gift of an HPLC column, Dr. C.E. Jeffree
for the scanning electron micrographs, and Mrs. J. Summers for typing
the manuscript.

REFERENCES
Eapen, S., Rangan, T.S., Chadha, M.S. & Heble, M.R. (1978a).
    Biosynthetic and cytological studies in tissue cultures
    and regenerated plants of haploid Atropa belladonna.
    Canadian Journal of Botany 56, 2781-4.
Eapen, S., Rangan, T.S., Chadha, M.S. & Heble, M.R. (1978b).
    Morphogentic and biosynthetic studies on tissue cultures
    of Atropa belladonna L. Plant Science Letters 13, 83-9.
Gosch, G., Bajaj, Y.P.S. & Reinert, J. (1975). Isolation, culture
    and induction of embryogenesis in protoplasts from cell
    suspensions of Atropa belladonna. Protoplasma 86, 405-
    10.
Khanna, P., Sharma, G.L. & Uddin, A. (1976). Atropine from Atropa
    belladonna Linn. tissue culture. Indian Journal of
    Experimental Biology 15, 323-4.
Konar, R.N., Thomas, E. & Street, H.E. (1972). Diversity of
    morphogensis in suspension cultures of Atropa belladonna
    L. Annals of Botany 36, 249-58.
Lindsey, K. & Yeoman, M.M. (1983a). The relationship between growth
    rate, differentiation and alkaloid accumulation in cell
    cultures. Journal of Experimental Botany 34, 1055-65.
Lindsey, K. & Yeoman, M.M. (1983b). Novel experimental systems for
    studying the production of secondary metabolites by plant
    tissue cultures. In Plant Biotechnology, eds. Mantell,
    S.H. & Smith, H. Society for Experimental Biology Seminar
    Series 18, pp. 39-66. Cambridge University Press.
Lindsey, K., Yeoman, M.M., Black, G.M. & Mavituna, F. (1983). A
    novel method for the immobilisation and culture of plant
    cells. FEBS Letters 155, 143-9.
Lindsey, K. & Yeoman, M.M. (1984). The viability and biosynthetic
    activity of cells of Capsicum frutescens Mill. c.v. annuum
    immobilised in reticulate polyurethane. Journal of
    Experimental Botany 35, 1684-96.
Morton, J.F. (1977). Major Medicinal Plants: Botany, Culture and
    Uses. Springfield, Illinois: Charles C. Thomas.
Phillipson, J.D. & Handa, S.S. (1975). N-oxides of hyoscyamine and
    hyoscine in the Solanaceae. Phytochemistry 14, 999-1003.
Phillipson, J.D. & Handa, S.S. (1976). Hyoscyamine-N-oxide in Atropa
    belladonna. Phytochemistry 15, 605-8.
Sharma, M. & Khanna, P. (1982). Role of hormones on atropine
    production in Atropa belladonna L. tissue cultures. Indian
    Journal of Pharmaceutical Sciences 43, no 5, 175-7.
Thomas, E. & Street, H.E. (1970). Organogenesis in cell suspension
    cultures of Atropa belladonna L. and var. Lutea. Annals
    of Botany 34, 657-69.

Thomas, E. & Street, H.E.   (1972).   Factors affecting morphogenesis in
        excised roots and suspension cultures of Atropa
        belladonna. Annals of Botany 36, 239–47.
West, F.R. & Mika, E.S.   (1957).   Synthesis of atropine by isolated
        roots and root-callus cultures of Belladonna.   Botanical
        Gazette 119, 50–4.
Yeoman, M.M., Miedzybrodzka, M.B., Lindsey, K. & McLauchlan, W.R.
        (1980).   The synthetic potential of cultured plant cells.
        In Plant Cell Cultures: Results and Perspectives.   Eds.
        F. Sala, B. Parisi, R. Cella, O. Ciferri, pp. 327–43.
        Elsevier: Biomedical Press.
Yeoman, M.M., Lindsey, K., Miedzybrodzka, M.B. & McLauchlan, W.R.
        (1982).   Accumulation of secondary products as a facet of
        differentiation in plant cell and tissue cultures.   In
        Differentiation in vitro: British Society for Cell Biology
        Symposium 4.   Eds. M.M. Yeoman & D.E.S. Truman, pp. 65–82.
        Cambridge: Cambridge University Press.

SECTION 3

BIOCHEMISTRY OF SECONDARY METABOLISM

11. MODULATION OF THE PHENYLPROPANOID PATHWAY IN BEAN
(PHASEOLUS VULGARIS) CELL SUSPENSION CULTURES

Richard A. Dixon and G. Paul Bolwell

Department of Biochemistry,
Royal Holloway and Bedford New College,
University of London,
Egham, Surrey. TW20 OEX. U.K.

INTRODUCTION
A major aim of plant biotechnology is to manipulate
cultured plant cells in order to increase the metabolic flux into
specific pathways of secondary product formation. In most cases this
involves alteration of physical and chemical components of the culture
environment; analysis of molecular events resulting from such changes
has generally been limited to measurements of end-product accumulation
only. While this approach is of undoubted value for preliminary
optimisation of culture conditions for secondary product formation,
its main disadvantage is that it may overlook the operation of
positive or negative endogenous biochemical regulatory mechanisms
which act to control the flux through the pathways under study and
which, if able to be triggered or circumvented, may result in
increased product yield. In short, studies on end product
accumulation alone may not necessarily indicate the total attainable
capacity for secondary product synthesis.

In order to assess the operation of endogenous regulatory mechanisms
controlling secondary product accumulation, it is necessary to have
knowledge of the enzymology of the biosynthetic pathways under
consideration and of the factors which might control enzymic capacity,
both in vitro (i.e. positive and negative effectors) and in vivo (i.e.
effects on transcription, translation and post-translational
modification including enzyme inactivation and/or degradation).
Although the end produce of the phenylpropanoid pathway from L-
phenylalanine (e.g., flavonoids, isoflavonoids, stilbenes, coumarins
and furanocoumarins) are not at present of great importance
commercially, sufficient information is now available on the
enzymology and control of their formation to make this pathway a
valuable model system for the definition of regulatory phenomena which
may be of general occurrence and importance.

In this chapter, we review the enzymology and control of the phenylpropanoid pathway in suspension cultured bean (Phaseolus vulgaris)
cells. These produce isoflavonoid-derived antifungal agents
(phytoalexins) in response to treatment with elicitor polysaccharides
heat-released from cell walls of the phytopathogenic fungus
Colletotrichum lindemuthianum. This system is highly amenable to
studies of molecular control mechanisms, as it is characterised by
rapid but transient enzyme induction phenomena; in addition,

mechanisms have been defined which result in the removal of active
enzyme molecules after their initial induction and which clearly,
therefore, may act to limit sustained flow from primary into secondary
metabolic pathways.   The implications of these findings to strategies
for the improvement of product yield in cultured plant cells are
briefly outlined.

ENZYMOLOGY OF THE PHENYLPROPANOID PATHWAY

The central phenylpropanoid pathway channels L-phenylalanine into the formation of hydroxycinnamoyl coenzyme A thiol esters;
these, and other intermediates of the pathway, serve as starting
points for branch pathways leading to the biosynthesis of a number of
classes of plant secondary product (Figure 1).   Many of the enzymes of
the central phenylpropanoid pathway and the lignin and flavonoid
branch pathways are now well characterised and their properties have
been reviewed (Hahlbrock & Grisebach, 1979; Grisebach, 1981; Dixon et
al., 1983a, Jones, 1984).   Some of these enzymes have not been studied
in detail from bean, and are better described from other plant
sources.

L-phenylalanine ammonia-lyase (PAL ; EC 4.3.1.5) is a tetrameric enzyme
which catalyses the removal of ammonia, by the elimination of the
amino group and the pro 3S hydrogen, from L-phenylalanine to form
trans-cinnamic acid; this is the first committed step in the
biosynthesis of phenylpropanoid compounds in higher plants.   The enzyme
is believed to possess two active sites, each containing a
dehydroalanine residue, per tetramer (Havir & Hanson, 1973).   The
subunits are considered to be of identical M_r.   However, the reported
subunit M_r varies between species (Nari et al., 1974; Havir & Hanson,

Figure 1.   Biosynthesis of plant secondary products via
the phenylpropanoid pathway.   Enzymes are: 1,
L-phenylalanine ammonia-lyase; 2, trans-cinnamic acid
4-hydroxylase; 3, 4-coumarate : CoA ligase; 4, chalcone
synthase; 5, stilbene synthase; 6, chalcone isomerase.

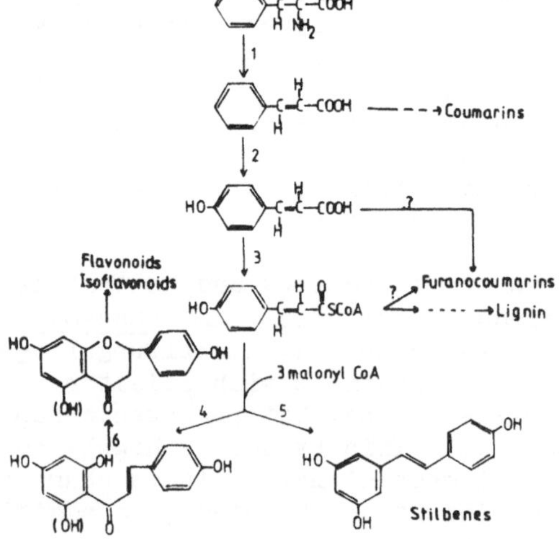

1973; Zimmerman & Hahlbrock, 1975; Gupta & Acton, 1979), although M_r
heterogeneity may result from partial degradation, as recently
demonstrated for the enzyme from bean (Bolwell et al., 1986b). The
apparent negative rate cooperativity with respect to the substrate
L-phenylalanine observed with partially purified preparations may
result from the presence of co-purified iso-enzymes with differing K_m
values (see below). The enzyme was purified to apparent homogeneity
from suspension cultured bean cells, which had been exposed to
Colletotrichum elicitor, by a combination of ammonium sulphate
fractionation, gel filtration, ion-exchange chromatography and two
cycles of chromatofocussing (Bolwell et al., 1985a) or by affinity
chromatography on L-α-aminooxy-p-hydroxyphenol-propionic acid linked
to epoxy-activated Sepharose 6B via the phenolic hydroxyl group
(Bolwell et al., 1986b). Purified enzyme preparations exhibited
heterogeneous subunit M_r values of 77,000, 70,000 and 53 000, the
relative proportions depending upon the enzyme source and the length
of time taken for purification. All these subunits were, however,
dissociated from active tetramers and were shown to be related to one
another on the basis of peptide mapping and interconversion studies
(Bolwell et al., 1986b). Four forms of the enzyme, differing in pI
and K_m values, were resolved by chromatofocussing, and these appear to
be differentially induced following elicitation (Bolwell et al.,
1985a).

The hydroxylation of trans-cinnamic acid to 4-coumaric acid is
catalysed by a microsomal cytochrome P-450 dependent monooxygenase
(EC. 1. 14.13.11). The activity of this membrane-bound enzyme system
has not yet been reconstituted in solubilised extracts from bean,
although the P-450 has been tentatively identified by indirect use of
a cross-reactive anti-(rat P-450) monoclonal antibody, as an elicitor-
inducible polypeptide of M_r 48,000 (Bolwell et al., 1985b; Bolwell &
Dixon, 1986). A second hydroxylation of 4-coumarate to caffeate can
be catalysed in vitro by 4-coumarate 3-hydroxylase, a phenolase of low
specificity, although there is some doubt as to whether this is the
function of the enzyme in vivo. Although neither this enzyme, nor the
subsequent caffeic acid O-methyltransferase involved in modification
of phenylpropanoid precursors of lignin, have been isolated from bean,
their activities are present, and all the hydroxy-cinnamic acid
intermediates of the phenylpropanoid pathway have been detected in
free and esterified forms, in cultured bean cells (Bolwell et al.,
1985b).

Prior to entry into the branch pathways of phenylpropanoid metabolism,
substituted cinnamic acids are activated to their corresponding
coenzyme A thiol esters. Such reactions are catalysed by multiple
forms of 4-coumarate : CoA ligase (EC 6.2.1.12). This enzyme has not
been purified from tissue cultured cells of bean owing to its inherent
instability (Dixon & Bendall, 1978), although the enzyme from parsley
is more stable and antibody and cDNA probes have been produced for it
(Schroder & Schafer, 1980; Hahlbrock et al, 1982). The hydroxycinnamoyl CoA esters have a number of species, tissue and cell-
specific fates. In the intact bean plant they may be channelled into

lignin, flavonoid or isoflavonoid production; in tissue cultures, the branch pathway(s) operating depend on the source and age of the cultures, and the specific stimulus applied.

Hydroxycinnamoyl CoA units are channelled into the formation of flavonoids and isoflavonoids via the enzyme chalcone synthase (CHS). This is a dimer of subunit M_r 46,000 which, like PAL, can be separated into active iso-forms by chromatofocussing (Hamdan & Dixon, unpublished results). The properties of the enzyme from bean have been studied (Whitehead & Dixon, 1983), although in less detail than for the apparently more stable enzyme from parsley. The necessity for a separate 6'-deoxychalcone synthase activity as a control site for the formation of 5-deoxy isoflavonoids, such as the phytoalexin phaseollin, has been demonstrate by ^{13}C-NMR experiments (Dewick et al., 1982). Further conversion of chalcones to flavonoids and isoflavonoids occurs via an isomerisation reaction catalysed by chalcone isomerase (CHI, EC 5.5.1.6). The isomerase from bean is an M_r 27,000 single subunit polypeptide whose properties have been studied (Dixon et al., 1982) and which has been purified to homo-geneity from suspension cultured cells (Robbins & Dixon, 1984). The subsequent pathways of isoflavonoid phytoalexin biosynthesis have been investigated in a number of plant species by radioisotope feeding experiments; in soybean, specific isoflavone synthetase, pterocarpan 6a-hydroxylase and dihydroxypterocarpan : dimethylallyl transferase enzymes have been identified (Hagmann & Grisebach, 1984; Hagmann et al., 1984; Zahringer et al., 1979). The two former enzymes appear to be microsomal, cytochrome P-450-dependent systems. The enzymology of the flavonoid branch pathway is now well documented for parsley (Hahlbrock et al, 1976), and the enzymic formation of stilbenes and furanocoumarins has also been the subject of recent studies (Rolfs & Kindl, 1984; Haufe et al., 1986).

EXPERIMENTAL MANIPULATION OF THE PHENYLPROPANOID PATHWAY IN CULTURED CELLS
As implied above, a number of different stimuli can induce rapid transient increases in the activities of some, if not all, of the enzymes of the central phenylpropanoid pathway. The subsequent fate of the resulting intermediates is dependent on the relative activities (constitutive or induced) of the branch pathway enzymes. Thus, in parsley cell cultures, the synthesis of flavonoids is stimulated by UV light or dilution (Hahlbrock et al., 1982), whereas fungal elicitor preparations induce the accumulation of furanocoumarin phytoalexins (Tietjen et al., 1983) and the suppression of UV-induced chalcone synthase activity (Hahlbrock et al., 1981). In bean, routine subculture into a maintenance medium leads to constitutive flavonoid production, whereas subculture into a medium whose growth regulator components induce xylogenesis results in induction of PAL and caffeic acid O-methyltransferase activities and consequent lignification (Haddon & Northcote, 1976). In contrast, treatment of bean cultures with elicitor macromolecules from the cell walls of Colletotrichum lindemuthianum results in rapid accumulation of the prenylated 5-hydroxy isoflavanone kievitone followed by later accumulation of the

5-deoxy-isoflavonoid-derived phytoalexin phaseollin (Robbins et al.,
1985).   In addition, wall bound phenolics are deposited.   The
accumulation of these products is preceded by rapid increases in the
extractable activities of PAL, cinnamic acid 4-hydroxylase, CHS, CHI
and 4-coumarate : CoA ligase, but not 4-coumarate hydroxylase or
caffeic acid 0-methyltransferase (Dixon & Bendall, 1978; Bolwell et
al., 1985b; Robbins et al., 1985).   These changes occur concomitantly
with increases in enzyme activities involved in the post-translational
modification of hydroxyproline-rich glycoproteins and their
accumulation in the cell wall (Bolwell et al., 1985b).

MOLECULAR MECHANISMS UNDERLYING THE INDUCTION OF
PHENYLPROPANOID BIOSYNTHETIC ENZYMES
Regulation of the phenylpropanoid pathway in bean cell
suspension cultures has been studied in particular detail with respect
to the elicitation response, which is both rapid and characterised by
high levels of induction; typically, 10-fold increases in PAL and CHS
activity would occur within 6 to 8h of elicitation.   This work has
involved the production of specific molecular probes and the
application of a variety of labelling, immunological and RNA/DNA
hybridisation techniques in order to answer the basic questions of
whether induced appearance of enzyme activities is a result of de novo
synthesis or activation and whether synthesis, where it can be
demonstrated, is the result of increased gene transcription.   The
general picture that has emerged from these studies is one of rapid,
transient increases in gene expression (eg., maximum levels of PAL and
CHS mRNA attained within 2-4h of elicitation) underlying most of the
elicitor-induced changes in enzyme levels; superimposed on this, in
some cases, are subsequent positive or negative post-translational
effects.

Evidence for rapid increases in the rate of synthesis of PAL protein
dependent upon increased gene transcription in elicitor-treated bean
cell suspension cultures has been obtained by a combination of the
following techniques:

(1)   Density labelling with ^{2}H from ^{2}H$_2$O, followed by analysis of
the equilibrium distribution of enzymes activity on CsCl and KBr
density gradients (Dixon & Lamb, 1979; Lawton et al., 1980); this
allows calculation of the rate constants for active enzyme synthesis
and removal.

(2)   In vivo labelling with [35S] methionine, followed by specific
immunoprecipitation of PAL using anti-(parsley PAL) or anti-(bean PAL)
polyclonal antisera (Lawton et al., 1983a; Bolwell et al., 1985a,
1986a); this measures the rate of synthesis of enzyme subunits.

(3)   Immune blotting to determine the actual levels of enzyme
protein (Bolwell et al., 1986a).

(4)   In vitro translation of total and polysomal mRNA, followed by
specific immunoprecipitation of [35S]-labelled PAL subunits (Lawton et

al., 1983b; Cramer et al., 1985; Bolwell et al., 1986a); this measures
the translational activity of extracted mRNA.

(5)   Northern blot hybridisation analysis of mRNA (and in
particular of newly-synthesised, thiouridine-labelled mRNA), using a
fully characterised cDNA probe complementary to bean PAL mRNA (Edwards
et al., 1985); this measures the actual levels of PAL mRNA, and an
increase in level in the newly-synthesised fraction points to
increased transcription.

(6)   Hybridisation of a specific, highly labelled bean genomic PAL
probe (intron : exon boundary sequence) to bean mRNA followed by S1
nuclease digestion and analysis of protected fragments (unpublished
results); this again measures mRNA level, but is considerably more
sensitive than Northern hybridisation.

(7)   Nuclear transcript run-off analysis (unpublished results, cf
Chappell and Hahlbrock, 1984); this measures the formation of
completed specific transcripts by isolated nuclei.

In addition to suggesting that increased transcription is the main
factor responsible for the initial elicitor-mediated increases in PAL
activity, these studies have also revealed that the multiple forms of
PAL observed at the enzyme activity level are also reflected by
multiplicity at the subunit and gene levels. Thus, immunoprecipitation experiments have demonstrated that although newly
synthesised PAL subunits in elicitor-treated cultures all have a
subunit M_r of 77,000, they exist in at least 10 charge iso-forms,
which can be resolved in the isoelectric focussing dimension on 2-D
IEF-SDS polyacrylamide gels (Bolwell et al., 1985a). The M_r 77,000
subunits newly synthesised in vitro by translation of mRNA are
resolved into 4 or 5 charge iso-forms by 2-D gel analysis (Bolwell et
al., 1985a), suggesting that some of the polymorphism seen in vivo
results from post-translational changes. Analysis of bean genomic DNA
by Southern hybridisation to bean PAL cDNA indicates the presence of a
multigene family. Comparative restriction analyses of genomic clones
hybridizing to the PAL cDNA, and genomic sequence data so far obtained
indicate a family of around 4 genes (K. Edwards & W. Schuch, personal
communication). Probes specific for each gene are currently under
development, and should facilitate studies of the differential
expression of PAL genes in response to environmental stimuli.

Use of one or more of the techniques described above has also
demonstrated rapidly increased de novo synthesis and mRNA appearance
underlying elicitor-induced changes in cytochrome P-450, CHS and CHI
in the bean cultures (Bolwell & Dixon, 1986; Dixon et al., 1986b).
Polymorphism similar to that seen with PAL has been demonstrated for
bean CHS, both at the level of subunits synthesised in vivo and in
vitro (Dixon et al., 1986a), and at the mRNA level by 2-D gel analysis
of hybrid-release-translated mRNAs selected by 3'-region probes
specific for different members of the CHS multigene family (T.B. Ryder
& C.J. Lamb, personal communication). In contrast, 2-D gel analysis

of CHI synthesised in vivo and in vitro from mRNA, and immuno-
precipitated using a polyclonal monospecific anti-(bean CHI) serum
(Robbins & Dixon, 1984), reveals the presence of only a single
polypeptide species (Dixon et al., 1986a).

Comparisons of the induction kinetics of mRNA species and the rates of
synthesis of their corresponding polypeptides in the above studies
suggests the absence of any major, intervening, rate-determining
control sites; following initiation of increased gene transcription,
elicitation is rapidly expressed as an increase in enzyme appearance.
This has also been indicated for the rising phase of the PAL response
during growth-regulator-induced differentiation in bean (Jones &
Northcote, 1981) and for the dilution and illumination responses in
parsley cells (Schroder et al., 1977; Kuhn et al., 1984; Chappell &
Hahlbrock, 1984). An extra factor appears to operate during the
elicitor-mediated induction of CHI in bean cultures, which has been
shown by density labelling (Dixon et al., 1983b) and immune blotting
(Robbins & Dixon, 1984) experiments to result from simultaneous de
novo synthesis and activation of pre-existing, and possibly newly
synthesised, enzyme.

As yet, the mechanism of response-coupling during elicitation is
unknown, as are the initial molecular sites of action of plant growth
regulators which modulate the phenylpropanoid pathway. Co-ordination
of the selective increased transcription of metabolically and/or
functionally related genes is presumably dependent upon specific
common regulatory sequences. Present indications suggest that this
expression is likely to be regulated by sequence-specific proteins
activated directly or indirectly as a result of specific signal
transmission (Dynam & Tjian, 1985). The extent of co-ordinate
induction of gene expression and enzyme appearance within the phenylpropanoid pathway has been discussed elsewhere (Dixon et al.,
1986,a,b; Robbins, this volume). The presence of multiple genes
encoding two key regulatory enzymes of phenylpropanoid metabolism may
hold exciting prospects for the future characterisation of specific
stimulus-related promoter sequences, their analysis by genetic
transformation techniques and the elucidation of the relationships
between specific genes and metabolic functions within a multigene
family.

MOLECULAR MECHANISMS UNDERLYING THE REMOVAL OF PHENYL-
PROPANOID BIOSYNTHETIC ENZYME ACTIVITY

As already described, the induction of the enzymes of the
phenylpropanoid pathway in cultured bean cells is of a transient
nature. If our ultimate aim is to maximise the biosynthetic potential
of the pathway for the production of secondary products from their
initial precursors, it is important to understand the processes
underlying the cessation of gene transcription/enzyme synthesis and
the mechanism of enzyme activity loss.

A number of studies have provided evidence which suggests that intracellular levels of trans-cinnamic acid, or some metabolite of it, may

act as a signal for the regulation of the flux through PAL.   The
effects of exogenous additions of cinnamic acid (which suppresses PAL
extractable activity in vivo) and L-α-aminooxy-β-phenylpropionic acid
(AOPP, a powerful competitive inhibitor of PAL whose application can
result in superinduction of PAL in vivo) can be interpreted in terms
of the involvement of cinnamic acid in a dual control mechanism
involving inhibition of the rate of PAL synthesis and stimulation of
the rate of PAL removal.   Both processes, whose importance was
initially revealed by density labelling studies (Lawton et al., 1980;
Shields et al., 1982) have recently been investigated in relation to
the effects of exogenous addition of cinnamic acid during the
elicitation response in bean cultures.   In terms of effects on enzyme
synthesis, Northern blotting, S1 nuclease-protection and nuclear
transcript run-off analyses have suggested a rapid switching off of
PAL and CHS transcription, and/or rapid removal of hybridisable mRNA,
following addition of cinnamic acid to elicitor-induced cells during
the initial period of increasing transcription; consequently, the mRNA
activities encoding PAL, CHS and CHI decline rapidly, resulting in
decreased synthesis of the corresponding enzyme subunits (G.P.
Bolwell, M.P. Robbins & R.A. Dixon, unpublished results).   These
changes occur against a background of no net quantitative effects on
mRNA and protein synthesis, although the overall patterns of newly
synthesised polypeptides differ qualitatively in the presence and
absence of cinnamate.   Furthermore, addition of cinnamate has little
effect on the translatable mRNA activity encoding cytochrome P-450
which may be involved in cinnamic acid hydroxylation.   It seems
possible, therefore, that cinnamate could act as a cellular modulator
for the "down-regulation" of expression of genes involved in phenylpropanoid metabolism.   Further work on the specificity of such effects
is in progress.

Unlike PAL, elicitor-induced CHS and cinnamic-acid 4-hydroxylase
extractable activities are relatively unaffected by exogenous
additions of trans-cinnamic acid to bean cells (G.P. Bolwell, M.P.
Robbins & R.A. Dixon, unpublished results), whereas CHI activity can
be markedly stimulated by such treatments (Gerrish et al., 1985).   The
effects of cinnamic acid on enzyme removal therefore appear to be
specific for PAL, and this effect has been investigated at the enzyme
activity and polypeptide subunit levels using antibody and active-
site-specific affinity probes (Bolwell et al., 1986a). Pulse-chase
experiments have revealed a pathway for PAL subunit turnover involving
M_r 70 000 and 53 000 partial degradation products in vivo (Bolwell et
al., 1986b).   However, further pulse-chase and immune blotting
experiments have indicated that cinnamic acid does not affect the rate
of turnover of PAL subunits in vivo, but rather mediates irreversible
inactivation of the enzyme.   A non-dialysble factor from cinnamate-
treated bean cells stimulates removal of PAL activity from enzyme
extracts in vitro; this effect is dependent upon the presence of
cinnamic acid, and is accompanied by an apparent loss or reduction of
the dehydroalanyl residue at the PAL active site (as detected by
active-site-specific tritiation) in the absence of an accompanying
loss in the levels of immunodetectable enzyme subunits.   Furthermore,

cinnamic acid-mediated loss of enzyme activity in vivo is accompanied,
in pulse-chase experiments, by a greater loss of [^{35}S]-labelled enzyme
subunits precipitated by an immobilized active site affinity ligand
than of subunits precipitated with anti-(PAL)IgG (Bolwell et al.,
1986a).   Taken together, this evidence strongly suggests that a likely
mechanism for cinnamic acid-mediated removal of PAL activity in vivo
may involve modification of the enzyme's active site.   This effect, in
addition to the effects on enzyme synthesis, could provide a rapid,
flexible mechanism for down-regulating PAL in response to an increased
cinnamate pool size.   If this hypothesis is correct, PAL can be seen
to be acting as a self-regulating valve controlling the entry of
phenylalanine (a primary metabolite required for protein synthesis)
into secondary pathways from which fixed carbon may not be readily
retrievable.   Such a mechanism is further tightened by the product
inhibition of PAL catalytic activity by cinnamic acid, especially in
view of the negatively rate cooperative kinetics of the enzyme (Lamb &
Rubery, 1976), whatever the molecular basis of this may be.

The overall interplay of factors which may regulate the in vivo level
of PAL in suspension cultured bean cells is summarised in Figure 2.

CONCLUDING REMARKS
          Manipulation of growth regulator levels in bean cell
cultures can lead to increased basal levels of phenylpropanoid
biosynthetic enzymes and consequent increased production of secondary
products such as isoflavonoid phytoalexins (e.g. Dixon & Fuller,
1978).   However, far greater production is obtained in response to
elicitation and prospects for the possible commercial application of
'biological stress agents' in the induction of useful plant secondary
products in culture systems have recently been discussed (Di Cosmo &
Tallevi, 1985).   It is now clear that the potential of such methods,
which must be as efficient as possible as they may involve losses in
cell viability which will preclude any form of continuous induction
regime, may be limited by endogenous regulatory mechanisms which
operate under natural conditions to protect the plant from excessive
flow from primary metabolism into secondary products.   These may of
course be by-passed by addition of precursors of the pathway beyond
the initial regulatory steps although, in the case of the phenylpropanoid pathway, cinnamic acid for example will also suppress the
overall induction of later regulatory enzymes such as CHS; addition of
later, more complex intermediates may not be cost effective.

It is not possible at present to suggest whether such negative
regulatory processes may easily be removed.   More information is
needed on their mode of operation; in addition to discovering the
nature of the regulatory proteins which may be involved in cinnamate-
mediated effects and assessing possibilities for their alteration or
inhibition, it will also be important to discover whether expression
of all PAL genes is switched off at the same rate in cinnamate-treated
cells, and whether all PAL subunit types containing catalytically
active residues are inactivated by cinnamate in the same manner.   If
not, prospects may emerge for the genetic manipulation of cells with

Figure 2.   Scheme for the possible control sites involved
in regulating the rapid transient induction of phenylalanine ammonia-lyase extractable activity in bean cell
cultures.   See text for details.

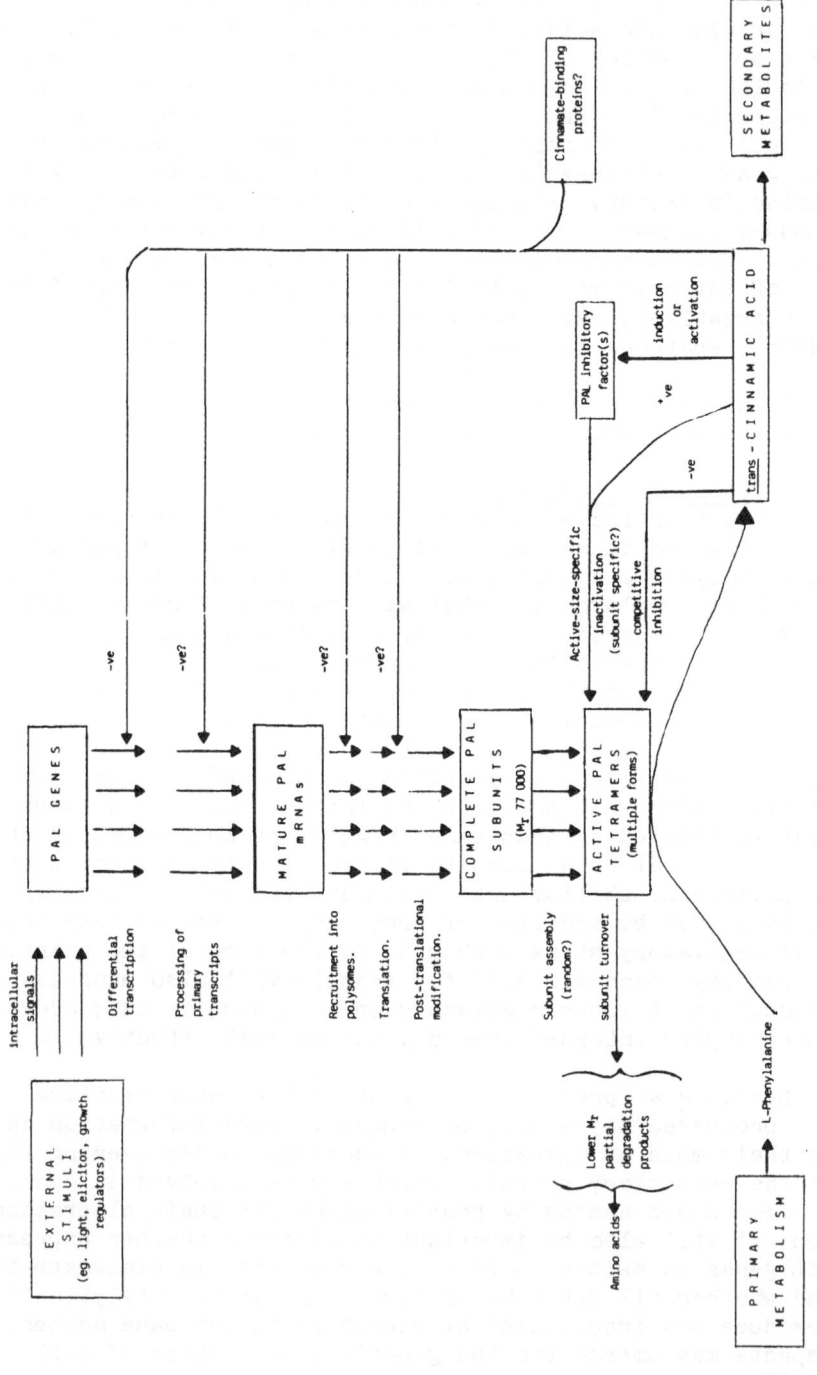

altered feedback regulatory mechanisms. In the meantime, the study of
the phenylpropanoid pathway is providing fascinating insights into the
complexities underlying the differential accumulation of gene products
in plant cells.

REFERENCES
Bolwell, G.P., Bell, J.N., Cramer, C.L., Schuch, W., Lamb, C.J. &
    Dixon, R.A. (1985a). L-Phenylalanine ammonia-lyase from
    Phaseolus vulgaris. Characterisation and differential
    induction of multiple forms from elicitor-treated cell
    suspension cultures. Eur. J. Biochem 149, 411-19.
Bolwell, G.P., Cramer, C.L., Lamb, C.J., Schuch, W. & Dixon, R.A.
    (1986a). L-Phenylalanine ammonia-lyase from Phaseolus
    vulgaris. Modulation of the levels of active enzyme by
    trans-cinnamic acid. Planta, in press.
Bolwell, G.P. & Dixon R.A. (1986). Membrane-bound hydroxylases in
    elicitor-treated bean cells. Rapid induction of prolyl
    hydroxylase and cytochrome P-450 synthesis and mRNA
    activities. Eur. J. Biochem., in press.
Bolwell, G.P., Robbins, M.P. & Dixon, R.A. (1985b). Metabolic changes
    in elicitor-treated bean cells. Enzymic responses in
    relation to rapid changes in cell wall composition. Eur.
    J. Biochem. 148, 571-78.
Bolwell, G.P., Sap, J., Cramer, C.L., Schuch, W., Lamb, C.J. &
    Dixon, R.A. (1986b). L-Phenylalanine ammonia-lyase from
    Phaseolus vulgaris. Partial degradation of enzyme
    subunits in vitro and in vivo. Biochim. Biophys. Acta,
    881, 210-21.
Chappell, J. & Hahlbrock, K. (1984). Transcription of plant defence
    genes in response to uv light or fungal elicitor. Nature
    311, 76-78.
Cramer, C.L., Bell, J.N., Ryder, T.B., Bailey, J.A., Schuch, W.,
    Bolwell, G.P., Robbins, M.P., Dixon, R.A. & Lamb, C.J.
    (1985). Co-ordinated synthesis of phytoalexin
    biosynthetic enzymes in biologically-stressed cells of
    bean. EMBO J. 4, 285-89.
Dewick, P.M., Steele, M.J., Dixon, R.A. & Whitehead, I.M. (1982).
    Biosynthesis of isoflavonoid phytoalexins ; incorporation
    of sodium [1,2-^{13}C$_2$] acetate into phaseollin and
    kievitone. Z. Naturforsch 37c, 363-68.
Di Cosmo, F. & Tallevi, S.G. (1985). Plant cell cultures and
    microbial insult: interactions with biotechnological
    potential. Trends in Biotechnol. 3, 110-11.
Dixon, R.A. & Bendall, D.S. (1978). Changes in the levels of enzymes
    of phenylpropanoid and flavonoid synthesis during
    phaseollin production in cell suspension cultures of
    Phaseolus vulgaris L. Physiol. Plant Pathol. 13, 295-306.
Dixon, R.A., Bolwell, G.P., Hamdan, M.A.M.S. & Robbins, M.P. (1986a).
    Molecular biology of induced resistance. In Genetics and
    Plant Pathogenesis, ed. P.R. Day & G.J. Jellis, Oxford :
    Blackwell Scientific Publications, in press.

Dixon, R.A., Bolwell, G.P., Robbins, M.P. & Hamdan, M.A.M.S. (1986b). Molecular targets for elicitor modulation in bean (Phaseolus vulgaris), In Biology and Molecular Biology of Plant-Pathogen Interactions, ed. J.A. Bailey, New York : Plenum, in press.

Dixon, R.A. Dey, P.M. & Lamb, C.J. (1983a). Phytoalexins : enzymology and molecular biology. Adv. Enzymol. 55, 1-136.

Dixon, R.A., Dey, P.M. & Whitehead, I.M. (1982). Purification and properties of chalcone isomerase from cell suspension cultures of Phaseolus vulgaris. Biochim. Biophys. Acta 715, 25-33.

Dixon, R.A. & Fuller, K.W. (1978). Effects of growth substances on non-induced and Botrytis cinerea culture filtrate-induced phaseollin production in Phaseolus vulgaris cell suspension cultures. Physiol. Plant Pathol. 12, 279-88.

Dixon, R.A., Gerrish, C., Lamb, C.J. & Robbins, M.P. (1983b). Elicitor-mediated induction of chalcone isomerase in Phaseolus vulgaris. Planta 159, 561-69.

Dixon, R.A. & Lamb, C.J. (1979). Stimulation of de novo synthesis of L-phenylalanine ammonia-lyase in relation to phytoalexin accumulation in Colletotrichum lindemuthianum elicitor-treated cell suspension cultures of French bean (Phaseolus vulgaris). Biochim. Biophys. Acta 586, 453-63.

Dynam, W.S. & Tjian, R. (1985). Control of eukaryotic messenger RNA synthesis by sequence-specific DNA-binding proteins. Nature 316, 774-78.

Edwards, K., Cramer, C.L., Bolwell, G.P., Dixon, R.A., Schuch, W. & Lamb, C.J. (1985). Rapid transient induction of phenylalanine ammonia-lyase mRNA in elicitor-treated bean cells. Proc. Natl. Acad. Sci. USA 82, 6731-35.

Gerrish, C., Robbins, M.P. & Dixon, R.A. (1985). Cinnamic acid as a modulator of chalcone isomerase in bean (Phaseolus vulgaris) cell suspension cultures. Plant Sci. Lett. 38, 23-27.

Grisebach, H. (1981). Lignins. In The Biochemistry of Plants, Vol. 7. Secondary Plant Products, E.C. Conn ed., Academic Press, New York; London; Toronto; Sydney; San Francisco.

Gupta, S. & Acton, G.J. (1979). Purification to homogeneity and some properties of L-phenylalanine ammonia-lyase of irradiated mustard (Sinapis alba L.) cotyledons. Biochim. Biophys. Acta. 570, 187-97.

Haddon, L. & Northcote, D.H. (1986). Correlation of the induction of various enzymes concerned with phenylpropanoid and lignin synthesis during differentiation of bean callus Phaseolus vulgaris L. Planta 128, 255-62.

Hagmann, M.L. & Grisebach, H. (1984). Enzymatic rearrangement of flavanone to isoflavone. FEBS Lett. 175, 199-202.

Hagmann, M.L., Heller, W. & Grisebach, H. (1984). Induction of phytoalexin synthesis in soybean. Stereospecific 3,9-dihydroxypterocarpan 6a-hydroxylase from elicitor-induced soybean cell cultures. Eur. J. Biochem. 142, 127-31.

Hahlbrock, K., Boudet, A.M., Chappell, J., Kreuzaler, F., Kuhn, D.N. & Ragg, H. (1982). Differential induction of mRNAs by light and elicitor in cultured plant cells, In Structure and Function of Plant Genomes, ed. O.Cifferi and L. Dure III, pp. 15-23, New York, Plenum.

Hahlbrock, K. & Grisebach, H. (1979). Enzymic controls in the biosynthesis of flavonoids and lignin. Annu. Rev. Plant Physiol. 30, 105-130.

Hahlbrock, K., Knobloch, K-H., Kreuzaler, F., Potts, J.R.M. & Wellmann, E. (1976). Coordinated induction and subsequent activity changes of two groups of metabolically inter-related enzymes. Light-induced synthesis of flavonoid glycosides in cell suspension cultures of Petroselenum hortense. Eur. J. Biochem. 61, 199-206.

Hahlbrock, K., Lamb, C.J., Purwin, C., Ebel, J., Fautz, E. & Schafer, E. (1981). Rapid response of suspension cultured parsley cells to the elicitor from Phytophthora megasperma var sojae. Induction of the enzymes of general phenyl-propanoid metaoolism. Plant Physiol. 67, 768-773.

Hauffe, K.D., Hahlbrock, K & Scheel, D. (1986). Elicitor-stimulated furanocoumarin biosynthesis in cultured parsley cells. : S-adenosyl-L-methionine : bergaptol and S-adenosyl-L-methionine : xanthotoxol O-methyltransferases. Z. Naturforsch, in press.

Havir, E.A. and Hanson, K.R. (1973). L-Phenylalanine ammonia-lyase (maize and potato). Evidence that the enzyme is composed of four subunits. Biochemistry 12, 1583-1591.

Jones, D.H. (1984). Phenylalanine ammonia-lyase : regulation of its induction, and its role in plant development. Phytochemistry 23, 1349-59.

Jones, D.H. & Northcote, D.H. (1981). Induction by hormones of phenylalanine ammonia-lyase in bean-cell suspension cultures. Inhibition and superinduction by actinomycin D. Eur. J. Biochem. 116, 117-125.

Kuhn, D., Chappell, J., Boudet, A. & Hahlbrock, K. (1984). Induction of phenylalanine ammonia-lyase and 4-coumarate CoA ligase mRNAs in cultured plant cells by UV light or fungal elicitor. Proc. Natl. Acad. Sci. USA 81, 1102-1106.

Lamb, C.J. & Rubery, P.H. (1976). Inhibition of co-operative enzymes by substrate-analogues : possible implications for the physiological significance of negative co-operativity illustrated by phenylalanine metabolism in higher plants. J. Theor. Biol. 60, 441-47.

Lawton, M.A., Dixon, R.A., Hahlbrock, K. & Lamb, C.J. (1983a). Rapid induction of the synthesis of phenylalanine ammonia-lyase and of chalcone synthase in elicitor-treated plant cells. Eur. J. Biochem. 129, 593-601.

Lawton, M.A., Dixon, R.A., Hahlbrock, K. & Lamb, C.J. (1983b). Elicitor induction of mRNA activity: rapid effects of elicitor on phenylalanine ammonia-lyase and chalcone synthase mRNA activities in bean cells. Eur. J. Biochem. 130, 131-139.

Lawton, M.A., Dixon, R.A. & Lamb, C.J. (1980). Elicitor modulation of
    L-phenylalanine ammonia-lyase in French bean cell
    suspension cultures. Biochim. Biophys. Acta 633,
    162-175.
Nari, J., Moutett, C., Fouchier, F. & Ricard, J. (1974). Some
    physico-chemical properties of L-phenylalanine ammonia-
    lyase of wheat seedlings. FEBS Lett. 23, 220-224.
Robbins, M.P., Bolwell, G.P. & Dixon, R.A. (1985). Metabolic changes
    in elicitor-treated bean cells: selectivity of enzyme
    induction in relation to phytoalexin accumulation. Eur.
    J. Biochem. 148, 563-569.
Robbins, M.P. & Dixon, R.A. (1984). Induction of chalcone isomerase
    in elicitor-treated bean cells. Comparison of rates of
    synthesis and appearance of immunodetectable enzyme. Eur.
    J. Biochem. 145, 195-202.
Rolfs, C-H. & Kindl, H. (1984). Stilbene synthase and chalcone
    synthase. Two different constitutive enzymes in cultured
    cells of Picea excelsa. Plant Physiol. 75, 489-492.
Shields, S.E., Wingate, V.P. & Lamb, C.J. (1982). Dual control of
    phenylalanine ammonia-lyase production and removal by its
    product cinnamic acid. Eur. J. Biochem. 123, 389-395.
Schroder, J., Betz, B. & Hahlbrock, K. (1977). Messenger RNA-
    controlled increase of phenylalanine ammonia-lyase
    activity in parsley. Light-independent induction by
    dilution of cell suspension cultures into water. Plant
    Physiol. 60, 440-445.
Schroder, J. & Schafer, E. (1980). Radioiodinated antibodies, a tool
    in studies on the presence and role of inactive enzyme
    forms: regulation of chalcone synthase in parsley cell
    suspension cultures. Arch. Biochem. Biophys. 203, 800-
    808.
Tietjen, K.G., Hunkler, D. & Matern, U. (1983). Differential response
    of cultured parsley cells to elicitors from two non-
    pathogenic strains of fungi. I. Identification of
    induced products as coumarin derivatives. Eur. J.
    Biochem. 131, 401-407.
Whitehead, I.M. & Dixon, R.A. (1983). Chalcone synthase from cell
    suspension cultures of Phaseolus vulgaris. Biochim.
    Biophys. Acta. 747, 298-303.
Zahringer, U., Ebel, J., Mulheirn, L.J., Lyne, R.L. & Grisebach, H.
    (1979). Induction of phytoalexin synthesis in soybean.
    Dimethylallylpyrophosphate: trihydroxypterocarpan
    dimethylallyl transferase from elicitor-induced
    cotyledons. FEBS Lett. 101, 90-92.
Zimmermann, A. & Hahlbrock, K. (1975). Light-induced changes of
    enzyme activities in parsley cell suspension cultures.
    Purification and some properties of phenylalanine ammonia-
    lyase. Arch. Biochem. Biophys. 166, 54-62.

12. INTERACTIONS AND INTERELATIONSHIPS BETWEEN PRIMARY AND SECONDARY METABOLISM.

M.W. Fowler

Wolfson Institute of Biotechnology,
The University of Sheffield,
Sheffield. S10 2TN.   U.K.

        While there is an already fairly extensive body of
information on the functioning of primary metabolic pathways in plant
cell culture, and information on secondary metabolism is beginning to
accumulate, little information is available on the interactions
between these two area. Yet the nature of events occurring in primary
metabolism is a key influence in determining the pattern and level of
secondary metabolite synthesis. This is not to say of course that it
is the only factor, the pattern of secondary metabolism in a cell
culture is a compound of many interacting factors, including the
physiological status of the culture, its origin, age, degree of
development, and so on. Primary metabolism underpins all this, and is
itself then dependent on the availability of a narrow range of
nutrients which provide the major support for all cellular activity.
These nutrients include a carbon source (typically glucose or
sucrose), a nitrogen source (nitrate or ammonia), molecular oxygen,
gaseous carbon dioxide or bicarbonate, and phosphate. These materials
of course not only feed the primary metabolic pathways, but in
modified form also serve as entry points into secondary metabolism, as
building blocks or core structures which are added to or much modified
through the highly versatile and flexible biosynthetic pathways of the
plant cell. Numerous entry points exist linking the primary metabolic
pathways of the plant cell with secondary metabolism. Figure 1 is
illustrative of such links, albeit in a much simplified form.

In following the flow of materials between primary and secondary
metabolism, the demands of secondary metabolite biosynthesis for
phosphorylative and redox energy donation should not be forgotten.
That there is a close association between those pathways concerned
with energy conservation and transfer, and biosynthesis is axiomatic;
many biosynthetic reactions also require a specific input of NADPH.
This latter is predominantly produced in the non-photosynthetic plant
cell through the oxidative pentose phosphate pathway. There are a
number of documented instances where high biosynthetic activity
correlates with an active participation of the pentose phosphate
pathway in carbohydrate oxidation and provision of NADPH (Stepan-
Sarkissian & Fowler, 1985).

The metabolism of the living cell is dynamic. There is no obvious
boundary between primary and secondary metabolism, the flow of
metabolite adjusting subtley to changes in cellular activity and

metabolic demand. Little is known of the control mechanisms which
integrate and maintain a balance between primary and secondary
metabolism, or what regulates the flow of metabolite into, perhaps
competing, secondary metabolic pathways. A number of gross influences
may be perceived, for instance quality and quantity of substrate
supply, or developmental state of the cells in the culture. The
precise way in which these influences are translated into controlled
responses by the cells is not understood.

Background information linking primary and secondary metabolism in
cell cultures is, as stated above, sparse. There are however in the
literature data indicative of the complex nature of the interactions
between the two areas. In some cases the data may be useful in
developing new approaches to product yield enhancement, a key area in
the development of plant cell cultures for industrial use. There are
precedents for such an approach from microbial physiology, where for
many years 'metabolite over-production' or 'overflow metabolism' based
on the manipulation of the major medium constituents has been an
important part of culture manipulation to improve product synthesis
(Neilssel & Tempest, 1974). Metabolite over-production or overflow
metabolism is based on the precept that the substrate is supplied at a
level sufficient not only to cope with the demands of primary
metabolic activity but has the capacity also to support secondary
metabolite synthesis. There are numerous examples from microbial
sources where increasing the level of one substrate or decreasing the
level of another key nutrient, thereby putting other nutrients into
overload, leads to increased secondary metabolite synthesis. There
are examples of a similar situation from a number of plant cell
cultures, although caution must be exercised in drawing too close a
correlation between the systems.

A particular situation with plant cell culture is quite often observed
with high levels of sucrose. Here while the level of biomass and the
growth rate achieved with sucrose is equivalent to that from equal
concentrations of glucose, the use of sucrose as the carbon source
often leads to enhanced levels of product synthesis (see Fowler &
Stepan-Sarkissian, 1986 for review). Whether this effect is caused
directly through sucrose as a metabolic precursor, or some less well
defined interaction as a control system, is impossible to say at this
juncture. Nonetheless, the observation is a consistent one with
important implications for the underlying metabolism. Reducing the
level of basal metabolites may also have major effects on secondary
metabolites. For instance, Wilson (1980) found that reducing the
level of inorganic phosphate available enhanced the synthesis of
anthocyanin in a number of cell cultures. Less clearly defined results
have also been obtained regarding alkaloid biosynthesis and the effect
of varying the quality and quantity of amino acids supplied to a
culture (Dougall, 1980). Unfortunately little detailed information is
available to enable us to unravel the interactions between the
'modulation' of the simple nutrient and the ultimate metabolic
response on the part of the culture.

Figure 1. INTERACTIONS BETWEEN SECONDARY AND PRIMARY
METABOLISM.

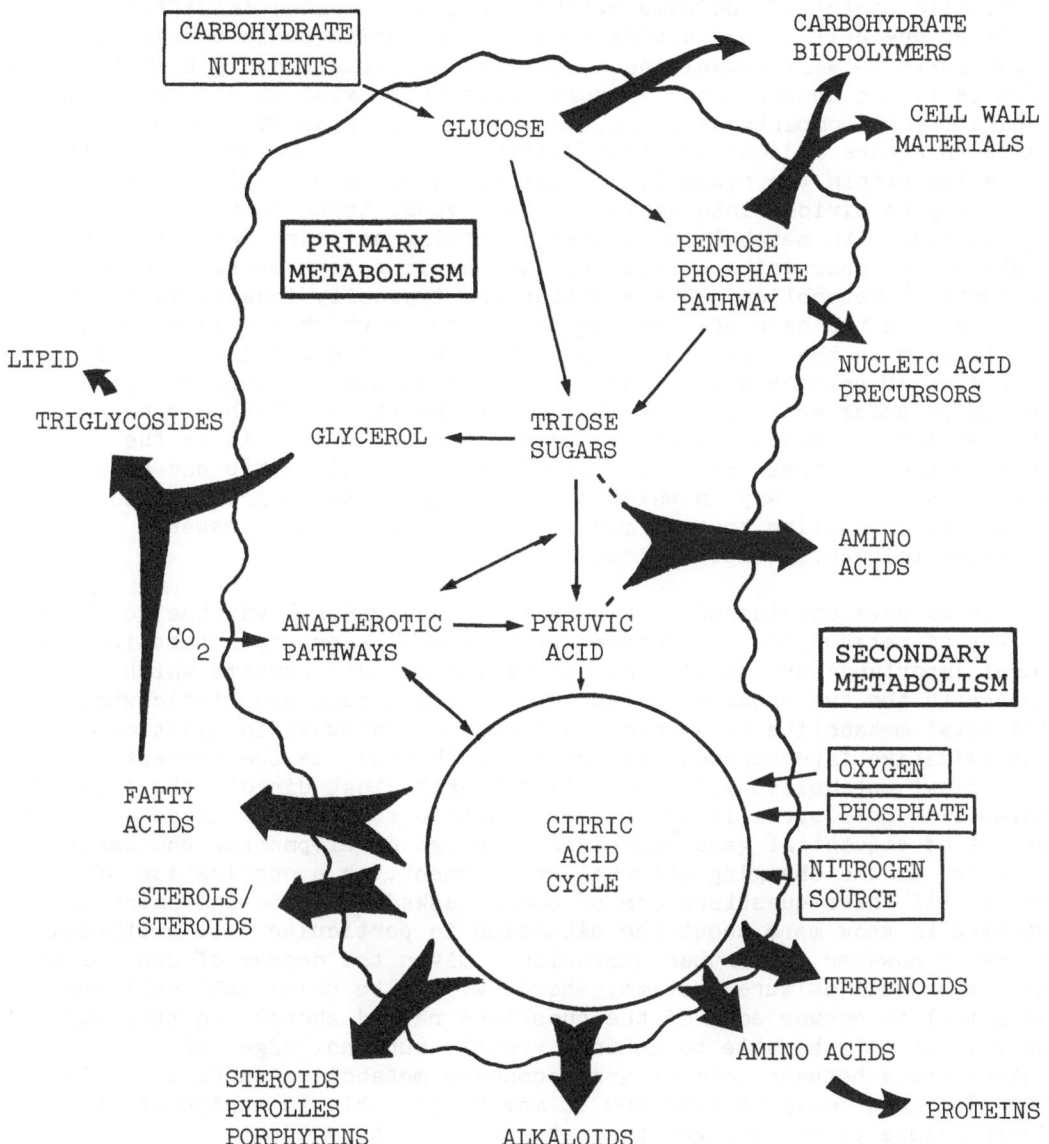

An additional complication of looking at interactions between primary and secondary metabolism in plant cells relates to the presence of 'metabolic pools' of the same metabolites, but located in different parts of the cell. That a wide range of metabolic pools exists in plant cells is well established. However the relationship between them is little understood. In some cases they exist in discrete and separate cell compartments, e.g. the vacuole or plastid. In other cases there are indications that distinct pools of the same metabolite may exist within the cytosol, presumably in close proximity. Metabolic pools may be divided into at least two types, those in dynamic equilibrium with metabolism, probably acting as short term storage systems and those which effectively act as long term stores for end-products of metabolism. These latter are typically located in the vacuole. In the case of those metabolic pools which are in dynamic equilibrium in the cell, they typically involve metabolites at the interface of primary and secondary metabolism e.g. C_4 organic acids, key amino acids such as glutamate, sugar phosphates which may be used in a variety of ways, and short chain fatty acids. It is in the regeneration of these metabolic pools, their relationship one with another and in the way in which they interact between primary and secondary metabolism that we may well find some of the answers relative to overflow metabolism.

So far we have considered in a very brief and general way the coupling by way of metabolite flow between primary and secondary metabolism. Of equal importance are considerations of the control systems which integrate the two areas of metabolism. At its most simplistic what 'directs' metabolite flow into the secondary pathways in addition to its oxidative fate through the primary pathways? Is the concept of 'overflow' metabolism valid in this context? What directs the flow of metabolite into specific secondary metabolic routes? Is it as a result of sequential gene expression through developmental phenomena, response to the changing external environment, or a combination of both? All these questions are of course asked from the viewpoint of wishing to know more about the situation in particular cell cultures, there is however, a further dimension. Given the degree of control we have over cell culture systems, should we not be using cell cultures as a tool to answer some of the questions raised above? In this way we may not only be able to enhance greatly our knowledge of interactions between primary and secondary metabolism in plant cells, but also gain insights into how it may be possible to manipulate such interactions to achieve greater natural product synthesis.

REFERENCES
Dougall, D.K. (1980). In: Plant Tissue Culture as a Source of
        Biochemicals (Ed. Staba, E.J.) pp.21-58. CRC Press, Boca
        Raton, Florida.
Fowler, M.W. & Stepan-Sarkissian, G. (1985). In: Primary and
        Secondary Metabolism of Plant Cell Cultures (Ed. Neumann,
        K.H., Barz, W. and Reinhard, E.). Springer-Verlag,
        Heidelberg, pp.66-73.

Neilssel, O.M. & Tempest, D.W. (1979). In: Microbial Technology;
          Current State, Future Prospects. (Eds. Bull, A.T.,
          Ellwood, D.C., Ratledge, C.). pp.53-82. Cambridge
          University Press, Cambridge.
Stepan-Sarkissian, G. & Fowler, M.W. (1984). In: Carbohydrate
          Metabolism in Cultured Cells. (Ed. Morgan, M.J.) Plenum
          Press, New York. (in press).
Wilson, G. (1980). Advances in Biochemical Engineering 16, 1-25.

13. CO-ORDINATED INDUCTION OF ENZYMES OF SECONDARY METABOLISM

M.P. Robbins

Department of Biochemistry, Royal Holloway and Bedford New College, University of London, Egham Hill, Egham, Surrey, TW20 0EX, U.K.

INTRODUCTION
        Cultured plant cells represent a potentially rich source of secondary metabolites of commercial importance. However, plant cells grown in culture may produce low or negligible quantities of the required secondary products. In recent times a number of workers have successfully used cell cultures to produce acceptable product levels, and much work has concentrated on the optimisation of growth and culture conditions (Morris et al., 1985; Fuller & Bartlett, 1985). It has also been suggested that a number of environmental stress conditions may result in an increase of secondary metabolite synthesis in plant cultures (reviewed by Di Cosmo & Towers, 1984). Recently there has been interest in the effects of the addition of biological chemicals to plant cell suspension cultures. In particular elicitor preparations derived from fungal and bacterial sources have been employed and it is possible that 'microbial abuse' of cell cultures may have a biotechnological potential for the production of natural compounds (Di Cosmo & Tallevi, 1985).

A well studied model system is the interaction between Phaseolus vulgaris cell suspensions and autoclaved cell wall preparations from the phytopathogenic fungus Colletotrichum lindemuthianum. Addition of elicitor to bean cell suspensions results in the accumulation of low molecular weight antibiotic phytoalexins (Dixon & Lamb, 1979) and recent work has shown that a number of different secondary metabolites are produced when bean cells are challenged with elicitor preparations from C. lindemuthianum. Two isoflavonoid phytoalexins, kievitone and phaseollin, show marked increases on elicitation (Robbins et al., 1985) and the levels of wall-bound phenolics and hydroxyproline also increase with this treatment (Bolwell et al., 1985). Table 1 details the peak levels of secondary products that accumulate in elicitor treated bean cells. Phytoalexin components increase from undetectable levels whereas wall-bound metabolites show more modest increases, with phenolic levels increasing nine-fold in a 24 hour period while hydroxyproline levels increase 2.8-fold in the first six hours after elicitation. Interestingly, the four different end points of secondary metabolism shown in Table 1 exhibit different timings with respect to the attainment of peak metabolite levels and it may be concluded that the accumulation of secondary products is not co-ordinate in plant cells treated with fungal elicitor preparations. The relative expression of different pathways of secondary metabolism deserves further study (Fowler, this volume; Brindle et al., this

volume), in order to understand the control processes involved in the
production of a range of secondary compounds by plant cells.

The appearance of secondary metabolites in elicited bean suspensions
is preceded by an increase in extractable activities of enzymes in
pathways leading to their production (Robbins et al., 1985; Bolwell et
al., 1985). Much work has concentrated on changes in phenylpropanoid
metabolism (Bolwell & Dixon, this volume) and a number of enzymes in
this pathway have been investigated (Robbins et al., 1985; Bolwell et
al., 1985; Dixon et al., 1986a). Table 2 outlines data representing
the induction of four enzymes involved in the biosynthesis of
isoflavonoid phytoalexins such as kievitone. The extractable
activities of all four enzymes increase on elicitor treatment and in
some cases the induction is quite dramatic, with 6'-hydroxy chalcone
synthase (CHS) activity increasing 47-fold while smaller proportional
changes are noted for phenylalanine ammonia lyase (PAL), cinnamic acid
4-hydroxylase (CA 4-H) and chalcone isomerase (CHI). Although the
increases in enzyme activities follow the addition of elicitor to the
cultures, the kinetics of the appearance of these different enzymes
are not identical (Dixon & Lamb, 1979) and one particularly striking
effect is the late induction of chalcone isomerase which clearly lags
behind the induction of earlier enzymes in the biosynthetic pathway
(Robbins & Dixon, 1984).

Current techniques permit the measurement of a number of parameters
underlying elicitor modulation of induced enzyme activities. Immune
blotting using antibody probes to elicited gene products can quantify
increases in amounts of immunodetectable protein, and in vivo and in
vitro labelling experiments allow the measurement of rates of
synthesis of individual components of an induced metabolic pathway.
In addition, cDNA hybridisation experiments can measure changes in
steady state message levels; analysis of recent data suggests that
many elicitor-mediated changes are characterised by rapid transient
changes in rates of enzyme synthesis and that this is consistent with
significant increases in mRNA levels encoding these enzymes (Dixon et

Table 1.   Peak levels of secondary metabolites following
elicitation of Phaseolus vulgaris cell suspension
cultures cv. Immuna.

Compound	Initial level/g. fresh weight	Maximum level/g. fresh weight	Time of peak level
Kievitone	Not detectable	42.0 nmol	12 hours
Phaseollin	Not detectable	537.0 nmol	48 hours
Wall-bound phenolics	4.8 A_{310} units	43.3 A_{310} units (24 hours)	>24 hours
Wall-bound hydroxyproline	0.76 µmol	2.14 µmol	6 hours

al., 1986b).   Table 3 shows data relating to changes in gene
expression of four components relating to isoflavonoid phytoalexin
accumulation.   PAL, CHS and CHI are early enzymes in the induced
pathways while cytochrome P_{450} containing enzymes are also members of
the pathway (Dixon et al., 1986a).   CA-4H is known to be a cytochrome
P_{450} enzyme and may be detected using a mammalian P_{450} monoclonal
antibody (Bolwell & Dixon, 1986).   Results in Table 3 indicate that
elicitation results in an increase in mRNA level, followed by an
increase in rates of protein synthesis which then precedes a peak
level of immunodetectable gene product.   It is interesting to note
that though PAL, cytochrome P_{450} proteins, CHS and CHI are all induced
by elicitor, CHI expression lags behind that of earlier members of the
pathway, which is consistent with the data for extractable enzyme
activities (Table 2).

Table 2.   Initial peak levels of four enzymes of
isoflavanoid metabolism in elicited cell suspension
cultures of Phaseolus vulgaris cv. Immuna.

Enzyme	Initial level/ Kg.fresh weight	Maximum level/ Kg.fresh weight	Time of peak level
Phenylalanine ammonia lyase	30 µkat	240 µkat	6-8 hours
Cinnamic acid 4-hydroxylase	5.0 µkat	14.6 µkat	6 hours
Chalcone synthase	0.2 µkat	8.4 µkat	6 hours
Chalcone isomerase	1.2 mkat	4.5 mkat	16-24 hours

Table 3.   Changes in gene expression of four components
of isoflavonoid metabolism in elicited cell suspension
cultures of Phaseolus vulgaris cv. Immuna (time of
attainment of peak levels).   ND. Not determined.

	Immunodetectable protein	Rate of protein synthesis	mRNA activity	mRNA level
Phenylalanine ammonia lyase	6-8 hours	4 hours	2-4 hours	2-4 hours
Cytochrome P_{450}	8 hours	2 hours	2 hours	ND
Chalcone synthase	ND	4 hours	2-4 hours	2-4 hours
Chalcone isomerase	12 hours	6-10 hours	6 hours	ND

The information presented in this paper suggests that in the pathway to phytoalexins there is not an absolute co-ordination of gene expression of different members of the pathway. However, under some culture conditions co-ordination of temporal kinetics may occur and this may explain the co-ordinate induction of PAL, CHS and CHI noted by Cramer et al., (1985) in P. vulgaris cultures challenged with C. lindemuthianum elicitor preparations. Non co-ordination of induction of enzymes of secondary metabolism is not restricted to elicitor perturbations of cell cultures and similar phenomena have been noted for light induction of flavonoids in parsley (Petroselinum hortense) cell cultures. Enzyme induction in the parsley flavonoid pathway appears to occur in two temporally distinct stages (Hahlbrock et al., 1976) and this timing appears to be rigorous even at the molecular level, where PAL and 4-coumarate : CoA ligase message level increases precede message level changes corresponding to CHS (Kuhn et al., 1984). The sequential induction of synthesis of enzymes of secondary metabolism noted by Hahlbrock and co-workers suggests a complexity in the processes underlying the production of low-molecular weight metabolites in cultured plant cells. Pathway intermediates may play a crucial role controlling the expression of particular pathways of secondary metabolism (Bolwell & Dixon, this volume) and other controlling factors may also exist. It is clear that if there is to be commercial exploitation of plant cell cultures, further studies are required in order to understand the control of metabolic pathways of interest.

REFERENCES

Bolwell, G.P., Robbins, M.P. & Dixon, R.A. (1985). Metabolic changes in elicitor-treated bean cells. Enzymic responses associated with rapid changes in cell wall components. Eur. J. Biochem., 148, 571-578.

Bolwell, G.P. & Dixon, R.A. (1986). Membrane-bound hydroxylases in elicitor-treated bean cells. Rapid induction of prolyl hydroxylase and cytochrome P_{450} synthesis and mRNA activities. Eur. J. Biochem. (In Press).

Cramer, C.L., Bell, J.N., Ryder, T.B. Bailey, J.A., Schuch, W., Bolwell, G.P., Robbins, M.P., Dixon, R.A. & Lamb, C.J. (1985). Co-ordinated synthesis of phytoalexin biosynthetic enzymes in biologically-stressed cells of bean (Phaseolus vulgaris L.) EMBO J., 285-289.

Di Cosmo, F. & Tallevi, S.G. (1985). Plant cell cultures and microbial insult : interactions with biotechnological potential. Trends in Biotechnology, 3, No. 5, 110-111.

Di Cosmo, F. & Tower, G.K.N. (1984) Stress and secondary metabolism. In Recent Advances in Phytochemistry, 18, ed. B.N. Timmerman, C. Steelink, F.A. Loewus, pp. 97-175. New York : Plenum Press.

Dixon, R.A. & Lamb, C.J. (1979) Stimulation of de novo synthesis of L-phenylalanine ammonia-lyase in relation to phytoalexin accumulation in Colletotrichum lindemuthianum elicitor-treated cell suspension cultures of French bean

(Phaseolus vulgaris). Biochimica and Biophysic. Acta.
586, 453-463.

Dixon, R.A., Bolwell, G.P., Robbins, M.P. & Hamdan, M.A.M.S. (1986a).
Molecular targets for elicitor modulation in bean. In
Biology and Molecular Biology of Plant : Pathogen
Interactions, ed. J.A. Bailey. New York, NATO ASL Series,
Plenum Press. (In Press).

Dixon, R.A., Bolwell, G.P., Hamdan, M.A.M.S. & Robbins, M.P. (1986b)
Molecular biology of induced resistance. In Genetics and
Plant Pathogenesis, ed. P.R. Day and G.J. Jellès. Oxford
: Blackwell Press. (In Press).

Fuller, K.W. & Bartlett, D.J. (1985). The chemosynthetic potential of
plants and its realisation by immobilized systems. Ann.
Proc. Phytochem. Soc. Eur., 26, 229-247.

Hahlbrock, K., Knobloch, K-H., Kreuzaler, F., Potts, J.R.M. &
Weltmann, F. (1976). Coordinated induction and subsequent
activity changes of two groups of metabolically
interrelated enzymes. Light induced synthesis of
flavanoid glycosides in cell suspension cultures of
Petroselinum hortense. Eur. J. Biochem., 61, 199-206.

Kuhn, D.N., Chappell, J. Boudet, A. & Hahlbrock, K. (1984). Induction
of phenylalanine ammonia-lyase and 4-coumarate : CoA
ligase mRNAs in cultured plant cells by uv light or fungal
elicitor. Proc. Natl. Acad. Sci. USA, 81, 1102-1106.

Morris, P., Scragg, A.H., Smart, N.J. & Stafford, A. (1985). Secondary
product formation by cell suspension cultures. In Plant
Cell Culture : a practical approach, ed. R.A. Dixon.
Oxford : IRL Press Ltd.

Robbins, M.P. & Dixon, R.A. (1984) Induction of chalcone isomerase in
elicitor-treated bean cells. Comparisons of rates of
synthesis and appearance of immunodetectable enzyme. Eur.
J. Biochem., 145, 195-202.

Robbins, M.P., Bolwell, G.P. & Dixon, R.A. (1985) Metabolic changes
in elicitor-treated bean cells. Selectivity of enzyme
induction in relation to phytoalexin accumulation. Eur.
J. Biochem., 148, 563-569.

14. THE BIOTRANSFORMATION OF MONOTERPENOIDS BY PLANT TISSUE CULTURE

G. Lappin, J. Tampion & J.D. Stride

School of Biotechnology,
The Polytechnic of Central London,
115 New Cavendish Street,
London W1M 8JS

Monoterpenoids are economically-important compounds, being the main constituents of essential oils which are used extensively by the perfumery industry. Natural oils are extracted from plants which have been grown in the field. The methods of extraction vary widely, for example, bergamot oil (from Citrus durdutium var bergamia) is removed by rasping the fruit under a stream of cold water. On the other hand, lavender oil (from Lavandula Sp.) is collected by steam distillation. Monoterpenoids are also chemically manufactured. The starting material is frequently ∝- or β- pinene, which is, itself, extracted from natural sources. The chemical manufacture can demand high temperatures and pressures and often involves industrially-secret catalysts (e.g. Theimer 1982). In addition, the products are nearly always racemic and minor impurities are almost inevitably present. For these reasons there has recently developed an interest in producing these valuable compounds by a biotechnological approach. Some micro-organisms are known to produce monoterpenoids (e.g. Schindlere & Schmid 1982) but they do not produce the same diversity of structural forms, nor the same quantities that are available from vascular plants. Efforts have been made to use micro-organisms to transform cheap and plentiful monoterpenoids into rare and expensive compounds. Although there have been successes (e.g. Rehm & Reed 1985), the results have been, on the whole, disappointing (Abraham et al. 1984). Efforts have been made to produce monoterpenoids from plant tissue cultures (e.g. Nickell 1980). The majority of undifferentiated cultures do not produce essential oils (with a few exceptions e.g. Watts et al. 1985), although there has been some success using suspension cultures consisting of aggregates which exhibit a differentiated or organised structure (e.g. Berlin et al. 1984). On the other hand, undifferentiated, unproductive, suspension cultures have been used to biotransform monoterpenoids. Cultures of Nicotiana tabacum have been used in the stereospecific hydroxylation of linalool, dihydrolinalool, linalyl acetate,∝- and β- terpineol and β-terpenyl acetate (Suga et al. 1980, Hirata et al. 1982) and 1-acetoxy-P-menth-4(8)ene (Lee et al. 1983). Furthermore, carvone and carvoxime were stereo-selectively reduced by tobacco cultures (Suga et al. 1984). Cannabis sativa suspension cultures oxidised and isomerised geraniol to geranial and neral but did not metabolise a number of other structurally-related substrates (Itokawa et al. 1977). Aviv et al. (1981) examined the biotransformation of menthone to neomenthol and pulegone to isomenthone by suspension cultures of

Mentha species. It was found that the neomenthol, once formed, accumulated to a maximum over twelve hours and then disappeared from the cultures over the following six hours. The disappearance of the neomenthol did not coincide with the formation of any other metabolites that could be detected by Gas Chromatography (GC).

We have established a rapidly-growing suspension culture of Lavandula angustifolia and have examined its ability to biotransform a number of monoterpenoids and structurally-related compounds. Myrcene, (+)-linalool, (-)-menthone, (-)-menthol, (±)-camphor, (±)-α-pinene and (S)-β-pinene were not metabolised, but the acyclic monoterpenoid aldehydes (±)-citronellal, citral and the cyclic monoterpenoid aldehyde, (S)-perillaldehyde were rapidly reduced to the corresponding alcohols (Fig. 1). Other, structurally-related substrates (benzaldehyde and (E)-cinnamaldehyde were also reduced but at a slower rate (Fig. 2). In contrast, octanal was not metabolised. The reductase activity was completely inhibited by heat treatment (70°C for 10 min) and by 10^{-2}M 4(hydroxymercuri)-benzoic acid (sodium salt). Geraniol and citronellol, once formed by the reduction of the corresponding aldehydes, were found to disappear from the L. angustifolia cultures (Fig. 3) without the formation of any other GC-detectable metabolites. In contrast, the aromatic primary alcohols, benzyl alcohol and (S)-cinnamalcohol, the cyclic monoterpenol primary alcohols, perillalcohol, linalool survived unchanged. This observation was similar to the one made by Aviv et al. (previously cited) when they noted the disappearance of neomenthol from suspension cultures of Mentha species. Neomenthol is known to be glucosylated in vivo (Croteau & Martinkus 1979) and many glucosylated monoterpenoids have been reported to exist in plants (e.g. Banthorpe & Mann 1971). It is possible that glucosylation was responsible for the disappearance of the products from the L. angustifolia cultures. This possibility was investigated by attempting to recover the parent alcohols by acid hydrolysis of the culture. It was found, however, that no parent alcohols were recoverable and, in addition, direct TLC analysis of the cultures failed to reveal the presence of any glucosides. From these results, therefore, it can be concluded that the monoterpenoids primary alcohols were being either incorporated into higher terpenoids or other secondary metabolites or they were being catabolised, possibly returning to the carbon pool. There is evidence that both types of activity may occur in plants (Balsevich et al 1982, Banthorpe & Osbourne 1984, Croteau 1984, Croteau & Sood 1985). In either case, such metabolic activity would be a serious barrier to the utilization of tissue cultures for the commercial biotransformation of monoterpenoids. Furthermore, this type of catabolic activity may explain the absence of monoterpenoids from undifferentiated cultures. The presence of monoterpenoids might occur in cultures when descrete glands are present, supplying a metabolic sink in which essential oils can accumulate. The composition of essential oils found in plants, and their seasonal variation, might depend upon the balance of catabolic and anabolic activity. These controls may not exist in culture and the monoterpenoids might be catabolised immediately following their formation.

Acknowledgements
This work was financially supported by a SERC grant (GL) and by Tate &
Lyle Ltd. We would also like to thank Drs. P. Rogers and C. Bucke, at
Tate & Lyle for their assistance.

Figure 1.   Time-courses for the reduction of citronellal
(O) and perillaldehyde (●). The time-course for the
reduction of citral was similar to that for citronellal.

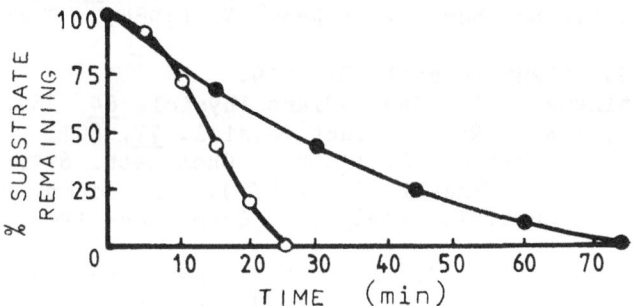

Figure 2.   Time-courses for the reduction of benzaldehyde
(●) and cinnamaldehyde (O).

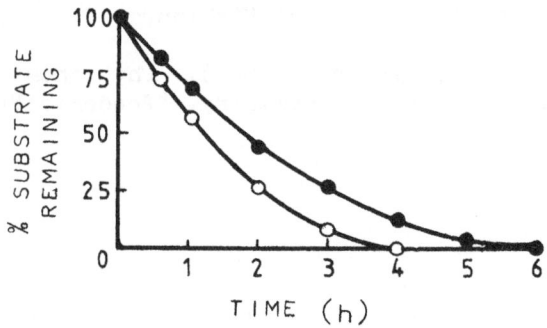

Figure 3.   Time-courses for the disappearance of
citronellol. The disappearance of geraniol followed a
similar time-course.

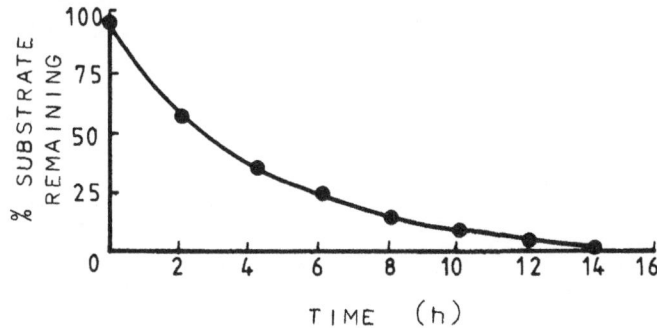

REFERENCES

Abraham, W.R., Kieslich, K., Reng, H. & Stumpf, B. (1984). 3rd.
    European Congress on Biotechnology. Vol. 1. 245-248.
    Verlag-Chemie Weinheim.
Aviv, D., Krochmal, E. & Galum, E. (1981). Planta Med. 42, 236.
Balsevich, J., Constabel, F. & Kurz, W.G. (1982). Planta Med. 44,
    231.
Banthorpe, D.V. & Mann, J. (1971). Phytochem. 11, 2589.
Banthorpe, D.V. & Osborne, M.J. (1984). Phytochem. 23, 905.
Berlin, J., Witte, L., Schubert, W. & Wray, V. (1984). Phytochem. 23,
    1277.
Croteau, R. (1977). Plant Physiol. 59, 519.
Croteau, R. & Mertinkus, C. (1979). Plant Physiol. 64, 169.
Croteau, R. & Sood, V.K. (1985). Plant Physiol. 77, 801.
Hirata, T., Lee, Y. & Takayuki, S. (1982). Chem Lett. 671.
Itokawa, H., Takeya, K. & Mihashi, S. (1977). Phytochem. 23, 1327.
Lee, Y., Hirata, T. & Suga, T. (1983). J. Chem. Soc. (Perkin Trans.
    I) 2475.
Nickell, L.G. (1980). Chap. 10. in Plant Tissue Culture as a Source
    of Biochemicals, Staba, E.J. ed. CRC Press Inc. Boca
    Raton.
Rehm, J-J. & Reed, G. (1985). Biotechnology (A Comprehensive
    Treatise in 8 vols) Vol. 6a. Verlag Chemie.
Schindler, J. & Schmid, R.D. (1982). Process Biochem. 17 (2), 2.
Suga, T., Aoki, T., Hirata, T., Lee, Y. & Nishimura, O. (1980). Chem.
    Lett. 229.
Suga, T., Hirata, T. and Futatsugi, M. (1984). Phytochem. 23, 1327.
Theimer, E.T. ed. (1982). Fragrance Chemistry. Academic Press.

15. BIOSYNTHESIS OF PHYTOALEXINS AND STEROLS IN POTATO CELL
SUSPENSION CULTURES

P.A. Brindle, P.J. Kuhn[*] & D.R. Threlfall

Department of Plant Biology & Genetics, University of
Hull, Hull   HU6 7RX, U.K.

[*] Shell Bioscience Laboratory, Sittingbourne Research
Centre, Sittingbourne, ME9 8AG, U.K.

INTRODUCTION
     Phytoalexins are low-M_r, antimicrobial compounds that are
synthesized by and accumulated in plants after exposure to micro-
organisms, to various fractions derived from pathogens and
non-pathogens (biotic elicitors) or to a wide variety of other agents
(abiotic elicitors). In phytochemical terms, phytoalexins are
secondary metabolites which are distinguished from chemically related
secondary metabolites solely by virtue of having been synthesized de
novo in an infected (or suitably treated) tissue. Phytoalexins with
terpenoid skeletons have been isolated from members of the
Convolvulaceae (sesquiterpenoids), Euphorbiaceae (diterpenoids),
Malvaceae (sesquiterpenoids) and Solanacea (sesquiterpenoids).

The carbocyclic sesquiterpenoid phytoalexins rishitin and lubimin are
accumulated rapidly in the upper cell layers of potato (Solanum
tuberosum L.) cv. Kennebec tuber discs which have been aged and then
the top surfaces either infected with an incompatible race (or an
appropriate compatible race) of the fungus Phytophthora infestans
(Mont.) de Bary or inoculated with a biotic elicitor from the fungus
(eg. cell-free mycelial preparations, arachidonic acid etc.) (see Kúc
& Rush 1985). Concomitant with the induction of phytoalexin
accumulation by these treatments, steroid glycoalkaloid (solanine and
chaconine) accumulation by the discs is inhibited (see Kúc & Rush
1985). It has been suggested that steroid glycoalkaloid accumulation,
which along which an increase in the sterol content of potato tuber
discs (Hartmann & Benveniste 1974) is part of the ageing process

Phytuberin

Solavetivone

Rishitin

Hydroxylubimin

Lubimin

(wounding reaction), is inhibited due to the diversion of farnesyl pyrophosphate away from steroid glycoalkaloid (and by implication sterol) biosynthesis and into phytoalexin biosynthesis, possibly as a result of the loss of squalene synthetase activity (see Kǔc & Rush, 1985). This simplistic interpretation, however, is open to criticism (Ishizaka & Tomiyama 1972).

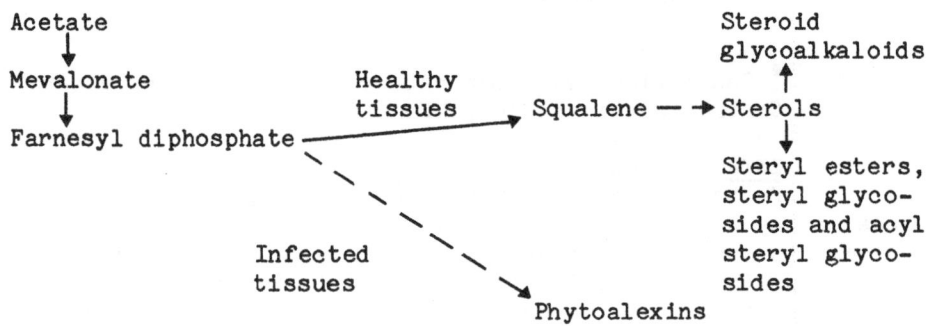

In order to study the biosynthesis and metabolism of potato phytoalexins and the interrelationship between sesquiterpenoid phytoalexin biosynthesis and triterpenoid biosynthesis in the absence of the wounding reaction, we have turned our attention to the use of potato cell-suspension cultures.

RESULTS AND DISCUSSION
In agreement with a previous study (Brindle et al. 1983), it was shown that phytoalexin accumulation in 21-day-old Kennebec (R_1) cell suspension cultures (ninth and tenth transfers) grown on MS medium supplemented with kinetin (0.1 mg.1) and NAA (1 mg/1) is induced by inoculation with sporangia (8 x 10^3/ml) of either an incompatible race (race 4) or a compatible race (complex race) of P. infestans. In addition, it was shown for the first time that phytoalexins are accumulated in Majestic (r)-cell suspension cultures (11-day old; media as before; fourth and fifth transfer) inoculated with the complex race (8 x 10^3 sporangia/ml) of P.infestans (Race 4 not tested). A noticeable browning of the white cultures became visible at two hours. Thereafter the browning became more intense and was maximal by 8-12 hours. Even after 24 hours only a small number of zoospores had germinated and there was no apparent invasion of the potato cells by zoospores. There was no evidence of lysis of the potato cells.

Phytoalexins were detectable by GC some three/four hours after inoculation, after which time the total phytoalexin content of the cultures rose in a linear fashion for some 24 hours and then declined (Table 1). The greater part (> 90%) of the phytoalexins in the cultures were recovered from the culture filtrates (and cell washings).

In the Kennebec cultures, the only phytoalexin produced was rishitin (Table 1). The Majestic cultures produced lubimin, 3-hydroxylubimin and rishitin and, in the presence of substrate amounts of MVA (see below), phytuberin (Table 1). The failure of the Kennebec cultures to accumulate large amounts of solavetivone and lubimin after the fifth transfer (Brindle et al. 1983 and this paper) and the eventual loss of their ability to produce rishitin show that the cultures were not stable. The patterns of accumulation and, in the case of lubimin, disappearance of phytoalexins from the Majestic cultures were consistent with the operation of the biosynthetic sequence: lubimin ⟶ 3-hydroxylubimin ⟶ rishitin, and the sequential induction of the enzymes concerned with the synthesis of lubimin from farnesyl diphosphate and its conversion to rishitin (Table 1). No metabolites of rishitin were isolated from the cultures, despite the fact that Kennebec-cell suspension cultures are known to metabolise exogenous rishitin to 13-hydroxyrishitin (Ward et al. 1977).

The accumulation of, and the incorporation of radioactivity into, phytoalexins, sterols and steryl glycosides (the cultures did not produce steroid alkaloids) were determined for 11-day-old Majestic (r) cultures (fourth and fifth transfers) which had been administered a single dose of either (R)-[2-^{14}C]MVA (3.3 mM) or [2-^{14}C]acetate (1 mM) at a saturating concentration for sterol synthesis and then, after 30 min, inoculated with sporangia (8 x 10^3/ml) of a compatible race of P. infestans. The results (Table 2) showed quite clearly that:

1. Added MVA acted as the major/sole carbon source for terpenoid synthesis in the cultures. It increased the rates of synthesis, and total amounts present, of squalene, sterols and steryl glycosides in healthy cultures and, in the short term, infected cultures and of phytoalexins in infected cultures.
2. Added acetate triggered an increase in phytoalexin and sterol (short term) production even though it was metabolized prior to

Table 1.   Amounts of phytoalexins at selected timepoints.   [a] Ca 25 g.fr.wt. of cells/100ml. Values relative to Me arachidate. [b] Similar amounts present in Kennebec/complex.   [c] Not continued beyond 24 hr.   [d] Includes 151 µg of phytuberin.

Interaction	Time (hr)			Amount (µg/100ml)[a]	
		Lubimin	3-Hydroxylubimin	Rishitin	Total
Kennebec/race 4[b]	24	0	0	543	543
	36	0	0	312	312
Majestic/complex	12	54	11	9	74
	24[c]	7	86	88	181
plus 3.3 mM R-MVA	24[d]	903	152	667	1873

the phase of phytoalexin production. The reason for this
non-specific enhancement of terpenoid synthesis is not clear. It
may be due to an enhancement of the general metabolic activity of
the cells brought about as the result of short term stimulation
of the TCA cycle by the added acetate.
3. Concomitant with the onset of phytoalexin accumulation, squalene
and cycloartenol synthesis stopped, whereas their utilization as
sterol intermediates was unaffected.

The MVA results showed that there was complete inhibition
of the flow of carbon from farnesyl diphosphate into sterol synthesis
at the start of phytoalexin accumulation, ie ca. four hours after
inoculation of the cultures with P. infestans. The onset of the
inhibition was extremely rapid and seemed to be due to the
inactivation or loss of squalene synthetase activity. The reason for
the inhibition is under investigation.

Table 2. Effect of exogenous R-[2-^{14}C]MVA. [a] Ca 1 g.dry wt. (25
g.fr.wt.)/100ml of culture. [b] Time after inoculation
with P. infestans ie T + 30 min after administration of
[2-^{14}C]MVA. [c] Rose to 48K cpm after 4 hr and then
declined. [d] This value reached after 4 hr.

Component	Amount (μg/g dry wt.)[a]		
	0 hr[b]	Control, 24 hr	Inoculated, 24 hr
Phytoalexins	0	0	1644
Squalene	(14K cpm)	(748K cpm)	(12K cpm)[c]
Cycloartenol	183	3773	513[d]
Monomethylsterols	13	114	48
Demethylsterols	340	440	384
Steryl glycosides	510	590	530
Lupeol	215	236	202

REFERENCES
Brindle, P.A., Kuhn, P.J. & Threlfall, D.R. (1983). Phytochemistry
22, 2719.
Hartmann, M.A. & Benveniste, P. (1974). Phytochemistry 13, 2667.
Ishizaka, N. & Tomiyama, K. (1972). Plant Cell Physiol. 13, 1053.
Kúc, J. & Rush, J.S. (1985). Arch. Biochem. Biophys. 236, 455.
Ward, E.W.B., Stoessl, A. & Stothers, J.B. (1977). Phytochemistry 16,
2024.

16. STUDIES ON THE REGULATION OF CAPSAICIN BIOSYNTHESIS IN
     IMMOBILIZED CELL CULTURES AND THE DEVELOPING FRUIT OF THE
     CHILLI PEPPER, CAPSICUM FRUTESCENS

R.D. Hall, M.A. Holden, M.M. Yeoman

Department of Botany, University of Edinburgh, Mayfield
Road, EDINBURGH, EH9 3JH

INTRODUCTION
     Much of the work carried out to study the accumulation of
secondary metabolites by cultured plant cells has had a strongly
empirical approach. This was necessitated by our very limited
knowledge of the processes of plant cell differentiation in general
and the control of the expression of secondary metabolic pathways in
particular. Although such investigations yield much important
information on the physiological and metabolic behaviour of cultured
cells they do not explain why the empirical procedures are successful.
Results obtained so far would suggest that it is only with a better
understanding of these processes that we are likely to be able to
manipulate cell cultures successfully to produce consistently high
metabolite yields. Consequently, work has begun in this laboratory to
investigate in detail the biosynthesis and accumulation of capsaicin
in cell culture systems and to compare this with accumulation in the
whole pepper fruit. In so doing it is envisaged that more detailed
information will be obtained as to the control mechanisms involved in,
and the possible limitations to capsaicin accumulation in vitro.
Treatments which have already been shown to enhance capsaicin yield in
pepper cultures (e.g. nutrient limitation, precursor feeding, cell
immobilization; Lindsey & Yeoman, 1985 and references therein) have
also proved to be successful in increasing secondary metabolite
accumulation in other systems (e.g. Knobloch & Berlin, 1981; Brodelius
et al. 1982). It is reasonable to assume therefore, that an
understanding of the regulation of capsaicin biosynthesis will allow
the application of a more directed approach to the biochemical/genetic
manipulation of a range of culture systems to improve product yield.

CAPSAICIN BIOSYNTHESIS IN WHOLE FRUIT AND CULTURED CELLS
     Capsaicin (8-methyl-N-vanillyl-6-nonenamide) is the
compound responsible for the hot 'taste' in the fruit of the chilli
pepper, C. frutescens. It is accumulated solely in the fruit of
mature pepper plants (Holden et al. 1986). Accumulation begins c. 20d
after fertilisation, is coincident with the decline in fruit growth
and continues for 10-15d (Hall et al. 1986). There is no evidence for
product turnover (authors, unpub. obs.; Lindsey, 1986). Capsaicin
accumulation in cell cultures has been studied in this laboratory for
a number of years. Results of these investigations have indicated
that cell suspension cultures of C. frutescens accumulate capsaicin in
the bathing medium although levels remain extremely low (Lindsey &

Yeoman, 1984). However, attempts to improve culture yield by
precursor feeding an/or nutrient limitation have been most successful,
particularly so if the cultured cells are also immobilized in an inert
support (reticulate polyurethane, Lindsey et al. 1983). Manipulation
of cultured pepper cells in this way can enhance capsaicin production
from ng to mg levels thus bringing culture yields on a par with whole
fruit (Lindsey & Yeoman, 1984).

STUDIES ON THE ACCUMULATION OF CAPSAICIN AND ITS PHENOLIC PRECURSORS

The proposed capsaicin biosynthetic pathway is presented
in Figure 1 in the context of general cell metabolism. In this
investigation standard HPLC techniques have been employed in
association with a Diode Array detector to separate and identify the
pathway components and to determine their intracellular pool sizes.
In addition, [^{14}C]-labelled precursors have been used to study the
metabolism of these intermediates within pepper cells. Work to date
has been concentrated on the phenolic side of the pathway as the
exogenous supply of precursors on this side of the pathway has been
shown to enhance capsaicin yield (Lindsey, this volume). Details of
the methodology may be found in Lindsey et al. (1983) and Hall et al.
1986.

In Table 1 are presented the data from a number of experiments where
[^{14}C]-phenylalanine and [^{14}C]-cinnamic acid were fed to whole pepper
fruit and immobilized cell cultures. Clearly both of these compounds
are capable of being taken up by pepper cells and in both whole fruit
and cultured cells become incorporated into capsaicin. In whole
fruit

Figure 1.   The proposed biosynthetic pathway of capsaicin
(8-methyl-N-vanillyl-6-nonenamide). * present as alkali-
labile esters in the cell wall of C. frutescens.

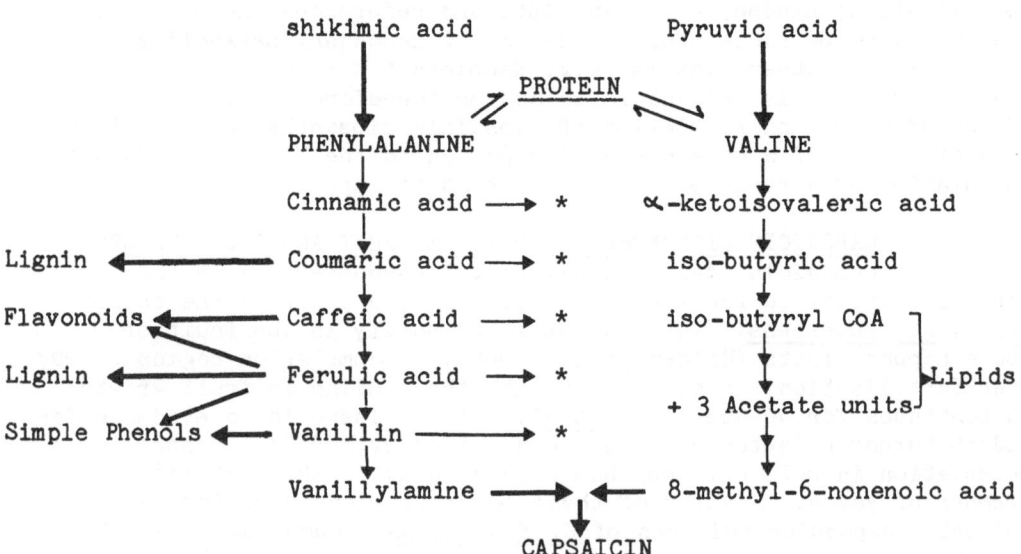

the level of incorporation of either precursor into capsaicin generally decreases with increasing fruit size (age) in accordance with the decrease in capsaicin biosynthesis associated with fruit maturation (Hall et al. 1986). A small amount (c. 1%) of the free [^{14}C]-phenylalanine added is always found to remain in both fruit and cultured cells even after 48h incubation (results not presented). Whether this level represents the labelling of an intracellular pool of phenylalanine with a very low turnover (e.g. in the vacuole) or represents an equilibrium level in the active pool is unclear. In this respect it is interesting to note that when [^{14}C]-cinnamic acid is fed to either fruit or immobilized cells, rather surprisingly, a certain amount of [^{14}C]-phenylalanine is accumulated. The very poor incorporation of this endogenously-produced amino acid into soluble protein, particularly in fruit tissue would strongly suggest that at least two phenylalanine pools occur in these tissues one of which appears not to be associated with protein synthesis but is likely to be associated with the synthesis of phenolic compounds. The existence of more than one intracellular pool of the same amino acid which are spatially or chemically isolated is now well established for plant cells (Oaks & Bidwell, 1970). Furthermore, labelling experiments using p-fluorophenylalanine (PFP)-sensitive and PFP-resistant cell lines have produced results which are also consistent with the proposal that separate phenylalanine pools for protein and phenol synthesis exist, in this case, in tobacco cells (Berlin & Widholm, 1978; Gilchrist & Kosuge, 1980). The possible regulatory role which

Table 1.   The incorporation of [^{14}C]-phenylalanine (A + C) and [^{14}C]-cinnamic acid (B + D) into soluble protein and capsaicin in attached pepper fruit (A + B) and 12d old immobilized cell cultures (C + D). The amount of activity remaining/incorporated into free-phenylalanine is also presented. All values in total dpm per culture or fruit. The specific activities of the stock solution were 2.07 GBq/mmol (Cinnamic acid) and 18.6 GBq/mmol (phenylalanine); 74 KBq were added to each culture/fruit. * mean ± se of 3 replicates.

	Fresh Weight (g)	Length (mm)	Soluble Protein (dpm)	Capsaicin (dpm)	Phenylalanine (dpm)	Label Time (h)
A	1.87	55	6.5×10^4	1.4×10^4	6.4×10^4	1
	2.65	68	3.0×10^4	1.3×10^3	4.9×10^4	1
	3.80	76	3.1×10^4	1.8×10^3	5.1×10^4	1
B	1.10	43	0	1.4×10^4	1.3×10^5	1
	1.60	50	10	6.5×10^3	6.0×10^4	1
	3.91	71	0	3.8×10^3	4.5×10^4	1
C*	-	-	$5.0 \pm 0.5 \times 10^5$	$8.3 \pm 0.5 \times 10^2$	$4.0 \pm 0.6 \times 10^4$	24
D*	-	-	$2.0 \pm 0.2 \times 10^3$	$1.6 \pm 0.2 \times 10^3$	$4.8 \pm 0.4 \times 10^4$	24

this multi-compartment system may have in determining the expression of metabolic pathways has been already considered (Oaks & Bidwell, 1970; Gilchrist & Kosuge, 1980) but clearly more work in this area is required before any detailed understanding is possible.

Investigations into the presence and size of the intracellular pools of some of the other, more immediate precursors of capsaicin have produced some interesting results. Detailed HPLC analysis of cultured cell extracts has revealed that all of the phenolic biosynthetic intermediates of capsaicin occur at extremely low levels, irrespective of the age of the tissue or the use of conditions which enhance capsaicin yield (Hall et al. 1986). A parallel investigation using whole fruit tissue at all stages of development produced similar results (Hall et al. 1986). From this it has been concluded that the metabolism of these compounds must be rapid in order to prevent a build-up within the cells to measurable levels as the input of metabolites into the pathway is considerable. Capsaicin precursors are however, detectable as radiolabelled compounds following incubation of the tissue in the presence of either $[^{14}C]$-phenylalanine or $[^{14}C]$-cinnamic acid (Hall et al. 1986). Interestingly, in the extracts of immobilized cells, in addition to the proposed phenolic precursors of capsaicin and capsaicin itself very few other compounds were labelled. This would provide further, albeit circumstantial, evidence that the pathway presented in Figure 1 is correct.

A number of experiments have been performed to determine the degree to which labelling of the capsaicin phenolic precursors occurs in immobilized cell cultures and whole fruit. It can be seen from the combined results presented in Table 2 that following feeding with $[^{14}C]$-phenylalanine the amount of label occurring in the phenolic intermediates in fruit tissue was generally less than the level in immobilized cells. This was accompanied, as might be expected, by a considerably greater incorporation of label into capsaicin, thus reflecting the more 'efficient' biosynthetic mechanism in vivo. It is particularly interesting to find that in response to the culture of cells in inductive conditions a general increase in the level of radioactivity in the phenolic intermediates results, in addition to the expected increase in incorporation into capsaicin. It has previously been demonstrated that the transfer of immobilized pepper cells to inductive conditions brings about a substantial diversion of radioactivity away from the methanol-insoluble and into the soluble phenolic fraction of the cells reflecting a major change in direction of cell metabolism. It is now clear that this change, in part at least, directly affects the metabolism of most of the intermediates in the capsaicin biosynthetic pathway. A proportionally higher increase was observed in the activity of free cinnamic and caffeic acids. Whether this increase reflects a restriction in the metabolic flux down the pathway at these points (i.e. a reduction in turnover) or a relative increase in the size of the intracellular pools of free cinnamic and caffeic acids is unclear. The ability to enhance capsaicin yield by the exogenous supply of ferulic acid (Lindsey, this volume) would tend to support the former possibility.

THE CONTROL OF CAPSAICIN BIOSYNTHESIS IN VITRO SYSTEMS

A considerable body of evidence has already been presented (Lindsey, this volume and refs. therein) which would indicate that an inverse correlation exists between culture growth and capsaicin yield in pepper cultures. Conditions which favour reduced culture growth are also found to be those which bring about increased capsaicin accumulation (Lindsey, 1985; Lindsey & Yeoman, 1984). It has been proposed that this phenomenon reflects, at least in part, a competition between protein synthesis and phenylpropanoid synthesis for the common primary metabolite phenylalanine (Lindsey, 1985). In support of this proposal it has been demonstrated that an induced reduction in the incorporation of [^{14}C]-phenylalanine into protein coincides with an increase in its incorporation into capsaicin (Lindsey 1986). In this article it has been demonstrated that at least two intracellular phenylalanine pools occur in these cells which have distinctly different roles in cell metabolism. It is possible therefore that one of the limitations to capsaicin biosynthesis is the transfer of phenylalanine between these pools and that under inductive conditions the increase in phenylalanine availability, resulting from a decrease in protein synthesis, brings about an increased transfer of this amino acid to the pool specifically associated with phenylpropanoid metabolism.

It has also become clear however, that in these cultures, in addition to the above-mentioned competition for phenylalanine, a second 'level' of competition occurs concerning its phenylpropanoid derivatives. Hall et al. (1986) have shown that under standard growth conditions the majority (60-80%) of the label from [^{14}C]-phenylalanine becomes

Table 2.   The level of radioactivity present in capsaicin and its phenolic precursors in 12d old immobilized cell cultures and attached fruit (2-4g) of C. frutescens after feeding with [^{14}C]-phenylalanine. (24h incubation period). All values are in dpm g^{-1} (dry weight). The cells were cultured in either full medium (non-induced) or in medium lacking growth substances and the major N sources (induced). np = not presented due to incomplete separation from unknowns after TLC.

| | Immobilized Cells | | |
Compound	Non-induced (dpm)	Induced (dpm)	Whole Fruit (dpm)
Phenylalanine	1.37×10^5	7.55×10^4	7.96×10^4
Cinnamic acid	2.64×10^4	2.33×10^5	2.89×10^3
Coumaric acid	2.14×10^3	3.01×10^3	8.88×10^2
Caffeic acid	3.95×10^2	1.73×10^3	np
Ferulic acid	2.19×10^3	5.01×10^3	4.44×10^2
Vanillin	1.79×10^3	1.71×10^3	2.26×10^3
Capsaicin	6.9×10^2	1.88×10^4	9.21×10^4

incorporated into the bound phenolic fraction of the cells whereas under inductive conditions this 'diversion' into the insoluble fraction is substantially reduced (to c. 30%). In a preliminary investigation it has been found that this fraction consists, in part, of cinnamic acid and some of its hydroxy-derivatives from the capsaicin pathway (unpub. obs.). These features, in addition to the findings reported here where an increase in radioactivity was observed in the free phenolic intermediates of the capsaicin biosynthetic pathway when cultures were grown in inductive conditions would strongly suggest that the bound phenolic fraction of the cells was also acting as a strong competitive sink for capsaicin precursors. The precise chemical nature of this sink has yet to be determined. It is considered that it consists mainly of the phenolic content typical of primary cell walls (see Fry, 1984) although the possible involvement of lignin cannot be excluded. Lignified cells are not however generally visible in these cultures.

CONCLUDING REMARKS
The synthesis of secondary metabolites and its control are clearly very complex and in relation to capsaicin production the problem of obtaining a detailed picture of events is particularly acute as the biosynthetic pathway shares intermediates not only with protein synthesis but also with a large range of other phenolic compounds (e.g. flavonoids, growth substances, pigments, polyphenols et.). The level of competition between different sinks for common precursors clearly plays an important regulatory role in determining the overall pattern of cell metabolism. It is evident that the yield of specific secondary metabolites from cell cultures can only be increased if the balance is tipped more in favour of secondary metabolic pathways at a time when the cells are able to respond to the modified conditions. The use of inhibitors to reduce the diversion of metabolites into 'undesirable' products (e.g. lignin, cell wall phenols) should prove of considerable value if suitable compounds are found.

REFERENCES
Berlin, J. & Widholm, J.M. (1978). Metabolism of phenylalanine and tyrosine in tobacco cell lines resistant and sensitive to p-fluorophenylalanine. Phytochemistry 17, 65-8.
Brodelius, P., Linse, L. & Nilsson, K. (1982). Viability and biosynthetic capacity of immobilised plant cells. In: Fujiwara, A. (ed.). Plant Tissue Culture 1982. Japanese Assn. Plant Tissue Culture, Tokyo. 371-2.
Fry, S.C. (1984). Incorporation of [14C]-cinnamate into hydrolase-resistant components of the primary cell wall of spinach. Phytochemistry 23, 59-64.
Gilchrist, D.G. & Kosuge, T. (1980). Aromatic amino acid biosynthesis and its regulation. In: Miflin, B.J. (ed.). The Biochemistry of Plants. Vol. 5. Acad. Press, London. 507-31.

Hall, R.D., Holden, M.A. & Yeoman, M.M. (1986). The accumulation of
    phenylpropanoid and capsaicinoid compounds in cell
    cultures and whole fruit of the chilli pepper Capsicum
    frutescens. Plant Cell, Tissue and Organ Culture. (In
    press).
Holden, M.A., Hall, R.D., Lindsey, K. & Yeoman, M.M. (1986).
    Capsaicin biosynthesis in cell cultures of Capsicum
    frutescens. In: Webb, C., Mavituna, F. and Faria, J.J.
    (eds.). Process Possibilities for Plant and Animal Cell
    Cultures. Institute of Chem. Engineers, London. (In
    press).
Knobloch, K.-H. & Berlin, J. (1981). Phosphate mediated regulation
    of cinnamoyl-putrescine biosynthesis in cell suspension
    cultures of Nicotiana tabacum. Planta Medica 42, 167-72.
Lindsey, K. (1985). Manipulation by nutrient limitation of the
    biosynthetic activity of immobilized cells of Capsicum
    frutescens Mill. cv. annuum. Planta 165, 126-33.
Lindsey, K. (1986). Incorporation of [^{14}C]-phenylalanine and [^{14}C]-
    cinnamic acid into capsaicin in cultured cells of Capsicum
    frutescens. Phytochemistry. (In press).
Lindsey, K. & Yeoman, M.M. (1984). The synthetic potential of
    immobilised cells of Capsicum frutescens Mill cv. annuum.
    Planta 162, 495-501.
Lindsey, K. & Yeoman, M.M. (1985). Immobilised plant cell culture
    systems. In: Neumann, K.-H., Barz, W. and Reinhard, E.
    (eds.). Primary and Secondary Metabolism of Plant Cell
    Cultures. Springer, Berlin. 304-15.
Lindsey, K., Yeoman, M.M., Black, S.M. & Mavituna, F. (1983). A
    novel method for the immobilization and culture of plant
    cells. FEBS Letters 155, 143-49.
Oaks, A. & Bidwell, R.G.S. (1970). Compartmentation of intermediary
    metabolites. Annual Review of Plant Physiology 21, 43-
    66.

17.   THE ELICITATION OF PHYTOALEXINS IN CULTURES OF CAPSICUM
       ANNUUM AND NICOTIANA TABACUM

D.G. Watson, C.J.W. Brooks & I.M. Freer

Chemistry Department,
University of Glasgow,
GLASGOW.
G12 8QQ U.K.

        The possibility that secondary product formation might be
induced in tissue cultures in a manner analogous to phytoalexins has
received recent consideration (Wolters & Eilert 1983; Heinstein 1985).
The link between phytoalexin elicitation and secondary product
formation per se may lie in the far-reaching hormonal effects exerted
by oligosaccharides in plants (Tran Thanh Van et al. 1985). The
elicitation of the antifungal sesquiterpenoid capsidiol in Capsicum
annuum cultures in response to a sterile extract from the spores and
mycelium of Gliocladium deliquescens can be seen in Figs. 1 and 2.

        Figure 1.   The accumulation of capsidol in Capsicum
        annuum cultures in response to a sterile extract from
        Gliocladium deliquescens: ●——● capsidiol present in the
        medium; ▲——▲   capsidiol present in the tissue.

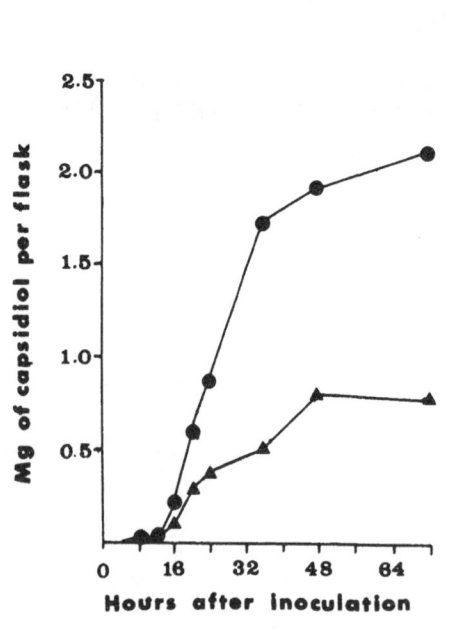
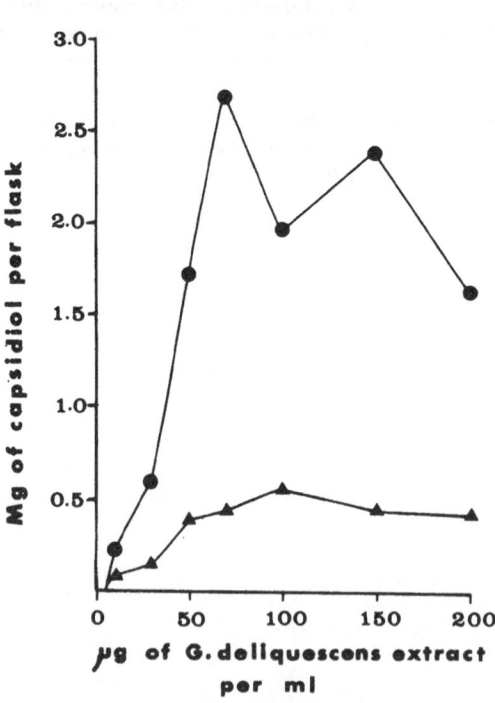

The response of the cultures to cellulase (ex Trichoderma viride) is similar but the maximal amounts of capsidiol accumulated are only ca. 30% of those produced in response to the G. deliquescens extract. (Brooks et al. 1986).

Fractionation of the G. deliquescens extract via gel filtration results in a marked decrease in its activity (this decline in elicitor power is more marked with respect to capsicum tissue cultures than to the fruits). The two most active fractions from gel filtration had molecular weights of ca. 2000 daltons and > 100,000 daltons respectively. The latter was almost completely composed of glucose, mannose and galactose (ca. 10:2:1) whereas the former contained large amounts of nitrogenous material and only ca. 20% of saccharide material, with the composition glucose, mannose and galactose (14:13:1). The decrease of activity in fractionated material may be due to the activity in the crude material being multicomponent. Glycosylase I, an enzyme isolated from soybean cultures (Cline & Albersheim 1981) has been shown to degrade elicitor β-glucans isolated from Phytophthora megasperma (Darvill & Albersheim 1984): a similar enzyme may perhaps exist in capsicum cultures and may be less effective in degrading glycan elicitors in the unfractionated G. deliquescens extract.

Hydrolysed citrus pectin has been shown to contain a dodecagalacturonic acid which elicits isoflavonoid phytoalexins in soybean (Nothnagel et al. 1983). Hydrolysed citrus pectin is also an elicitor of capsidiol in capsicum cultures. Fractionation via gel filtration reduces the activity of the hydrolysate, whereas ion-exchange chromatography concentrates the elicitor activity, largely into the neutral sugar fraction and into a fraction co-eluting with galacturonic acid. The latter fraction is the more active, remaining effective down to a concentration of ca. 0.2 mg/ml in the culture medium; these elicitor fractions are composed of a range of polysaccharides of between 1000 and 4000 daltons. The neutral sugar fraction is composed of galactose with ca. 3% of galacturonic acid (probably passing through the column in the form of its lactone) and the fraction co-eluting with galacturonic acid is composed of galactose and galacturonic acid in a 3:1 ratio. Thus it would appear that fragments from the neutral side chains of citrus pectin as well as fragments from the galacturonic

Capsidiol

Debneyol

Phytuberol

Phytuberin

acid backbone of the polymer have significant elicitor action.   Some
of the problems encountered in studying the action of fractionated
elicitors in tissue cultures may be due to the fact that the cultures
are not injured prior to the application of the elicitor.   In studies
with whole plants, excised cotyledons or tissue slices are used and in
a number of instances injury has been found to elicit phytoalexins
(Hargreaves & Bailey 1978; Whitehead et al. 1982) or to stimulate some
of the enzymes involved in phytoalexin synthesis without leading to
accumulation (Oba & Uritani 1981).   Thus there may be a requirement
for a potentiator of the phytoalexin response as well as an elicitor
in a tissue culture system; unfractionated extracts may contain such a
combination of factors which could become separated upon
fractionation.

In Nicotiana tabacum cultures cellulase ex T. viride elicits a number
of sesquiterpenoids as indicated in Fig. 3 (Watson et al. 1985).

Fig. 3 indicates the usefulness of butaneboronic acid in selectively
derivatising vicinal diols in the extracts without suppressing the
peak due to capsidiol (in contrast to methane boronic and benzene-
boronic acid).   Peaks 1, 2, 3 and 4 also correspond to cylic boronates
of sesquiterpenoid vicinal or 1,3-diols.

The extract from G. deliquescens is a less active elicitor in N.
tabacum cultures than cellulase and elicits a different range of
metabolites.

It is quite possible to envisage that secondary metabolism in plants
might be controlled by carbohydrate hormones which are dependent on
seasonal factors for their release rather than upon microbial attack.
These types of compounds might also hold the key to the control of
secondary metabolism in plant tissue cultures.

Figure 3.   GLC trace of an extract from the medium of
cellulase-treated Nicotiana tabacum cultures.   The extract
was sublimed then treated with butaneboronic acid.

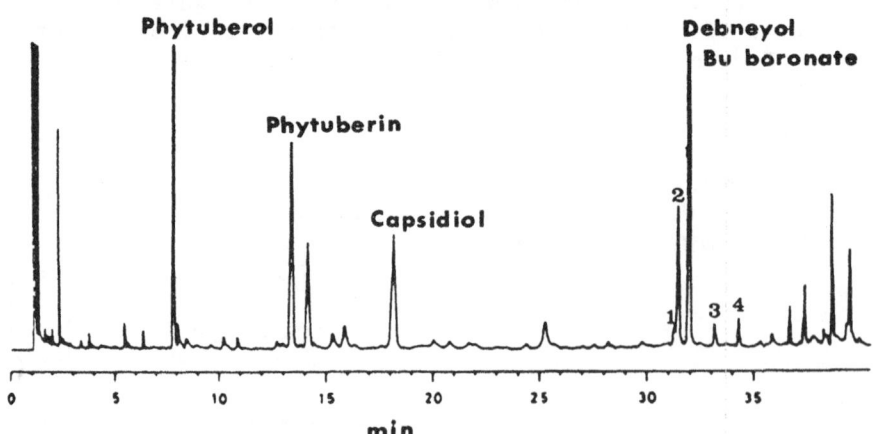

Grateful acknowledgement is made to the SERC for a project grant, and to Professor K.H. Overton for providing the facilities of the Plant Tissue Culture Unit.

REFERENCES

Brooks, C.J.W., Watson, D.G. & Freer, I.M. (1986). The elicitation of capsidiol accumulation in suspended callus cultures of Capsicum annuum. Phytochemistry,25, 1089-92.

Cline, K. & Albersheim, P. (1981). Host-pathogen interactions XVII. Hydrolysis of biologically active fungal glucans by enzymes isolated from soybean cells. Plant Physiology, 68, 221-8.

Darvill, A.G. & Albersheim, P. (1984). Phytoalexins and their elicitors. A defense against microbial infection in plants. Annual Review of Plant Physiology, 35, 243-75.

Hargreaves, J.A. & Bailey, J.A. (1978). Phytoalexin production of hypocotyls of Phaseolus vulgaris in response to constitutive metabolites released by damaged bean cells. Physiological Plant Pathology, 13, 89-100.

Heinstein, P.F. (1985). Future approaches to the formation of secondary natural products in cell suspension cultures. J. Natural Products, 48, 1-9.

Nothnagel, E.A., McNeil, M., Albersheim, P. & Dell, A. (1983). Host-pathogen interactions XXII. A galacturonic acid oligosaccharide from plant cell walls elicits phytoalexins. Plant Physiology, 71, 916-26.

Oba, K. & Uritani, I. (1981). Mechanism of furanoterpene production in sweet potato root tissue injured by chemical agents. Agricultural & Biological Chemistry, 45, 1635-39.

Tran Thanh Van, K. Toubart, P., Coussan, A., Darvill, A.G., Gollin, D.J., Chelf, P. & Albersheim, P. (1985). Manipulation of the morphogenetic pathways of tobacco explants by oligo-saccharides. Nature, 314, 615-7.

Watson, D.G., Rycroft, D.S., Freer, I.M. & Brooks, C.J.W. (1985). Sesquiterpenoid phytoalexins from suspended callus cultures of Nicotiana tabacum. Phytochemistry, 24, 2195-2200.

Whitehead, I.M., Dey, P.M. & Dixon, R.A. (1982). Differential patterns of phytoalexin accumulation and enzymic induction in wounded and elicitor treated tissues of Phaseolus vulgaris. Planta, 154, 156-64.

Wolters, B. & Eilert, U. (1983). Elicitoren-Ausloser der Akkumulation von Pflanzenstoffen. Deutsche Apotheker Zeitung, 123, 659-67.

18. STUDIES ON ALLIINASE FROM ONION (ALLIUM CEPA) AND THE
    POTENTIAL USE OF ONION CELL CULTURES AS AN ENZYME SOURCE
    IN ONION FLAVOUR PRODUCTION

S.B. Wright, G. Stepan-Sarkissian & M.W. Fowler

Wolfson Institute of Biotechnology,
University of Sheffield,
Sheffield. S10 2TN. U.K.

INTRODUCTION

The characteristic flavour of onion (Allium cepa) is due
to a complex mixture consisting mainly of sulphur-containing
compounds. These arise following rupturing of the onion tissue
resulting in an initial interaction between three flavour precursors
(S-methyl-, S-propyl-, and S-1-propenyl-L-cysteine sulphoxide) and an
enzyme commonly known as alliinase (E.C.4.4.1.4.).

The wide use of onion flavour products in processed foods means that
their production on a commercial scale is of some importance to the
food industry. Consequently, a more detailed knowledge of the factors
controlling the key enzyme process is likely to be of interest,
particularly if it could lead to enhanced production. Our initial
work has been to determine the stability and activity of partially
purified alliinase preparations under different conditions of
temperature, pH and co-factor concentration.

Parallel to the enzyme studies we are also interested in whether onion
cell cultures could be of value as a means of improving onion flavour
production. The production of onion flavour solely from cell cultures
has not so far proved possible owing to the very low levels or total
absence of the flavour precursors in the callus cultures studied.
However, Selby & Collin (1976) showed that significant alliinase
activity can be obtained from onion callus cultures and we have set up
a series of callus cultures to study their potential as a source of
alliinase.

Partially purified alliinase preparations were obtained from onion
bulbs as follows: Approximately 600g bulb tissue was homogenised in
300ml extraction buffer (0.2M potassium phosphate, pH 6.8, containing
0.3M sucrose and 0.05% (v/v) mercaptoethanol (as stabilising agents).
Two $(NH_4)_2SO_4$ cuts of 0-30% and 30-80% were then carried out, followed
by gel filtration of the 80% $(NH_4)_2SO_4$ pellet on a Sephadex G-25
column (30cm x 3cm). All the steps were performed at 5°C. Alliinase
activity was usually assayed by coupling activity to that of lactate
dehydrogenase (LDH). However, where the presence of LDH would have
been undesirable (ie. in the determination of pH and temperature
optima) a total hydrazone method was used (Schwimmer & Mazelis, 1963).
The substrate for alliinase was synthetic (+) S-methyl-L-cysteine
sulphoxide (Synge & Wood, 1956).

Alliinase preparations had a temperature optimum of approx. 45°C. The stability of the enzyme rapidly decreased above this temperature: At 50°C, 50% of activity was lost in 12 minutes whilst at 30°, 14°C and 4°C, 18 hours, 3.5 days and 11 days respectively were required for a similar loss of activity. In addition, alliinase preparations stored at -20°C showed an average _increase_ in specific activity of 200% after 36 weeks.

Fig. 1 shows the pH stability of a single preparation which is representative of the broad pH stability profile generally observed. The pH optimum obtained in 0.1M Tris-HCl buffer was approx 7.0 (Fig. 2). This is slightly lower than the values of 7.4 and 8.0-8.2 previously reported by Kupiecki & Virtanen (1960) and by Schwimmer & Mazelis (1963) and Tobkin & Mazelis (1979) respectively.

The effect of exogenous pyridoxal phosphate on alliinase activity is shown in Table 1. Although we have observed that 0.05×10^{-3} μM NH_2OH causes a 69% inhibition of activity, thus confirming the enzyme's requirement for pyridoxal phosphate, the addition of pyridoxal phosphate to the assay mixture clearly tends to inhibit alliinase activity. This effect is currently being investigated further.

Onion callus cultures were initiated from root tissue on Dunstan and Short's (1977) BDS medium and on a modification of Gamborg et al's (1968) B5 medium. After a three month initiation period callus was subcultured approximately every 4 weeks. Callus tissue samples were homogenised in extraction buffer and the crude extracts assayed for alliinase activity. All 10 callus lines assayed yielded alliinase activity to varying levels. Five lines gave specific activities comparable to those of "fresh" medium strength bulbs (Table 2).

Table 1   Effect of exogenous pyridoxal phosphate on alliinase
           activity.   mean ± S.E. represent 4 determinations on
           each of 3 different extracts.

conc. of exogenous pyridoxal phosphate (μM)	specific activity as % specific activity of control
0 (control)	100
25	88
40	81
50	76
75	64
100	60
200	50
300	39
500	18

Figure 1.   Effect of pH on alliinase stability.   mean
values + S.E. represent four replicate determinations.

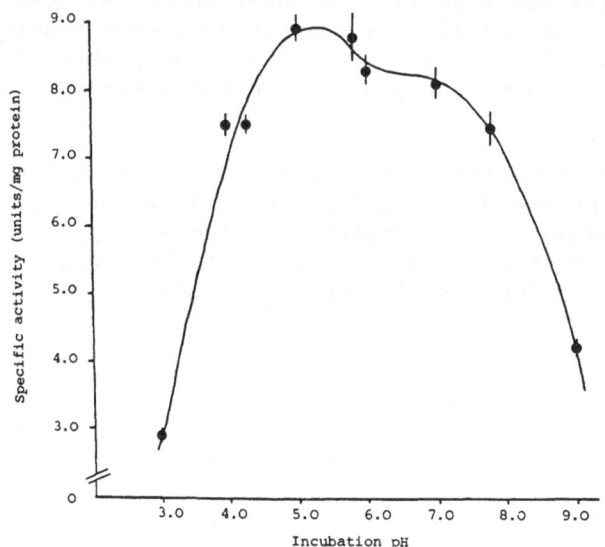

Figure 2.   Effect of pH on alliinase activity in the
presence of Tris-HCl buffer.   mean values + S.E.
represent triplicate determinations on each of four
extracts.

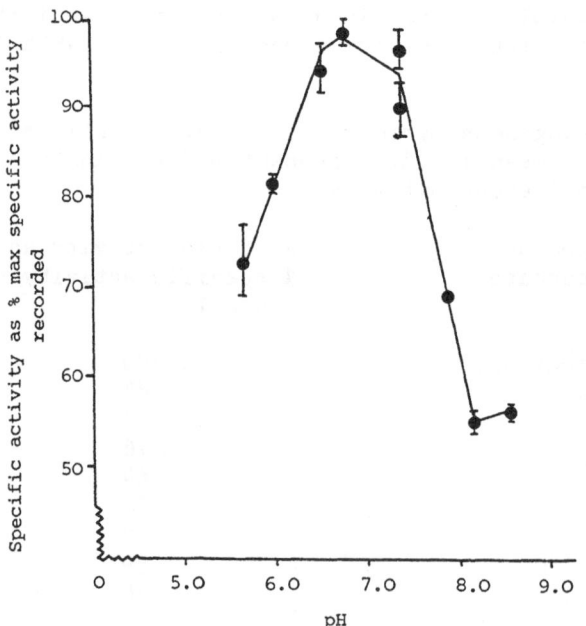

In the course of the basic enzyme studies it was also observed that
significant variations in alliinase activity occur both between
different bulb varieties and during bulb storage (Table 3).   Alliinase
preparations from English and Spanish bulbs, which were purchased in
early autumn and assayed within a few days (ie. minimal known
storage), showed a marked difference in specific activity.   The
specific activity of alliinase obtained from White French bulbs after
10 weeks storage ($6^{o}C$, 45% humidity) was only 15% of the mean specific
activity obtained during the initial storage period of 7 weeks.

Such observed variability in levels of extractable alliinase activity,
combined with the variability in flavour precursor levels which occurs
in bulbs of different ages and varieties (Lancaster & Kelly, 1983),
may be sufficient to affect the quality of the ultimate flavour
product.   However, if consistently high levels of alliinase were
produced by an onion cell culture, the culture-derived enzyme might be
used in a commercial process to compensate for the inconsistency in
the bulbs.   Further work is therefore now being undertaken to
determine whether the alliinase levels already observed in culture are
maintained through repeated subcultures and to compare the properties
of culture-derived enzyme to that obtained from bulbs.

Table 2:   Alliinase activity in onion callus cultures.   Varieties
Ailsa Craig, AC; Rijinsburger, R; Bedfordshire Champion, BC;
Paris Silverskin, PS.   BDS = Dunstan & Short's medium with
0.25 mg/l 2,4-D   B5' = Gamborg et al's medium with 0.5
mg/l 2,4-D   Mean values $\pm$ S.E. for duplicate extracts.

Callus line	Medium	Subculture	Specific Activity
AC20	B5'	S7	1.75 $\pm$ 0.00
AC21	BDS	S7	1.63 $\pm$ 0.07
R00	B5'	S5	4.31 $\pm$ 2.89
BC10	BDS	S5	2.47 $\pm$ 0.03
PS10	BDS	S5	1.56 $\pm$ 0.00

Table 3   Variations in alliinase activity in bulbs of different
varieties and following varying periods of bulb storage.
[a]mean $\pm$ S.E. for duplicate extractions.   [b]mean $\pm$ S.E.
for six extractions.

Bulb Variety	Known period of storage (weeks)	Mean specific activity (units/mg protein)
English	0	4.80 $\pm$ 0.58[a]
Spanish	0	0.26 $\pm$ 0.17[a]
White French	7	3.17 $\pm$ 0.61[a]
White French	10	0.48 $\pm$ 0.06[b]

Acknowledgements

S. B. Wright wishes to thank the S.E.R.C. for a CASE studentship to carry out this work. We also wish to thank the cooperating body, Leatherhead Food R.A. for their support.

REFERENCES

Dunstan, D.I. & Short, K.C. (1977). Improved growth of tissue cultures in onion, Allium cepa. Physiol Plant. 41, 70-72.

Gamborg, O.L., Miller, R.A., & Ojima, K. (1968). Nutrient requirements of suspension cultures of soybean root cells. Exp. Cell Res. 50, 151-158.

Kupiecki, F.P. & Virtanen, A.I. (1960). Cleavage of alkyl cysteine sulphoxides by an enzyme in onion (Allium cepa). Acta Chem Scand. 14, 1913-1918.

Lancaster, J.E. & Kelly, K.E. (1983). Quantitative analysis of the S-alk(en)yl-L-cysteine S-oxides in onion. J. Sci Fd. Agric 34 (11), 1229-1235.

Schwimmer, S. & Mazelis, M. (1983). Characterization of alliinase of Allium cepa (onion). Arch. Biochem. Biophys. 100, 66-73.

Selby, C. & Collin, H.A. (1976). Clonal variation in onion tissue cultures. Ann. Bot. 40, 911-918.

Synge, R.L.M. & Wood, J.C. (1956). (+)-S-methyl-L-cysteine S- oxide in cabbage. Biochem J. 64, 252-253.

Tobkin, H.E. Jr. & Mazelis, M. (1979). Allium lyase: Preparations and characterization of the homogenous enzyme from onion bulbs. Arch. Biochem. Biophys. 193, 150-157.

19.   UPTAKE, BINDING AND METABOLISM OF MORPHINAN ALKALOIDS BY
       CELL CULTURES OF PAPAVER SOMNIFERUM

P. Morris and S. Gibbs.

Wolfson Institute of Biotechnology
University of Sheffield
Sheffield   S10 2TN. U.K.

INTRODUCTION
The general inability of cell suspension cultures of P.
somniferum to accumulate free morphinan alkaloids (see Constabel 1985)
may be due in part to rapid metabolism of these alkaloids as found in
the whole plant. However, recent evidence suggests that morphine and
codeine may be present in some cell cultures in bound forms (Hutin et
al, 1983), similar to that found in capsules (Fairbairn and Steele,
1980). The recent finding that thebaine and sanguinarine were not
accumulated in the same compartments, or even in the same cell types,
in cell cultures of P. bracteatum (Kutchan et al, 1985) also suggests
that morphinan alkaloids may not be stable when stored in the cell
vacuole and that special sites of storage of these alkaloids in cytoplasmic vesicles or in laticifer cells may be required.

In order to ascertain if rapid metabolism of soluble alkaloids might
be a limiting factor in their accumulation, studies of the uptake,
binding and rate of turnover of morphinan alkaloids supplied
exogenously to cell cultures of P. somniferum were undertaken.

Uptake and Binding of ^{14}C-Morphine
Stable three year old cell suspension cultures of P.
somniferum which accumulated benzophenanthridine alkaloids but not
free or bound morphinan alkaloids were supplied with N-methyl ^{14}C-
morphine. Experiments were performed on 14 day old cell cultures grown
on M&S medium containing 2% sucrose, 1 mg/l 2,4-D and 0.1 mg/l kinetin
pH 5.8.

A steady state equilibrium for soluble morphine was reached within 1
hour of adding ^{14}C-morphine to the cells. Soluble ^{14}C-morphine had an
accumulation ratio of 1.0 at pH 5.8 (pKa morphine = 8.2) indicating
passive equilibration of the charged species, however, the total ^{14}C-
morphine content of the cells did not reach equilibrium. Binding of
^{14}C-morphine to the cell wall fraction accounted for these
differences. This binding was linear over the initial uptake period
of 2 hours and was not saturated by 24 hours (Fig. 1).

These uptake kinetics closely resembled the computer simulation of
uptake incorporating the formation of type II complexes (slow binding
to cell components in an irreversible or slowly reversible manner)
described by Renaudin and Guern, (1982). Acid hydrolysis may

solublize such type II complexes while not releasing attached morphine
and this differential may account for the large amount of radioactivity (84%) remaining at the origin after TLC of the hydrolysed
fraction.

Metabolism of ^{14}C-Morphine

After 24 hours incubation with ^{14}C-morphine 38% of the
activity in the cells was found in the MeOH soluble fraction and 59%
was released from the cell wall fraction by acid hydrolysis, 3%
remained in the final pellet (Table 1). Of the MeOH soluble fraction
84% was alkaloid in nature with 16% non-alkaloid, and in the
hydrolysed fraction 75% was alkaloid and 22% non-alkaloid (Table 1).
Some turnover or degradation of the applied ^{14}C-morphine during the
incubation period or during extraction therefore occurred.
N-demethylation to normorphine as monitored by the release of $^{14}CO_2$
amounted to only 0.02% of the applied activity. Thin layer chromatography showed that the majority of the activity in the soluble
alkaloid fraction (70%) remained as morphine but morphine constituted
a much smaller proportion of the alkaloid fraction of the hydrolysed
cell extract (16%).

Fig. 1 Uptake and binding of ^{14}C-morphine by cell
suspension cultures of P.somniferum. Cells incubated at
25°C, 70 rpm in medium containing 1.3 mM morphine and 0.1
µCi/ml N-methyl ^{14}C-morphine.

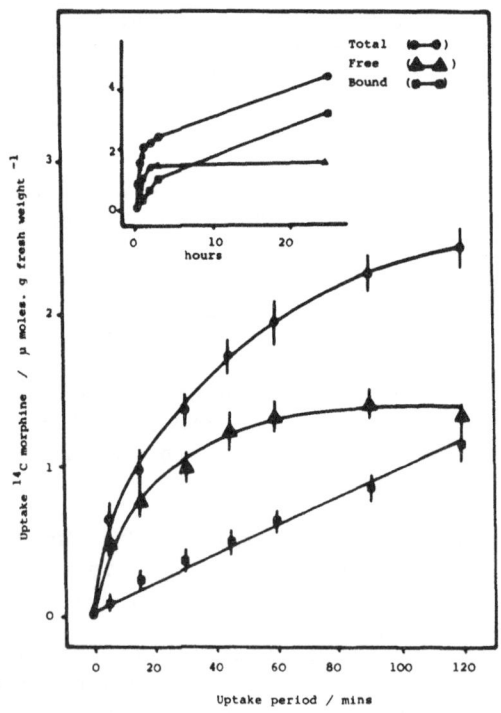

Table 1   Metabolism of applied ^{14}C-morphine. Cells incubated for 24 hours at 25°C with 1.3 mM morphine and 0.25 μCi/ml N-methyl ^{14}C-morphine pH5.8. Flasks contained a CO_2 trap (1 ml 6M KOH) (* assuming all activity remains as morphine).

Sample	Total Activity dpm	% Total Activity	% Activity in cells	μg/g fresh* wt morphine in cells
MEDIUM				
Alkaloid fraction	2.4×10^7	88.9	-	-
Non-alkaloid fraction	1.9×10^6	6.9	-	-
$^{14}CO_2$	4.0×10^2	0.02	-	-
<u>CELLS</u> Total	1.15×10^6	4.2	100	86.3
<u>MeOH extract</u>				
Alkaloid fraction	3.7×10^5	1.34	32.4	28.0
Non-alkaloid fraction	6.9×10^4	0.25	6.0	5.0
<u>Cell pellet</u> (acid hydrolysed)				
Alkaloid fraction	5.3×10^5	1.95	46.2	40.0
Non-alkaloid fraction	1.5×10^5	0.54	12.7	11.0
<u>Final cell pellet</u>	3.0×10^4	0.11	2.7	2.3

Table 2   Nature of binding of ^{14}C-morphine to cell walls. Cells were loaded with ^{14}C-morphine for 18 hours then exhaustively extracted with MeOH. Separate samples of the pellet were then treated as listed below and then the residue acid hydrolysed.

Treatment	% activity recovered in each fraction		
	Treated fraction	Hydrolysed fraction	Final pellet
Hexane extraction	1.0	30.4	68.6
MeOH (100° 5 min)	2.0	29.6	68.4
Morphine wash (10mM 1hr)	3.0	30.0	67.0
$CHCl_3$ extraction	5.0	32.3	62.7
Water wash (100°C 10 min)	6.5	28.0	65.5
KCl wash (0.2M 30 min)	8.5	35.2	56.3
Triton 100 digest (1% 1hr)	10.0	25.4	64.6
Dil HCl wash (pH2 30 min)	15.4	21.3	63.3
Dil NaOH wash (pH12 30 min)	26.4	12.7	60.9
Alkali hydrolysis (1N NH_4Cl 100°C 1hr)	33.0	16.5	50.5
Formic acid hydrolysis (3N 100°C 4min)	34.0	15.6	50.4
Acid hydrolysis (1N HCl 1hr 100°C)	49.6	10.7	39.7

Table 3   Binding of ^{14}C-morphine to cell walls in cultured cells. Cell walls were exhaustively extracted with MeOH and dil NaOH after ^{14}C-morphine loading for 18 h

Treatment	% activity recovered from cell walls
Final NaOH wash	2.8
1st hydrolysis (1N HCl 1hr 100°C)	49.6
2nd hydrolysis (1N HCl 1hr 100°C)	10.7
MeOH wash	7.7
Final pellet	31.9

Nature of Binding of ^{14}C-Morphine

Bound ^{14}C-morphine could not be totally released from the
cell wall fraction either by washing with a variety of solvent systems
or by hydrolysis. It is evident (Table 2) that both adsorption and
covalent binding of ^{14}C-morphine occurs. After removal of adsorbed
activity with a dilute NaOH wash, 70% of the bound activity could be
released by acid hydrolysis (Table 3). The nature of the activity
remaining in the pellet is not known but may be the labelled N-methyl
group. Incubation of isolated cell wall fragments with ^{14}C-morphine
showed some adsorption, but most of this activity could be removed
with a dilute alkali wash (Table 4).

The mechanism by which morphine binds to cell wall components and the
factors responsible for the rapid metabolism of free alkaloids calls
for some speculations. The formation of alkaloid-N-oxides and their
binding to polysaccharide cell wall components and the involvement of
peroxidase, polyphenol oxidases and H_2O_2 in binding and degradation
are currently under investigation. These factors have been shown to
be involved in the modification of morphinan alkaloids (Hsu et al,
1984) and in the binding of other alkaloids (Meyer 1983).

Uptake and Metabolism of Unlabelled Morphinan Alkaloids

The uptake and turnover of unlabelled morphine, codeine
and thebaine is shown in Fig. 2. All the alkaloids were rapidly taken
up and rapidly degraded by the cells, the amount of alkaloid found in
the cells at any one time being the difference between uptake and
turnover. In cells fed with these alkaloids no alkaloid biotransformation products were identified.

Codeinone fed to cell cultures of P. somniferum was rapidly degraded
to non-alkaloid material extracellularly but a small amounts of
morphine and codeine were found in the medium.   Codeinone was also
rapidly taken up by the cells and biotransformed to morphine and
codeine (Fig. 2). However these alkaloids were also rapidly
metabolized by the cells and after 5 days incubation no morphine,
codeine or codeinone was found in the MeOH soluble cell fraction.

Biotransformation of codeinone to codeine by cell cultures of P.
somniferum has been reported previously (Furuya et al, 1978). In our
cells morphine and codeine, whether supplied exogenously or derived by
metabolism from exogenously supplied codeinone, were rapidly degraded.

Table 4   Binding of ^{14}C-morphine to extracted cell walls.
Cell walls extracted and then incubated with ^{14}C-morphine
for 18 hours.

Treatment	% activity recovered from cell walls
Final MeOH wash	6.8
NaOH (pH 12) wash 1	60.7
NaOH (pH 12) wash 1	19.0
Acid hydrolysis (1N HCl 1hr 100oC)	8.2
Final pellet	5.2

Fig. 2   Uptake and metabolism of unlabelled morphinan
alkaloids by cell suspension cultures of P. somniferum
cells (4g fresh wt) suspended in 100 ml medium and
morphine (15 mg), codeine (12 mg), codeinone (10 mg) or
thebaine (10 mg) added to triplicate flasks.   Alkaloids
were extracted and quantified by reverse phase ion-pair
HPLC and identity confirmed by TLC.

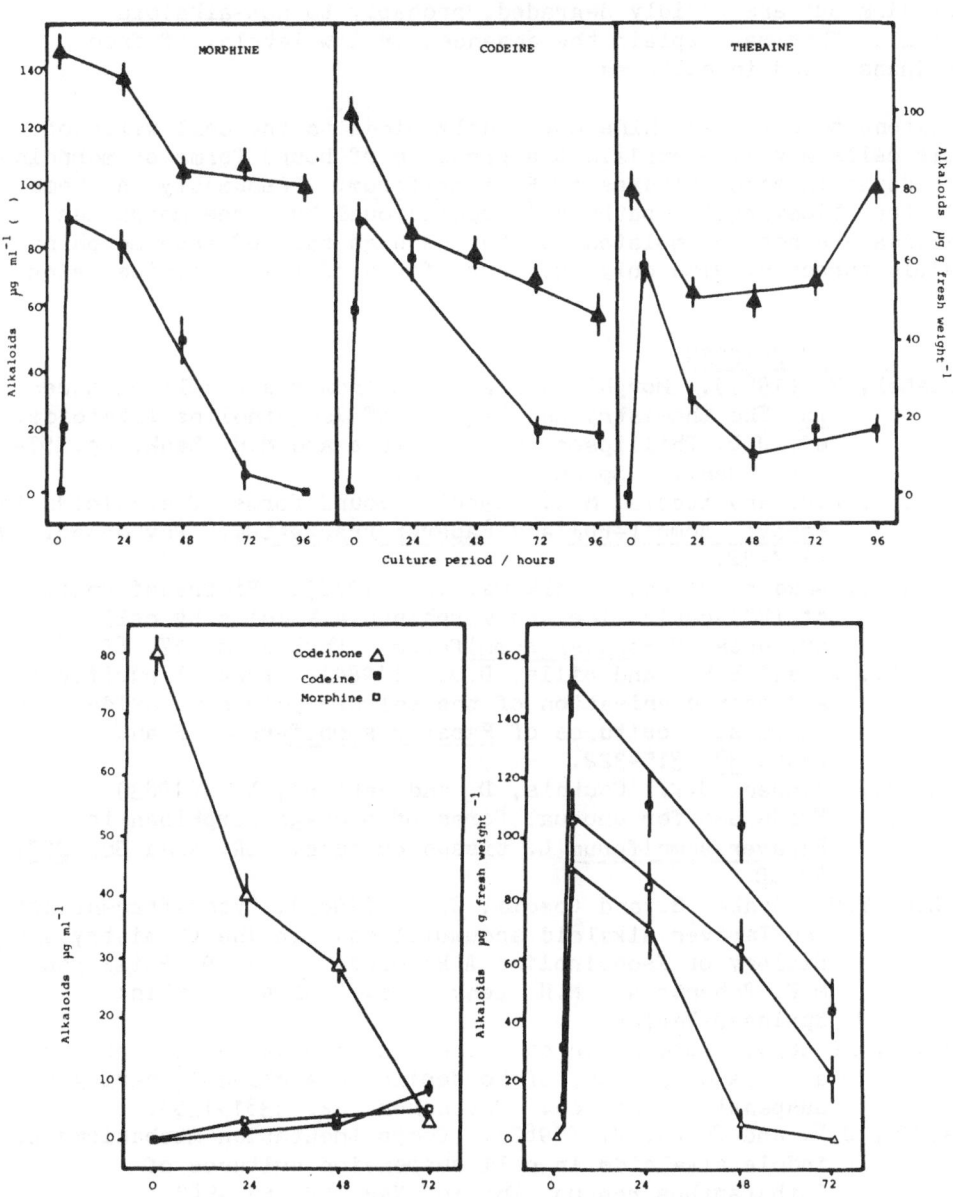

However it would appear that some discrimination between exogenously
supplied and endogenously produced codeine is apparent as morphine was
not detected when cultures were fed codeine but was produced,
presumably via codeine, when cultures were fed codeinone.

CONCLUSIONS
It may be concluded that in cell suspension cultures of P.
somniferum, morphine, codeine and thebaine are not end products of
metabolism but are rapidly degraded, probably to non-alkaloid
material. This may explain the absence, or low levels, of free
morphinans found in cultures.

The extent to which morphine covalently binds to the cell walls of
viable cells may also explain the presence of bound forms of morphine
and codeine in some cultures of P. somniferum. Presumably in these
cells the biosynthetic pathway is operational but free morphinan
alkaloids are not accumulated due to the high rate of free morphinan
alkaloid turnover, with only the bound forms of the alkaloids being
stable.

REFERENCES
Constabel, F. (1985). Morphinan alkaloids from plant cell cultures.
          In The Chemistry and Biology of Isoquinoline Alkaloids,
          ed. J.D. Phillipson, M.F. Roberts and M.H. Zenk, pp.257-
          264. Berlin: Springer-Verlag.
Fairbairn, J.W. and Steele, M.J. (1980). Bound forms of alkaloids in
          Papaver somniferum and Papaver bracteatum. Phytochem, 19,
          2317-22.
Furuya, T., Nakano, M. and Yoshikawa, T. (1978). Biotransformation
          of (RS)-reticuline and morphinan alkaloids by cell
          cultures of Papaver somniferum. Phytochem, 17, 891-893.
Hsu, A-F., Kalan, E.B., and Bills, D.D. (1984). Partial purification
          and characterisation of the soluble polyphenoloxidase from
          suspension cultures of Papaver somniferum. Plant Sci
          Lett, 34, 315-322.
Hutin, M., Foucher, J.P., Coutois, D. and Petiard, V. (1983).
          Evidences for unusual forms of storage morphinan in a
          Papaver somniferum L. tissue culture. CR. Acad Sci 297,
          47-50.
Kutchan, T.M., Ayabe, S. and Coscia, C.J. (1985). Cytodifferentiation
          and Papaver alkaloid accumulation. In The Chemistry and
          Biology of Isoquinoline Alkaloids. ed. J.D. Phillipson,
          M.F. Roberts and M.H. Zenk. pp.281-294. Berlin:
          Springer-Verlag.
Meyer, E. (1983). Peroxidase catalysed incorporation into polymers
          as a major pathway of hordenine metabolism in barley cell
          suspension cultures. Phytochem, 22, 1381-1385.
Renaudin, J.P. and Guern, J. (1982). Compartmentation mechanisms of
          indole alkaloids in cell suspension cultures of
          Catharanthus roseus. Physiol Veg, 20, 533-547.

SECTION 4

MASS CELL CULTURE AND IMMOBILISED PLANT
CELL SYSTEMS

20. THE PRODUCTION OF SECONDARY METABOLITES BY IMMOBILISED
PLANT CELLS

K. Lindsey

Department of Botany,
University of Edinburgh
Mayfield Road
Edinburgh.   EH9 3JH   U.K.

INTRODUCTION
The use of cultured plant cells for the large scale
production of specific secondary metabolites such as drugs, flavours,
perfumes, pigments and agrochemicals has been viewed with optimism as
our knowledge of the factors influencing secondary metabolic activity
in a number of experimental systems accumulates. The effects of a
range of nutritional and 'hormonal' conditions in particular on the
synthesis and accumulation of secondary compounds in vitro have been
the subject of intense investigation and have been well reviewed (e.g.
Mantell & Smith 1983). It is possible to enhance the productivity of
a particular species in culture by variously adopting techniques of
cell line selection (Ogino et al. 1978), precursor feeding,
manipulation of the composition of the nutrient medium supplied
(Mantell & Smith 1983) and of immobilisation (Lindsey & Yeoman 1985).
These approaches together provide the basis for the development of a
commerical process (Yeoman & Lindsey 1985), in which the emphasis is
switched, in the cultures, from primary to secondary metabolic
activity. Much of the work on the application of plant cell
immobilisation techniques performed to date suggests that their use
would be beneficial to the efficiency of such a production process for
a wide range of useful chemicals, synthesised by either one- or two-
step biotransformations or by more complex multi-enzyme process
(Lindsey & Yeoman 1985).

Although these advancements are some cause for optimism, it is a fact
that, with only a handful of exceptions at most, plant cell cultures
still fail to produce specific compounds in yields of a commercial
size. It is the object of this article, therefore, to consider the
nature of some of the limitations to the synthesis and accumulation of
secondary metabolites, to describe the experimental approach developed
at Edinburgh to study the regulation of secondary metabolism, and to
speculate on the extent to which one can expect to overcome the
limitation in the system. So, one can rephrase the question, 'How can
we enhance the productivity of plant cell cultures?', to ask instead
'What goes wrong with the expression of secondary metabolism in
cultured plant cells?', and some possibilities immediately spring to
mind. Is it simply that in the majority of cases the culture
conditions are not optimal for secondary metabolite formation, and
that the main problem lies with finding the appropriate chemical
signal to trigger the biosynthetic machinery? Or is the problem

concerned with the inability of the cultures to respond to the
environmental signals provided, perhaps due to genetic aberrations
within the culture or to the disorganised growth habit exhibited by
cell and tissue cultures? Available evidence points to the
involvement of a combination of these deficiencies, resulting in non-
producing or low-producing cultures. In an attempt to clarify the
situation by identifying limitations to production, we will now
consider the experimental approaches which have been adopted to
investigate the regulation of secondary metabolism in plant cell
cultures, and their relevance to the development of immobilisation
techniques.

The experimental approaches
Two types of experimental approach in particular have been
important in the evolution of immobilisation techniques in our
laboratory. These are, firstly, time course studies, in which the
kinetics of the synthesis and accumulation of a specific secondary
metabolite or of a group of compounds are determined over a growth
cycle; and secondly, studies in which the primary metabolic activity
of the cultures is manipulated by altering the composition of the
nutrient medium or by the application of inhibitors. Let us firstly
consider some of the important features to arise from the first type
of experiment. In 1972 Davies showed (Davies 1972) that the
accumulation of polyphenolic metabolites by cell cultures of Paul's
Scarlet Rose (Rosa sp.) declined as the cells entered a period of
rapid division after a lag phase following subculture, accumulation
increasing only in the post-exponential period as, with time, an
increasing proportion of the cells stopped dividing. A similar
pattern has been observed for the accumulation of some alkaloids such
as nicotine (Speake et al. 1964), trigonelline (Khanna & Jain 1972),
and tropane alkaloids (Lindsey & Yeoman 1983a), steroids such as
diosgenin (Yeoman et al. 1982) and pigments such as anthocyanins (Hall
& Yeoman 1986) and anthraquinones (Wijnsma et al. 1985). In each of
these examples, it is the case that, during the period of active cell
division, the accumulated level of these compounds either declines, as
for polyphenols (Davies 1972), anthocyanins (Hall & Yeoman 1986) and
anthraquinones (Wijnsma et al. 1985), due to degradation or remains
steady at a relatively low level on a per culture basis, if the
product is not degraded, as is the case for capsaicin (P.A. Aitchison
& M.M. Yeoman, unpublished observations; Lindsey 1986). Once
division of the cell population begins to stop, then accumulation of
the products is initiated. This pattern of accumulation is not
universal, for a small number of secondary metabolites appear to be
accumulated in a growth-dependent manner (Fowler 1983), but the
regulatory processes in such systems have not been studied in detail.
In some solanaceous species in particular, the stationary phase of the
growth cycle is associated with the development of organised
structures, ranging from compact aggregates and embryoids to shoot-
and root- like organs which accumulate significantly higher levels of
tropane or sterodial glycoalkaloids than do the rapidly-dividing and
structurally disorganised cells (e.g. Thomas & Street 1970; Lindsey &
Yeoman 1983b). This important observation, which has similarly been

made for cardenolide accumulation in <u>Digitalis</u> sp. cultures has given
rise to the view that secondary metabolite accumulation can be
considered to be an aspect of differentiation (Luckner 1980; Yeoman <u>et</u>
<u>al</u>. 1982). This concept has been integral in our development of cell
immobilisation systems, and this will be considered in more detail
below.

Let us now consider the second group of experiments which have had a
bearing on our development of immobilisation methods. These are
studies in which the growth rate (a combination of cell division and
expansion) and the rate of protein synthesis were manipulated by
altering the concentrations of supplied nutrients (particularly
phosphate and/or nitrate) or by supplying inhibitors of transcription
or translation. The effects of a range of nutrients on the
accumulation of specific secondary metabolites has been reviewed by
Mantell & Smith (1983), and in some cases, namely in those cases in
which the nutrient of interest was growth-limiting, the effect on
product accumulation was observed to be associated with an altered
growth rate of the cultures, although the precise relationship between
the two phenomena was unclear. For example, Yeoman <u>et al</u>. (1980) have
reported that by restricting the growth of callus cultures of <u>Capsicum</u>
<u>frutescens</u> by reducing the supply of nitrogen and sucrose, it is
possible to enhance the incorporation of [^{14}C]phenylalanine and
[^{14}C]valine into capsaicin. Furthermore, Lindsey (1985) has shown
that the incorporation of [^{14}C]phenylalanine into capsaicin is
inversely related to both the growth index and the accumulation of
radioactive protein in cultured pepper cells grown in a range of
concentrations of nitrate. Similarly, Knobloch <u>et al</u>. (1981) have
observed an inverse relationship between the growth rate of tobacco
cell cultures, manipulated by altering the phosphate supply, and the
accumulation of cinnamoyl putrescines. Other results in agreement
with this general observation have been discussed by Lindsey (1985).
The effects of inhibitors of protein synthesis or transcription are
also illuminating. Cycloheximide, for example, causes increases in
the accumulation of capsaicin in cultures of <u>Capsicum frutescens</u>,
D-threo-chloramphenicol stimulates the accumulation of tobacco
alkaloids (Neumann & Mueller 1971), Actinomycin D stimulates the
accumulation of <u>Macleaya</u> alkaloids (Neumann & Mueller 1974) and the
addition of streptomycin sulphate enhanced the accumulation of
shikonin derivatives in cultures of <u>Lithospermum erythrorhizon</u>
(Mizukami <u>et al</u>. 1977). In all cases protein synthesis and growth
were blocked. The implications fo these results are discussed in
detail by Yeomann <u>et al</u>. (1982), and are further evidence that cell
division and protein synthesis can be separated from secondary
metabolite synthesis and accumulation.

Taken together, the evidence presented so far indicates that 1)
secondary metabolite production is in many cases growth-independent,
and can be manipulated by experimentally altering the rate of cell
division of the culture, and 2) accumulation is enhanced in organised
tissues, compared with 'undifferentiated' cell cultures. These
observations and concepts therefore encouraged the development of

culture systems in which both the growth rate and the degree of organisation of the cells could be manipulated. We will now discuss these, and then go on to discuss the effects on secondary metabolic activity.

The flatbed immobilisation system
The flatbed immobilisation system (Yeoman et al. 1978; Lindsey & Yeoman 1983b) comprises two essential elements: a culture vessel, in which cells are fixed to a substratum of polypropylene fabric matting, and a separate reservoir of nutrient medium. The medium is supplied to the cultured cells by dripping from the reservoir onto the polypropylene fabric and moving across it by capillary action, and is returned to the reservoir by means of a peristaltic pump. The rate of supply of the medium is controlled by the rate at which it is pumped back to the reservoir. This simple apparatus therefore has two major advantages over the use of suspension cultures for the production of secondary metabolites: the circulating loop system of medium supply allows a facilitated control of the composition of the nutrient medium, and so of the rate of cell division of the cultured cells; and cell-cell contact in increased.

A modification of the design of this culture system resulted in the development of column bioreactors (Lindsey & Yeoman 1983b), which, it was felt, afforded better control over the supply of nutrients, a saving of space, and greater suitability for prospective industrial scale-up. Here the flat culture vessel was replaced by a vertical column, and this format necessitated the support of the cells along the length of the column in an inert matrix. Three principal entrapment methods were employed: in an agar gel supported by nylon netting (Lindsey & Yeoman 1983b); in a calcium alginate gel supported by nylon netting (Lindsey & Yeoman 1983b); and in reticulate polyurethane foam particles (Lindsey et al. 1983). These and other methods for the entrapment of plant cells have been described and discussed in detail by Lindsey & Yeoman (1985). The column culture system retains the advantages of the flatbed associated with the sequential chemical treatment of the cells, but has greater advantages with respect to the manipulation of the aggregation of the cells, both in terms of the density of the aggregates and their size.

For the reasons given, these features provide a rationale for the immobilisation of plant cells which is of a physiological nature, in the sense that the aim is to mimic those conditions which favour high accumulated yields of secondary metabolites (i.e. slow-growing and organised cultures). There are, however, advantages to immobilisation which are of a chemical engineering nature, and these are summarised in Table 1. In this sense, but not so obviously in the physiological sense, the immobilisation of plant cells can be considered to be an extension of the technology of the immobilisation of microbial cells. Let us now examine the secondary metabolic activity of immobilised plant cells, and consider what this tells us of the regulation of that activity.

The secondary metabolic biosyntheses which have been studied for
immobilised plant cells can be described as falling into one of two
general categories: biotransformations, in which a substrate is
metabolised in one or two reactions to a desired product, or multi-
enzyme syntheses, in which a precursor, which may be metabolically
distant from the designated product, is catabolised in several
reactions.

One or two-step biotransformations
Cultured plant cells can perform stereospecific biotransformations which involve the addition or removal of single chemical
groups - hydroxylations, glycosylations, acetylations and
methylations. The first report of such activity in immobilised cells
was the 12,β-hydroxylation of digitoxin to digoxin by cells of
Digitalis lanata entrapped in a gel of calcium alginate (Brodelius et
al. 1979). The activity of this reaction was maintained for 33 days,
and the efficiency of conversion was similar to that of
freely-suspended cells.

Alfermann et al. (1980) have similarly investigated the biotransformation of digitoxin and β-methyldigitoxin by calcium alginate-
immobilised cells of Digitalis lanata. Qualitatively, the reactions
were identical in immobilised and in freely-suspended cells. The
supplied digitoxin was converted primarily to purpurea glycoside A,
with the formation of small amounts of digoxin and deacetyllanatoside.
β-methyldigitoxin was hydroxylated primarily to β-methyldigoxin.
Although qualitatively the transformations were the same in free and
immobilised cells, the hydroxylation activity of the latter was only
50% that of the suspended cells, although it was maintained for a
longer period (61 days).

Immobilised carrot cells (Daucus carota) can also perform steroid
transformations (Jones & Veliky 1981). Although freely suspended
cells were found to be able to hydroxylate digitoxigenin to
periplogenin at a faster rate than cells entrapped in calcium
alginate, the rate could be manipulated by permeabilising the cells,
and by altering the relationship between substrate concentration and
cell concentration. By supplying fresh medium at 48h intervals it was
possible to obtain relatively high rates of biotransformation for five
consecutive batch periods, i.e. for ten days, after which time the
rate declined to an almost negligible level.

Table 1:   Some process engineering advantages of
immobilisation

Reusability of biomass
High biomass concentrations
Facilitated biomass recovery and cell handling
Resistance to dilution of endogenous metabolites
Resistance to shear damage
Facilitated sequential chemical treatments
Facilitated product recover
Reduced clogging of hardware

Hydroxylation of gitoxigenin to 5 -hydroxygitoxigenin was retained for 30 days by carrot cells immobilised in calcium alginate, and the rate could be manipulated by the aeration method (Veliky & Jones, 1981).

More recently, Furuya et al. (1984) have demonstrated that opium poppy cells (Papaver somniferum), immobilised in calcium alginate, could biotransform (-) codeinone to (-) codeine, and immobilised cells were more efficient at performing the reaction than were freely suspended cells, in batch cultures.

It would seem from these results that biotransformations are performed in a qualitatively similar way in both immobilised and freely-suspended cells, and though the rate of activity of immobilized cells may be reduced in some cases, productivity may be retained for a longer period of time with immobilised cells.

Complex multi-enzyme syntheses
Considering the potential commercial importance of this type of reactions, there have been relatively few reports of complex biosyntheses of specific secondary metabolites by cultures of immobilised plant cells and even less consideration of possible regulatory effects of immobilisation. The first of such reports was of the synthesis of the indole alkaloid ajmalicine from exogenously supplied precursors tryptamine and secologanin in cells of Catharanthus roseus, immobilised in calcium alginate gel (Brodelius et al. 1979). Indeed, the efficiency of conversion by the cells entrapped in either calcium alginate or agarose was greater than in cells immobilised in agar or carrageenan, and, importantly, greater than in freely-suspended cells (Brodelius & Nilsson 1980). It was also found (Brodelius et al. 1979; Brodelius & Nilsson 1980) that alginate-entrapped cells of Morinda citrifolia accumulated higher levels of anthraquinone pigments than did freely-suspended cells under the same, growth-limiting, culture conditions.

The accumulation of alkaloids by immobilised cells of Solanum nigrum and Datura innoxia has similarly been found to be greater than in rapidly-dividing freely-suspended cells. Solanum nigrum cells, either immobilised on the flatbed bioreactor or entrapped in gels of calcium alginate or agar accumulated enhanced levels of the steroidal glycoalkaloids α-solanine and α-chaconine (Lindsey 1982). Datura innoxia cells, immobilised in calcium alginate, accumulated higher levels of the tropane alkaloids scopolamine and atropine than did freely-suspended cells, but similar levels to callus (Lindsey 1982). Secondary metabolite activity has been reported to be retained in cells of Solanum aviculare immobilised on polyphenyleneoxide (Jirku et al. 1981) and of Catharanthus roseus in a gel mixture of polyacrylamide and alginate (Lambe & Rosevear 1982), and these examples have been reviewed in more detail elsewhere (Lindsey & Yeoman 1985).

The most detailed comparison of the biosynthetic activity of immobilised and freely-suspended cells has been made studying the synthesis and accumulation of capsaicin, a pungent acid amine derived from phenylalanine and valine, by cells of Capsicum frutescens (the Chilli pepper) immobilised in reticulate polyurethane foam (Lindsey et al. 1983b). In this experimental system it was found that, on a dry weight basis, cultures of cells immobilised in polyurethane accumulated between two and three orders of magnitude higher levels of capsaicin than did cultures of freely-suspended cells, a result reflected by an increased incorporation of radioactivity into capsaicin from [14C]phenylalanine (Lindsey & Yeoman 1984). The capsaicin accumulates predominantly in the medium rather than in the cells (Lindsey et al. 1983b; Lindsey 1986). If capsaicin synthesis and accumulation in vitro is subject to similar general regulatory control mechanisms to some of the other secondary metabolites considered above, then it is possible that this effect of immobilisation may be dependent on the two, apparently critical and related parameters, namely a reduced rate of cell division and associated metabolic processes, and an increased degree of aggregation of the cells. We will now consider some of the evidence supporting this hypothesis, and discuss what this can tell us about the level(s) at which capsaicin synthesis and accumulation is modulated in cell cultures.

The rate of increase in fresh and dry weight of immobilised cells is less than that of freely-suspended cells in culture conditions favouring cell division (Lindsey 1986), and this is reflected in the rate of protein synthesis, measured as the incorporation of [14C] phenylalanine into soluble protein. Furthermore, in those suspended cell cultures in which the incorporation of [14C]phenylalanine into protein was relatively high, incorporation into capsaicin was relatively low, i.e. compared with immobilised cell cultures (Table 2). This evidence is supportive of the concept of an inverse relationship between the synthesis of protein and of secondary products derived from amino acids, as suggested by Phillips & Henshaw (1977) and Lindsey & Yeoman (1983a). This possibility was investigated further by experimentally manipulating the rate of protein synthesis by supplying immobilised cells with different concentrations of nitrate (Lindsey 1985).

Table 2:   The incorporation of L-[U-14C]phenylalanine into capsaicin and soluble protein in immobilised and freely-suspended pepper cells, on d5 and d12 of culture.

mean dpm g^{-1} DW

	Suspended	Immobilised	Labelling
capsaicin (d5)	996 ± 226	7189 ± 765	2μCi, 24h
protein (d5)	1853 ± 473	901 ± 133	0.2μCi, 5 min
protein (d12)	594 ± 98	-	0.2μCi, 5 min
ratio, protein: capsaicin (d5)	2:1	1:8	

Results demonstrated that, when determined after 9 days of culture, the incorporation of [14C]phenylalanine into soluble protein was directly related to the initial nitrate concentration and to the growth index of the cells, and when supplied with less than 10mM nitrate, compared with 25mM in controls, there was a between two- and three-fold increase in the incorporation of [14C]phenylalanine into capsaicin. Moreover, an increase in the nitrate supply over that present in the standard culture medium resulted in an increased incorporation of radioactivity into soluble protein and a decreased incorporation into capsaicin. Uptake of radioactivity was unaffected by nitrate supply, and the results support the view that a high rate of cell division may limit secondary metabolite accumulation by providing a sink for common precursors in protein synthesis.

But is this model likely to hold for other secondary metabolites derived from precursors which are not utilised in protein synthesis? This can be tested indirectly in the Chilli pepper culture system, by following the fate of cinnamic acid, the immediate product of phenylalanine, which is not involved in protein synthesis (Lindsey 1986). The pattern of incorporation of [14C]cinnamic acid into capsaicin in immobilised and freely-suspended cells was determined over a culture period of 27d in growth medium, and the results are presented in Fig. 1. In the cultures of suspended cells it was found that during the lag and exponential phases (d. 3-9), the level of radioactivity in capsaicin declined to reach a level of 4000-8000 dpm g^{-1}DW during the linear phase of growth (d. 9-18). During the stationary phase (d. 18-27) the incorporation increased to approx. 30000 dpm g^{-1}DW.

In the cultures of immobilised cells the incorporation of [14C]cinnamic acid into capsaicin was greater than in suspended cell cultures throughout the greater part of the growth cycle. The incorporation declined during the immobilised cell exponential phase (d. 9-15) and increased thereafter. These results indicate that

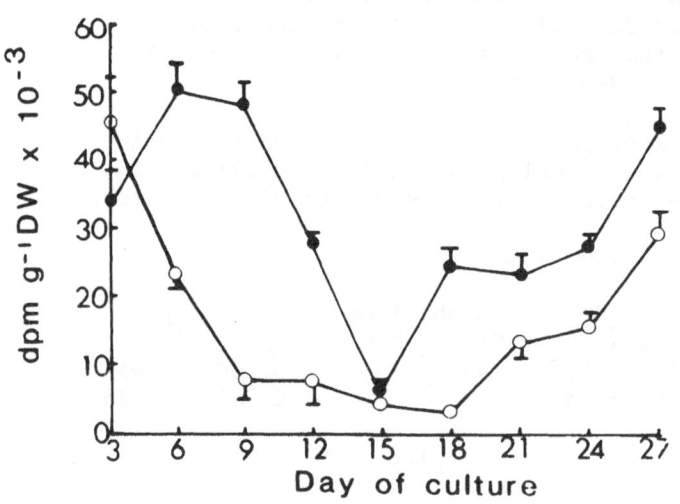

Incorporation of [14C] cinnamic acid into capsaicin in immobilised (●) and freely suspended cells (O) of C. frutescens

cinnamic acid is, like phenylalanine, incorporated into capsaicin to
the greatest extent when cell division is limited, and the explanation
would appear to be, in part, that the derivatives of cinnamic acid are
esterified to the cell wall in rapidly-dividing cells and are
consequently of limited availability for capsaicin synthesis (Lindsey
1986; see also Holden et al. this volume). If this is the case, then
the cell wall can be considered to be, like protein, a sink for
intermediates in capsaicin biosynthesis, and the synthesis and
accumulation of capsaicin in rapidly-dividing cells may be limited by
substrate supply. Further evidence in support of this view is
provided by experiments in which phenolic precursors to capsaicin were
supplied to cell cultures. The supply of 1mM phenylalanine (Lindsey &
Yeoman 1984) and 1mM ferulic acid (Lindsey, unpublished observations)
to immobilised pepper cell cultures both resulted in enhanced yields
of capsaicin.

Taken together, these results indicate that, in cultured pepper cells,
a major limitation to capsaicin accumulation is the availability of
substrate for the enzymes of the biosynthetic pathway. The reduced
yield in cultures of freely-suspended cells is due, it is suggested,
largely to the preferential utilisation of precursors in the growth-
associated processes of protein and cell wall synthesis, which are
limited in immobilised cells due to the altered intercellular
environment in which oxygen and perhaps other nutrients and growth
regulators are of reduce availability. The dispersed nature of
suspended cells might also result in a washing out of intracellular
precursors into the medium, again limiting their availability for
capsaicin biosynthesis; and this dilution effect may be less
pronounced in densely immobilised cells. There is some evidence
(Lindsey, unpublished observations), that a relatively large ratio of
medium volume to cell weight results in a relatively low accumulation
of capsaicin in column cultures of immobilised pepper cells, and vice
versa.

One other possible difference in the metabolic nature of immobilised
cells can be considered, but at present is purely speculative; and
this is that there may be enzymological differences between
immobilised and suspended cultured cells. There is some
circumstantial evidence which may have a bearing on this question, and
concerns the ability of the two types of culture to respond to
treatments designed to enhance the accumulation of capsaicin. Firstly,
the incorporation of [14C]phenylalanine and [14C]cinnamic acid, and
the total accumulation of capsaicin, has been found to be
significantly greater in immobilised than suspended cell cultures,
even though both cultures were subjected to identical, growth-
inhibiting, nutrient stress conditions (Lindsey 1986; Table 3).
Furthermore, immobilised cell cultures accumulate up to 50 fold higher
levels of capsaicin when supplied with 5mM isocapric acid, a precursor
of capsaicin, than do suspended cell cultures under identical
conditions (Lindsey & Yeoman 1984). It is, however, possible that a
dilution of intracellular isocapric acid in the suspended cells would
account for this result. It is also conceivable that the nutrient

stress results could be explained by a lower concentration of the valine-derived intermediates, necessary for capsaicin accumulation, in suspended cells. Evidence, however, for abnormal patterns of enzyme activity in suspended cells does exist (e.g. Zeleneva and Khavkin 1980), suggesting that, by their very nature, structurally disorganised cultured cells fail to express the totipotency with which they are often accredited.

CONCLUSIONS
It has been the intention of this article, not to provide a list of secondary metabolic reactions which immobilised plant cells are capable of, but to consider what we can learn of the regulation of secondary metabolism in cultured plant cells by manipulating, in a very simple way, the degree of structural organisation of those cells. Such an approach can be used to mimic biochemical aspects of differentiation and development in vitro, and can complement developmental studies on somatic embryos. What this tells us of the regulation of secondary metabolism in the intact plant, however, is a separate question, for we are studying undifferentiated or partially differentiated tisues which, all too evidently, fail to express their full potential. Immobilisation appears to go some way to restoring metabolic activity, but it is important to realise that in exploiting cultured cells for the production of a specific product, we are trying to manipulate, in a very crude manner, a highly complex process which requires that the cells are competent to respond. In this sense plant cells cannot be treated as microbes, for they depend on intercellular communication to provide the appropriate microenvironment through which competence is generated. It is hoped that experimentation with immobilisation techniques will provide some insight into the role of multicellularity in differentiation.

Table 3: Capsaicin contents of cultures incubated for 15d in the presence and absence of nitrate. Mean of 3 replicates $\pm$ SE.

mean mg g^{-1} DW l^{-1}

	suspended	immobilised
controls (+ 25mM-NO_3')	0.004 $\pm$ 0.000	1.007 $\pm$ 0.242
NO_3'-free	0.052 $\pm$ 0.012	2.497 $\pm$ 0.495

REFERENCES
Alfermann, A.W., Schuller, I. & Reinhard, E. (1980). Biotransformation of cardiac glycosides by immobilised cells of Digitalis lanata. Planta Med. 40, 218-23.
Brodelius, P., Deus, B., Mosbach, K. & Zenk, M.H. (1979). Immobilised plant cells for the production and transformation of natural products. FEBS Lett. 103, 93-7.
Brodelius, P. and Nilsson, K. (1980). Entrapment of plant cells in different matrices. FEBS Lett. 122, 312-16.

Davies, M.E. (1972). Polyphenol synthesis in cell suspension cultures of Paul's Scarlet Rose. Planta 104, 50-65.

Fowler, M.W. (1983). Commercial applications and economic aspects of mass plant cell culture. In: Plant Biotechnology, ed. S.H. Mantell & H. Smith, pp.3-37. Cambridge: Cambridge University Press.

Furuya, T., Yoshikawa, T. & Taira, M. (1984). Biotransformation of codeinone to codeine by immobilised cells of Papaver somniferum. Phytochemistry 23, 999-1002.

Hall, R.D. & Yeoman, M.M. (1986). Temporal and spatial heterogeneity in the accumulation of anthocyanins in cell cultures of Catharanthus roseus (L.) G. Don. J. Exp. Bot. 37, 48-60.

Jirku, V., Macek, T., Vanek, T., Krumphanzl, V. & Kubanek, V. (1981). Continuous production of steroid glycoalkaloids by immobilised plant cells. Biotechnol. Lett. 3, 447-50.

Jones, A. & Veliky, I.A. (1981). Examination of parameters affecting the 5-β-hydroxylation of digitoxigenin by immobilised cells of Daucus carota. Eur. J. Appl. Microbiol. Biotechnol. 13, 84-9.

Khanna, P. & Jain, S.C. (1972). Effect of nicotinic acid on growth and production of trigonelline by Trigonella foenum-graecum tissue cultures. Indian J. Exp. Biol. 10, 248-9.

Knobloch, K.H., Beutnagel, G. & Berlin, J. (1981). Influence of accumulated phosphate on culture growth and formation of cinnamoyl putrescines in medium-induced cell suspension cultures of Nicotiana tabacum. Planta 153, 582-5.

Lambe, C.A. & Rosevear, A. (1982). Production of chemical compounds from viable cells. UK Patent No. GB2 096 169A.

Lindsey, K. (1982). The growth and metabolism of plant cells cultured on fixed-bed reactors. Ph.D. thesis, University of Edinburgh.

Lindsey, K. (1985). Manipulation, by nutrient limitation, of the biosynthetic activity of immobilised cells of Capsicum frutescens Mill. cv. annuum. Planta 165, 126-33.

Lindsey, K. (1986). Incorporation of [^{14}C]phenylalanine and [^{14}C]cinnamic acid into capsaicin in cultured cells of Capsicum frutescens. Phytochemistry, in press.

Lindsey, K & Yeoman, M.M. (1983a). The relationship between growth rate, differentiation and alkaloid accumulation in cell cultures. J. Exp. Bot. 34, 1055-65.

Lindsey, K. & Yeoman, M.M. (1983b). Novel experimental systems for studying the production of secondary plant metabolites by plant tissue cultures. In: Plant Biotechnology, ed. S.H. Mantell & H. Smith, pp.39-66. Cambridge: Cambridge University Press.

Lindsey, K. & Yeoman, M.M. (1984). The synthetic potential of immobilised cells of Capsicum frutescens Mill. cv. annuum. Planta 162, 495-501.

Lindsey, K. and Yeomann, M.M. (1985). Immobilised plant cells. In: Plant Cell Culture Technology, ed. M.M. Yeoman, pp. 226-65. Oxford: Blackwell Scientific Publications.

Lindsey, K., Yeoman, M.M., Black, G.M. & Mavituna, F. (1983). A novel
        method for the immobilisation and culture of plant cells.
        FEBS Lett. 155, 143-9.
Luckner, M. (1980). Expression and control of secondary metabolism.
        In: Secondary Plant Products, ed. E.A. Bell & B.V.
        Charlwood, pp.23-63. Berlin, Heidelberg, New York:
        Springer-Verlag.
Mantell, S.H. & Smith, H. (1983). Cultural factors that influence
        secondary metabolite accumulation in plant cell and tissue
        cultures. In: Plant Biotechnology, ed. S.H. Mantell & H.
        Smith, pp.75-108. Cambridge: Cambridge University Press.
Mizukami, H., Konoshima, M. & Tabata, M. (1977). Effect of
        nutritional factors on shikonin derivative formation in
        Lithospermum callus cultures. Phytochemistry 16, 1183-6.
Neumann, D. & Mueller, E. (1971). Beitrage zur Physiologie der
        Alkaloide. Alkaloidbildung in Kallus and
        Suspensionskulturen von Nicotiana tabacum L. Biochem.
        Physiol. Pflanzen. 162, 503-13.
Neumann, D. & Mueller, E. (1974). Beitage zur Physiologie der
        Alkaloide. Alkaloidbildung in Kalluskulturen von
        Macleaya. Biochem. Physiol. Pflanzen. 165, 271-82.
Ogino, T., Hiraoka, N. & Tabata, M. (1978). Selection of high
        nicotine-producing cell lines of tobacco callus by single
        cell cloning. Phytochemistry 17, 1907-10.
Phillips, R. and Henshaw, G.G. (1977). The regulation of synthesis of
        phenolics in stationary phase cell cultures of Acer
        pseudoplatanus L. J. Exp. Bot. 78, 785-94.
Speake, T., McCloskey, P., Smith, W., Scott, T. & Hussey, H. (1964).
        Isolation of nicotine from cell cultures of Nicotiana
        tabacum. Nature 210, 614-15.
Thomas, E. & Street, H.E. (1970). Organogenesis in cell suspension
        cultures of Atropa belladonna L. and Atropa belladonna
        cultivar lutea Doll. Ann. Bot. 34, 657-69.
Veliky, I.A. & Jones, A. (1981). Bioconversion of gitoxigenin by
        immobilised plant cells in a column bioreactor.
        Biotechnol. Lett. 3, 551-4.
Wijnsma, R., Go, J.T.K.A., van Weerden, I.N., Haarkes, P.A.A.,
        Verpoorte, R. & Baerheim-Svendsen, A. (1985).
        Anthraquinones as phytoalexins in cell and tissue cultures
        of Cinchona spec. Plant Cell Rep. 4, 241-4.
Yeoman, M.M., Fagandini, D.A.A. & Childs, A.F. (1978). Callus culture
        in nutrient flow. British patent application no.
        32185/78.
Yeoman, M.M. & Lindsey, K. (1985). The scientific and commercial
        potential of immobilised plant cells. Process Biochem
        (suppl.): Advances in Fermentation II, pp.87-92. London:
        Turret-Wheatland.
Yeoman, M.M., Lindsey, K., Miedzybrodzka, M.B. & McLauchlan, W.R.
        (1982). Accumulation of secondary products as a facet of
        differentiation in plant cell and tissue cultures. In:
        Differentiation in vitro, ed. M.M. Yeoman & D.E.S. Truman,
        pp.65-82. Cambridge: Cambridge University Press.

Yeoman, M.M., Miedzybrodzka, M.B., Lindsey, K. and McLauchlan, W.R.
        (1980). The synthetic potential of cultured plant cells.
        In: Plant Cell Cultures - Results and Perspectives, ed. F.
        Sala, B. Parisi, R. Cella & O. Cifferi, pp. 327-43.
        Amsterdam: Elsevier/North Holland.
Zeleneva, I.V. & Khavkin, E.E. (1980). Rearrangement of enzyme
        patterns in maize callus and suspension cultures: is it
        relevant to changes in the growing cells of the intact
        plant? Planta 148, 108-15.

21.   IMMOBILIZATION TECHNOLOGY

A. Rosevear and C.A. Lambe

Biotechnology Group
Bldg. 353
Harwell,
Oxfordshire.   OX11 ORA.   U.K.

INTRODUCTION
Traditionally, valuable phytochemicals have been obtained
by extracting dead botanical tissue. This necessitates either
destroying the whole plant or selectively harvesting specialised
organs. The availability of tissue cultured cells presents new
opportunities to synthesize such chemicals in a controlled
environment. However normal culture techniques inevitably lead to the
sacrifice of the whole biomass in order to recover the product stored
within individual cells. Immobilization offers a method of avoiding
this problem by maintaining the biosynthetic machinery of a cell and
using this repeatedly over a prolonged period. Nevertheless it is
only applicable where a product is released under conditions which do
not disrupt the primary metabolism of the cell. It was thought that
tissue cultured cells had lost the ability to excrete many products.
However, it now appears that homogeneous, suspension culture itself
may create conditions which cause products to be retained
preferentially inside the cell. Immobilisation not only recreates an
environment where increased secondary metabolite synthesis is promoted
but it also facilitates the removal of these products on a continuous
basis.

STRUCTURE
The full potential of an immobilized system can only be
realized if a minimum level of control and organisation is imposed on
the aggregated cells. Bound cells are held fixed in a closed reactor,
so that the mobile aqueous phase is the principal agent for
environmental control. Consequently efficient contact between this
medium and the immobilized cell mass is essential at all times. In
the growing plant, this organisation is self-generated. However this
vascularisation and differentiation cannot be formed quickly or
efficiently in tissue cultured cells. Immobilization provides an
opportunity to re-impose a degree of spatial stability on the cells
but this must be done efficiently in order to meet the minimum
requirements for mass transfer within the biomass. No part of the
cell mass ought to be more than a critical distance (about 1mm) from a
well-mixed region of the mobile phase but at the same time, the
packing density of the cells should be as high as possible. Dense
cell culture is not only necessary to maximize the productivity per
unit volume of reactor, but also to simulate those conditions which
are generally associated with secondary metabolism. These include the

restriction of cell division, proximity to similar cells which may synthesize control phytochemicals and the presence of cellular structures in which the newly made compounds might be stored.

Consequently, the immobilization process should aim to pack as many cells as possible within a dimensionally stable structure. Since it is currently impossible to direct the growth of dedifferentiated cells, it is preferable to aggregate the biomass in a single event at the desired final packing density. Any subsequent growth is likely to reduce mass transfer in the matrix and, as a result of outgrowth, impair the free flow of fluid through the whole reactor. Physical restraint placed on the cells by the matrix may reduce the rate of subsequent cell division. Nevertheless, ommission of key nutrients (e.g. phosphate) or phytohormones from the medium in which the immobilized cells are maintained is probably the most effective method of restricting further cell growth. As well as reducing the cost of the medium these deletions may also promote product synthesis.

IMMOBILIZATION METHODS
Two general approaches have proved successful in immobilization of tissue cultured plant cells. The first exploits the natural tendency of the individual cells to adhere to many surfaces under conditions of low shear. The second is to take advantage of the relatively large size of plant cells (0.1mm) and of microcallus aggregates, by entrapping the biomass within or behind a macroporous structure.

Adhesion
Unlike many other biocatalysts, plant cells will spontaneously bind to suitable surfaces. The task of the technologist is to control this property so that the biomass is deployed in an open and stable manner. A callus is the simplest example of this approach. The mutual adhesion of polysaccharides in the cell walls binds adjacent cells into a multicellular clump which resists disruption by liquid shear forces in the reactor.

If these aggregates are sufficiently large, they can be retained within a stirred tank reactor merely by sedimentation. Such an approach was used by Sahai & Shuler (1984) and Pareilleu & Vinas (1984) who incorporated decoupler regions on the fluid outlets so that aggregates were not carried out of the reactors. Filters may in principle offer an alternative barrier but such devices are susceptible to blockage by cell debris and viscous carbohydrates.

Cells which slough off from the periphery of the callus may be swept out of these reactors with a resultant loss in catalytic activity. The effective size of the callus is limited by poor diffusion of compounds through the dense biomass and the scavenging of nutrients by the outer layer of cells. Thus microcallus culture (Dainty et al. 1985) is probably only suited to short term exploitation of biomass, e.g. biotransformation, where the cells are used mainly as a crude multienzymes system and cell growth can be fully arrested.

An alternative approach is to provide a secondary matrix to which the cells will adhere preferentially. Macroreticular, polyurethane foams provide an excellent environment in which cell clumps develop. Many plant cells, when grown in the presence of 1cm^3 blocks of foam, infiltrate the open structure within a day or so (Lindsey & Yeoman, 1984). The foam must first be washed free of toxic contaminants and preconditioned with medium and the size of the pores is critical (18 pores/in.). The main problem with this approach is to control growth in and around these large blocks but the technique is reported to be suitable for large scale operation (Yeoman et al. 1982). Fibrous materials have been used as alternatives to foam blocks (Rhodes & Kirsop, 1982) and the present authors have had success with other thin section supports.

The methods considered above rely on interactions between cells. Specific binding to the support can be achieved by coating the surface with a ligand which has an affinity for the cell wall. Lectins have the potential to achieve this. Bornmann & Zachrisson (1982) illustrated that plant protoplasts changed their shape to fit around a Cytodex 1 microcarrier which had been pre-coated with concanavalin A. Con A which is directly bound to Sepharose, fails to adsorb proto- plasts (Warren & Fallon, 1984) but if the lectin is bound through a flexible link, the cells adhere strongly. Despite the elegance of such methods, their application is limited by cost to fundamental studies of cell surfaces. Less specific binding is provided by polyphenylene oxide beads which, after treatment with glutaraldehyde, were found to bind Solanum cells covalently (Jirku et al. 1981). However the turnover of cell surface polymers is likely to release cells held in this way, unless they generate adhesive polysaccharides.

In general, methods involving adhesion are simple, inexpensive and subject the cells to minimum stress during immobilization. In principle no additional barrier to diffusion is created save that imposed by adjacent cells. Methods involving selective binding are impractical at present while those relying on non-specific binding require surface roughness or open porosity to ensure reliable anchorage of the biomass. Nevertheless a number of inert non-toxic matrices have been identified and.provide a convenient method of immobilization where cellular control is not critical.

Entrappment
This approach relies on the enormous size differential between the plant cell and the solutes which form its nutrients and products. The usual approach is to distribute a cell suspension in a polymer solution which is subsequently gelled. This traps the cells throughout a three dimensional, porous matrix. Plant cells have proved less tolerant of gelling reagents than microbial cells. Brodelius & Nilsson (1980) concluded that the polymerization of acrylamide, or the cross-linking of proteins by glutaraldehyde, killed C. roseus cells. Only the gelation of marine polysaccharides such as agarose, carrageenan and especially calcium alginate were sufficiently mild to immobilize plant cells without loss of viability. They

recommended dropping a suspension of cells in sodium alginate into a 50mM solution of calcium chloride to give 3.5mm beads, suitable for use in shake flasks or packed beds. This method has since been adopted by many other workers (see reviews by Rosevear & Lambe, 1985; Brodelius, 1985). The size of the beads formed from drops of viscous polymer is often rather large and cell aggregates may block the orifice. Nilsson et al. (1983) described an improved method of forming beads in which the pre-gel is emulsified in a water-immiscible silicone or soya bean oil and gelation of the droplets then promoted. This method is best suited to polymers which gel on cooling e.g. agarose and carrageenan.

Despite the ease with which cells can be entrapped in these gels, there are serious limitations to their prolonged use. Calcium alginate gel in particular, can break up as a result of calcium loss through chelation by phosphate or organic acids. Barium ions form stronger complexes but in our experience are toxic to plant cells. We have investigated methods of cross-linking alginate to provide long term stability. The inclusion of small amounts of acrylamides in the pre-gel makes it possible to initiate graft polymerization of the alginate to give a stable gel without causing significant loss of cell viability (Rosevear, 1980). High molecular weight water soluble pre-polymers are unlikely to enter the cell and so may not be toxic to cells. This has been illustrated by Galun et al. (1983) who cross-linked the hydrazide of polyacrylamide with glyoxal to entrap cells.

Despite the availability of non-toxic gels, entrappment of plant cells is relatively complex, difficult to perform in situ and can result in some loss of cell viability. However the distribution of cells in the matrix can be well defined and the microenvironment closely controlled.

Membranes
Semi-permeable membranes in the form of fibres or sheets have been used extensively to immobilize other biocatalysts but there are only a few reports of their use for plant cells. Shuler (1981) used an Amicon hollow fibre unit to immobilize soyabean cells while Prenosil & Pederson (1983) entrapped carrot cells in a similar device and followed the production of phenolic compounds over a period of 19 days.

Most important plant products diffuse freely through cellulosic or polysulphone membranes but biomass build-up may reduce mass transfer. However the major disadvantage of these devices is that the most costly element in the system, the fibre bundle, cannot be recovered after use.

EXPLOITING IMMOBILIZED CELLS
The aim of immobilization is to make full use of plant biomass by facilitating contact between non-growing cells and a defined medium. It is essentially a two stage process in which

biomass is first grown rapidly in suspension culture, and the
biosynthetic machinery then used over an extended period. The growth-
independent product must be liberated into the medium and swept out of
the reactor by a mobile phase (liquid or gas).

Although suspension cultured cells appear to make only intracellular
products, the equilibrium can be displaced in favour of an external
product by altering environmental conditions (Vinas & Pareilleux
1982). Immobilization may fortuitously create these conditions
(Lindsey et al. 1983) or may permit the removal of barriers to
equilibration (e.g. by treatment of cells with solvents such as
dimethyl sulphoxide (Brodelius & Nilsson 1983). An alternative
approach is to continuously remove the small amounts of product from
the extracellular medium. This was strikingly illustrated by Becker
et al. (1984) who found that adding an immiscible solvent to a
suspension culture resulted in the accumulation of compounds not
previously detected in the cells. Knoop & Beiderbeck (1983) have also
increased the yield of coniferyl aldehyde from a suspension culture
60-fold by including a solid adsorbent in the medium, thus providing a
high affinity "sink" for the product.

Immobilization provides the best way of exploiting these phenomena
since the adsorption system can be kept separate from the biocatalytic
unit and product transferred to the former in the mobile phase. We
have found this concept of an infinite, rechargeable "sink" to
displace unfavourable product equilibria very effective for recovering
secondary metabolites, without disrupting the cells.

This more efficient integration of synthesis with product recovery is
probably the most important aspect of plant cell immobilization (Sahai
& Knuth 1985) and may be the factor which makes the production of
phytochemicals by cultured cells economic.

REFERENCES
Becker, H., Chavadej, S., Baumer, J. & Stoeck, M. (1984). Two-Phase
        culture; a new method to yield secondary products. Proc.
        3rd European Cong. Biotechnol., Munich., 1-209.
Bornmann, C.H. & Zachrisson, A. (1982).   Attachment of plant
        protoplasts to microcarriers.   Plant Cell Rpts, 1,
        151-3.
Brodelius, P. & Nilsson, K. (1980). Entrapment of plant cells in
        different matrixes: a comparative study.   FEBS Lett. 122,
        312-16.
Brodelius, P. & Nilsson, K. (1983). Permeabilization of immobilized
        plant cells, resulting in release of intracellularly
        stored products with preserved cell viability. Eur. J.
        Appl. Microbiol. Biotech., 17, 275-80.
Brodelius, P. (1985). The potential role of immobilization in plant
        cell biotechnology. Trends Biotechnol., 3, 280-5.
Dainty, A.L., Goulding, K.H., Robinson, P.K., Simpkins, I. &
        Trevan, M.D. (1985).   Effect of immobilisation on plant
        physiology-real or imagined? Trends Biotechnol., 3, 59.

Galun, E., Avid, D., Dantes, A. & Freeman, A. (1983).
    Biotransformation by plant cell immobilized in crosslinked
    polyacrylamide-hydrazide. Monoterpene reduction by
    entrapped Mentha cells.Planta Med., 49, 9-13.
Jirku, V., Macek, T., Vanek, T., Krumphanzyl, V. & Kubanek, V. (1981).
    Continuous production of steroid glycoalkaloids by
    immobilised plant cells. Biotechnol. Lett. 3, 447-50.
Knoop, P. & Beiderbeck, R. (1983). Adsorbent culture. Method for the
    enhanced production of secondary substances in plant
    suspension cultures. Z. Naturforsch. 38C, 484-6.
Lindsey, K. & Yeoman, M.M. (1984). The synthetic potential of
    immobilized cells of Capsicum frutescens.   J. Exp. Bot.,
    35, 1684-96.
Nilsson, K., Birnbaum, S., Flygare, S., Linse, L., Schroder, U.,
    Jeppsson, U., Larsson, P-O., Mosbach, K. & Brodelius, P.
    (1983). A general method for the immobilization of cells
    with preserved viability.   Eur. J. Appl. Microbiol.
    Biotech., 17, 319-26.
Pareilleux, A. & Vinas, R. (1984). A study on the alkaloid production
    by resting cell suspensions of Catharanthus roseus in a
    continuous flow reactor. Appl. Microbiol. Biotech. 19,
    316-20.
Prenosil, J.E. & Pederson, H. (1983). Immobilized plant cell
    reactors. Enzyme Microb. Technol. 5, 323-31.
Rhodes, M.J.C. & Kirsop, B.H. (1982). Plant cell culture as sources
    of valuable secondary products. Biologist 29, 134-40.
Rosevear, A. (1980). Immobilized cells.   GB 2083827B.
Rosevear, A. & Lambe, C.A. (1985). Immobilized plant cells.   Adv.
    Biochem. Eng., 31, 37-58.
Sahai, O.P & Shuler, M.L. (1984). Multistage continuous culture to
    examine secondary metabolism in plant cells. Phenolics
    from Nicotiana tabacum.   Biotechnol. Bioeng., 26, 27-36.
Sahai, O.P. & Knuth, M. (1985). Commercialising plant tissue culture
    processes: economics, problems and prospects. Biotechnol.
    Progress 1, 1-9.
Shuler, M.L. (1981). Production of secondary metabolites from plant
    tissue culture - problems and prospects. Ann. N.Y. Acad.
    Sci. 369, 65-79.
Vinas, R. & Pareilleux, A. (1982). Production of alkaloids by cell
    suspension cultures of Catharanthus roseus cultured in
    vitro. Physiol. Veg. 20, 219-25.
Warren, G.S. & Fallon, R. (1984). Reversible, lectin mediated
    immobilization of plant protoplasts on agarose beads.
    Planta, 161, 201-6.
Yeoman, M.M., Fagandini, D.A.A. & Childs, A.F. (1978). Callus
    Culture in nutrient flow. British Patent No. BP
    32185/78.

22. STUDIES OF ENVIRONMENTAL FEATURES OF IMMOBILISED PLANT
     CELLS

R.J. Robins, A.J. Parr, S.R. Richards and M.J.C. Rhodes

Plant Cell Culture Group,
AFRC Institute of Food Research
(Norwich Laboratory),
Colney Lane,
Norwich, NR4 7UA, U.K.

INTRODUCTION

The opportunities for the production of secondary products
from suspensions of plant cells in batch culture are probably limited
to products of high value (> US\$ 500/kg). A possible way of extending
the technology to products of lower value is to use immobilised plant
cells held in the productive state for extended periods in a
continuous culture system. Immobilisation involves providing a
support in which cells are held in particles of uniform size and are
thus separated from the bulk liquid phase of the fermenter.

Approaches to immobilisation

Immobilisation can be achieved by three approaches. Two
involve loading cells at high density from suspension culture into a
support matrix. This may be either gel beads, such as calcium
alginate, the cells being entrapped in a uniform polymeric matrix
(Brodelius 1982), or particles of a semi-rigid material, such as
polyurethane foam, which is mixed with a high density suspension
culture. Plant cells become lodged within the foam, filling much of
the void volume available (Lindsey et al. 1983). In both these cases
there is little possibility of cell growth and of establishing
cell-to-cell inter-connections. The third alternative is to seed a
semi-rigid matrix, such as foam or nylon particles with a low density
of cells and promote these to grow, filling the whole volume of the
particle. This enables intimate inter-cellular contact to be
established and maintained and for cells to become conditioned to the
micro-environment in which they are to be exploited.

Advantages and disadvantages of immobilisation

The exploitation of cells in the immobilised state offers
a number of potential bioengineering (1-5) and biological (6-8)
advantages:
1. The use of continuous culture systems with slow or non-growing
    cells.
2. Separation of cells and medium is simply achieved. Thus, product
    can be harvested without damage to the cells.
3. Improved mixing due to low viscosity at high cell densities.
4. Fermenters need not be tailored to the needs of individual plant
    cell lines but can be optimised for a specific type of particle.
5. Product is readily harvested in a concentrated semi-purified form
    (Robins & Rhodes 1986).

6. High concentrations of biomass are possible.
7. Stabilisation of catalytic activity can be achieved.
8. The micro-environment created within the particle may modify the catalytic activity of the culture beneficially (Brodelius 1982). Predominant aspects of the micro-environment are (a) its low shear compared to free suspensions (Scragg et al. 1986), (b) the establishment of gradients across the depth of the particle and (c) the possibility of maintaining intimate cell-to cell connections.

Against these possible advantages, which may not all be exploited simultaneously, a number of disadvantages must be balanced:

9. Product must be released to the medium, either by a natural transport process, which may be influenced by the physico-chemical environment of the cells, or by altering the properties of the cellular membranes, provided this does not lead to general cellular damage. The efficiency of the process is more critically dependent on the rate of release of product than on the rate of biosynthesis.
10. Catalytic activity may be lost on immobilisation.
11. The rate of growth may be lower in situ than in free suspension.
12. Loss of free cells from the particles may cause downstream problems but these are unlikely to be as serious as with a free-cell system.
13. The micro-environment created within the particles may be detrimental to secondary product synthesis (Dainty 1985; Berlin 1985).

Strategy
The full potential of immobilisation is most readily realised in a process in which a few cells are used to seed a large void volume and biomass is generated in situ. This avoids the need for large first-stage free-cell fermentations and encompases the special advantages envisaged in establishing specific micro-environments. In addition, the gradients that exist during the growth phase could lead to limited differentiation in a controlled manner which can be beneficial to secondary product formation (Lindsey & Yeoman 1985). It is this approach that we have adopted at Norwich, a portion of the work having been generously assisted by the British Technology Group. Initially, we considered the factors involved in the efficiency of utilisation of the innoculum (i.e. the percentage of cells that become immobilised in a defined period); the development of a minimum effective inoculum (i.e. how small a seeding is required to achieve rapid growth to fill the void volume in a defined period); and the uniformity with which particles are seeded. Substantial progress in these areas has been made in both shake flasks and fermenters (Rhodes et al. 1986; unpublished results).

In this paper we consider aspects of the micro-environment of immobilised cells grown in situ. In particular, the surface features of cells grown on semi-rigid matrices and the internal gradients of oxygen concentration generated by cells which have been grown in situ in foam particles or entrapped at high densities in alginate beads.

SCANNING ELECTRON MICROSCOPE STUDIES OF IMMOBILISED PLANT CELLS
Daucus carota in alginate beads

Spherical beads (3.3-3.8mm dia) were made by mixing cultures with 3% sodium alginate (1:2) and extruding through an orifice 0.5mm diameter into 50mM $CaCl_2$. After incubation in growth medium, beads were prepared for low·temperature SEM and viewed in the partially-hydrated state, as described by Rhodes et al. (1985). By 11 days (Fig. 1a) cells are seen erupting from the otherwise smooth surface of the beads; by 14 days the beads are undergoing extensive disintegration. Sections taken at different ages and viewed in the LM reveal that micro-colonies within the first 0.5mm from the surface tend to grow rapidly but those towards the interior grow much less during the same period (Fig. 1b).

Plant cells grown in situ in semi-rigid support particles

Figure 1c-e show cells grown in situ in polyurethane foam particles or in nylon matrix (Fig. 1f). Figure 1c and 1d are of cells of Beta vulgaris viewed as critical-point dried or frozen-hydrated material respectively. Figure 1d shows a film of material covering he cells and support matrix, which when freeze-dried in the microscope is damaged, cracking and contracting until an image equivalent to that seen in freshly viewed critical-point dried material is generated (Fig. 1c). These properties have previously been observed in hydrated mucilage layers (Read et al. 1983), leading us to suggest that the film we observe is a layer of hydrated mucilage (Rhodes et al. 1985). Cinchona pubescens on foam (Fig. 1e) and Humulus lupulus on nylon (Fig. 1f) are similarly covered in a film, indicating that this general phenomenon occurs with several plant species immobilised on different matrices and may be analogous to the films observed with immobilised cyanobacteria (Robins et al. 1986).

Figure 1a and 1d illustrate the differences between gel-entrapped and foam-seeded particles. In beads little of the surface is composed of exposed cells; in foams much of the surface is cells. A mucilaginous layer is apparent over the cells in both environments: the extent to which this determines the nature of the micro-environment is not clear. Cells at the surfaces of the two immobilised states may experience comparable environments, but internally the two systems probably differ considerably as the inter-cellular spaces are filled with free liquid in the foams but by gel in the beads.

The mucilages produced by B. vulgaris and C. pubescens, currently under investigation by Dr. R. Selvendren (IFRN), contain carbohydrate, protein and phenolics. Preliminary carbohydrate analyses indicate that cells immobilised in foam particles produce greater amounts of pectic arabinogalactans and xylo-glucans than in free solution and that these are particularly retained in the mucilage layer. Such polymers are typical of mucilages and hemicelluloses. The

Figure 1.   Micrographs of immobilised plant cells. (a)
Low temperature SEM of Daucus carota entrapped in an
alginate bead (x20); (b) LM section through an alginate
bead containing D. carota cells (x95); (c) SEM of
critical-point-dried Beta vulgaris cells immobilised on
polyurethane foam (x130); (d) SEM of frozen-hydrated B.
vulgaris cells on polyurethane foam (x35); (e) SEM of
frozen-hydrated Cinchona pubescens cells on polyurethane
foam (x130); (f) SEM of frozen-hydrated Humulus
lupulus grown on nylon matrix (x145).   SEMs taken by
R.J. Turner (IFRN); full details in Rhodes et al.
(1985).

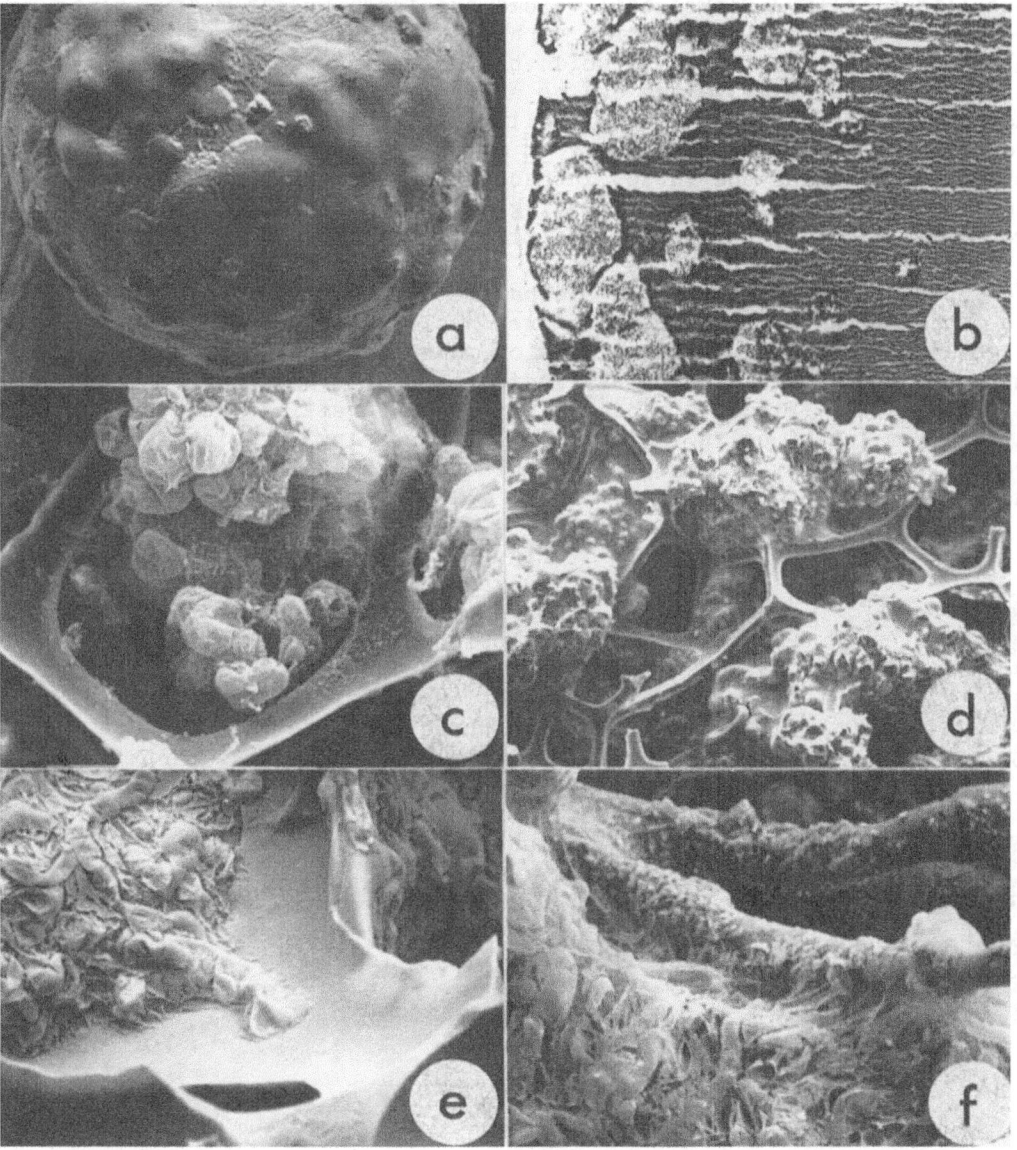

mucilagenous layers surrounding some cyanobacteria (Bar-Or et al. 1985), which form a film in the frozen-hydrated state (Robins et al. 1986), are necessary for floculation (Fatton & Shilo 1984) and are also composed of carbohydrate and protein (Bar-Or et al. 1985). As we have recently shown that cyanobacteria appear to adhere to the foam support matrices by this mucilage, we have suggested that the carbohydrate-rich mucilage of the plant cells is performing a comparable role (Robins et al. 1986).

GROWTH AND PRODUCTION BY IMMOBILISED CELLS

In any cell culture there are problems associated with the parameters used to define growth. This is particularly so with immobilised cells. Packed cell volume and fresh weight measurements are hindered by liquid retention within the matrix. Dry weights depend on having particles in which the mass of support material falls within a narrow range, which can be achieved, but still suffer from the problem of retention of dead cell matter. To circumvent such problems, we have developed (Parr et al. 1984) a non-destructive method to determine the intact intracellular space within a support particle. This relies on determining the differential dilutions of two molecules, one excluded by the plasmalemma, the other able to diffuse freely throughout the cell. For the former we use a ^{14}C-labelled non-metabolisable saccharide, such as mannitol, and for the latter ^{3}H-water. The method gives a close correlation with dry weight during the growth and expansion phase of a Beta vulgaris culture but much more accurately indicates necrosis within the culture in later phases (Fig. 2). The ability to differentiate dead cellular matter from intact cells is important in assessing the viability and productive capacity of immobilised cultures over a long period (see Rhodes et al. 1986).

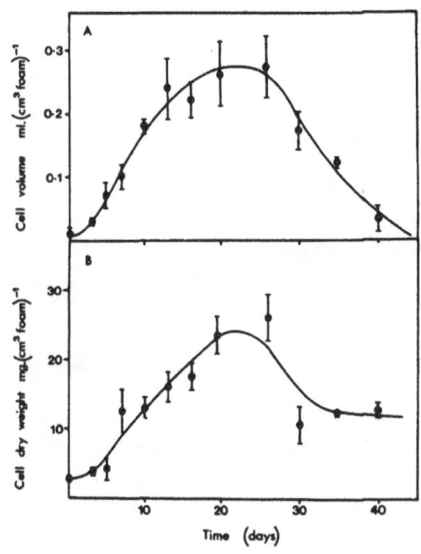

Figure 2. The growth of Beta vulgaris immobilised on foam particles determined by (a) isotope dilution method (Parr et al. 1984) and (b) dry weight. Each determination represents the mean ± S.E. for three separate flasks. The "viable volume" is that part (expressed as a percentage) of a defined space which excludes mannitol but is accessible to water.

With D. carota in alginate, it was found that the loss of cells from gel beads became a serious problem when the cells had grown to occupy only 30-40% of the bead volume. At higher levels disintegration occurred. With foams, however, high cell densities of 80-90% viable volume is attained before significant levels (< 5%) of free cells appear in the medium. This may reflect the more even distribution of growth within the foams and demonstrates an advantage of a support matrix composed of void volume rather than a space-filling gel polymer.

THE GASEOUS ENVIRONMENT OF IMMOBILISED CELLS
Oxygen utilisation rates

Entrapping microbial organisms within a matrix leads to a decreased mean oxygen utilisation rate (OUR) (Tramper et al. 1983). Despite the much lower respiratory rates shown by suspension cultures of plant cells (0.03-0.7 mmol/g.dry wt/h) compared to free-living aerobic microbes (3-5 mmol/g.dry wt/h) similar restrictions are found with plant cells, whether entrapped in alginate beads or grown in foam particles.

Figure 3 shows that, while the OUR for free cells is constant until very low levels (< 3%) of oxygen are reached, with immobilised cells the OUR deviates from linearity at much higher concentrations. As shown in Figure 4, the initial OUR increases with increasing concentration of cells within the bead or foam, but less rapidly than would be anticipated for the same mass of free cells, as demonstrated by the inverse relationship between the specific OUR and the cell mass.

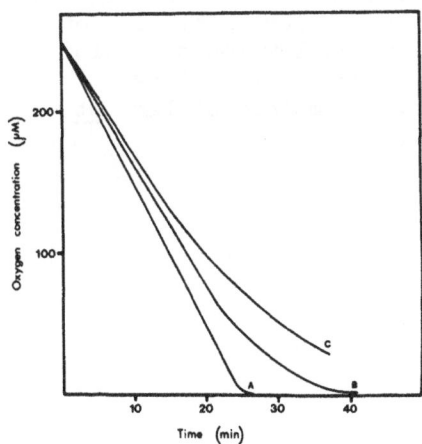

Figure 3. Oxygen utilisation rate (OUR) curves as determined with a Clarke-type electrode (10ml capacity) at 25°C. Curves show (a) Daucus carota free cells, 49mg dry weight; (b) Beta vulgaris immobilised in foam (50mg dry weight, 66% viable volume); (c) Daucus carota entrapped in alginate beads 3.5mm diameter: 10 beads (4.2mg dry weight/bead)

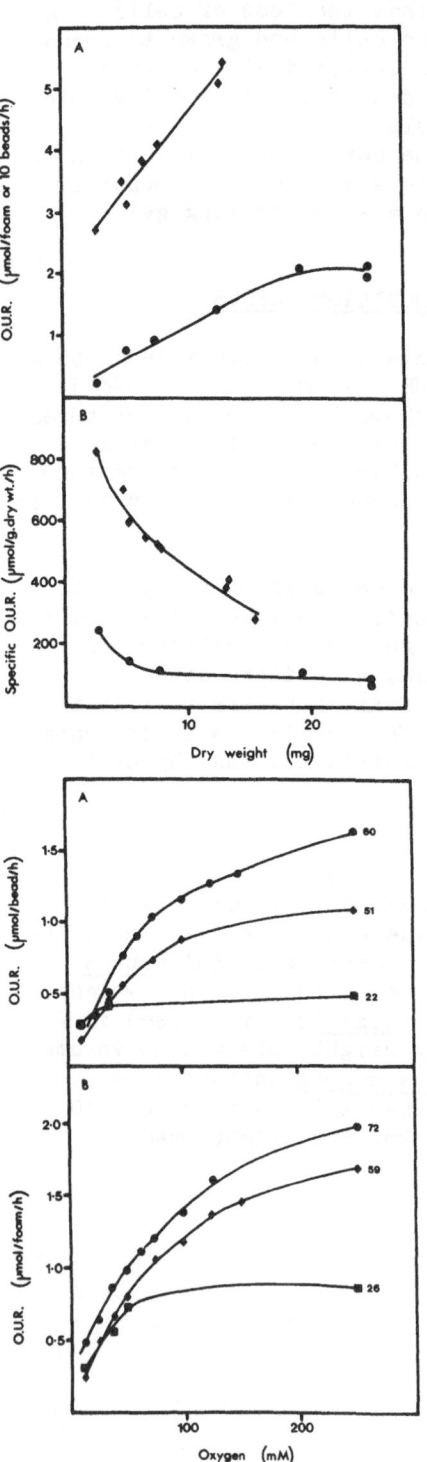

Figure 4.   The effect on initial OUR of varying the concentration of cells immobilised in particles. (a) Initial OUR on the basis of per particle; (b) specific initial OUR, on the basis of the mass of cells. ● = C. pubescens immobilised on foam particles, various ages; (◆) D. carota entrapped in alginate beads, all same age cells. All rates are those initially obtained with a saturated solution of 25°C.

Figure 5.   Progress curves for the OUR of (a) alginate beads and (b) foam particles as the oxygen concentration decreases (cf. Fig. 3). Each frame shows curves typical of high, medium and low loads. Value given is percentage viability obtained by the method of Parr et al. (1984).

From tangents to curves of the type in Figure 3, the relationship between oxygen concentration and OUR can be assessed (Fig. 5). It is satisfactorily described by a rectangular hyperbola, giving linear double-reciprocal plots, from which an apparent $K_{0.5}$ can be determined – i.e. the concentration of oxygen giving half the maximal OUR. The maximal OUR and the $K_{0.5}$ are dependent on the mass of cells present, whether in gel beads or in foam particles (Fig. 5). From numerous such analyses it is found that the apparent $K_{0.5}$ bears an approximately constant relationship to the mass of cells present (Fig. 6). There are two possible explanation for such a relationship. One is that there is more than one terminal oxidase, of varying affinities for oxygen, present, which operate differentially at different oxygen tensions. A much more likely explanation is that there are oxygen limitation within the immobilised particles, causing aberent estimates of the affinity for oxygen because the concentration at the active site is not that measured in the free liquid phase.

Oxygen micro-environment within foam particles

To test for such a limitation within alginate beads, beads were prepared, cut into quarters and finally disintegrated with EDTA/phosphate, the OUR and $K_{0.5}$ being determined at each stage. As shown in Table 1, this treatment caused a marked alteration to the $K_{0.5}$ values with little coincident change in the OURs. That is, the cells have the same ability to utilise oxygen, but the supply of substrate is restricted in full-sized particles. A similar conclusion was reached by Nilsson et al. (1980) who found comparable changes in the apparent K_m for nitrate reductase in immobilised Pseudomonas.

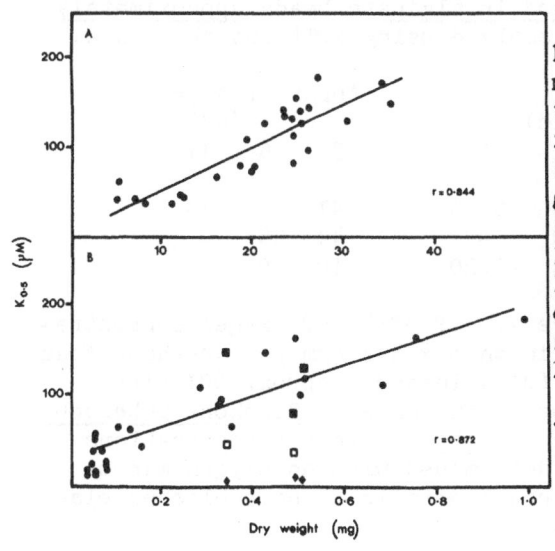

Figure 6. Relationship between mass per particle and the concentration giving half the maximal initial OUR (designated $K_{0.5}$) for (a) C. pubescens immobilised and grown in situ in polyurethane and (b) Daucus carota immobilised in 3% calcium alginate gel. By linear regression analysis the data is satisfactorily and significantly (p < 0.01) fitted by a linear relationship. In (b) ■ , □ and ◆ show gel beads that were quartered and disrupted (see Table 1).

Using an oxygen micro-electrode, we have measured the oxygen gradient within foams (Fig. 7). The oxygen concentration in a fully loaded particle diminishes rapidly to a value of > 5% saturated within the outermost 1mm of the cube. Thus, cells different distances from the surface experience different oxygen concentrations, those at the centre experiencing a semi-anaerobic environment. Comparable profiles occur in fungal pellets (Whittler et al. 1984). Probably, similar gradients exist for other metabolites, making a highly complex micro-environment. When, however, the bulk liquid phase at the surface of the particles is of constant composition, such as in a continuous flow fermenter, then these gradients will exist in a meta-stable state.

Although cells towards the centre of the particle are only exposed to a low oxygen level this need not imply that they are metabolically inactive. Using the method of Parr et al. (1984), we have shown that even in densely loaded foams 80-90% of the void volume contains intact cells. Yet only 60% of the volume of the cube (8 x 8 x 8mm) is contained in the outermost 1mm box. Possibly cells at the centre are supplied with their metabolic requirements via the symplastic connections between cells grown in situ, a situation that will not exist in gel beads or foams loaded at high density. Only after extended periods of culture in the same medium (10-12 weeks) do necrotic regions appear at the centre.

Table 1. Showing the effect of decreasing bead size and dissolution on the oxygen gaseous exchange parameters for Daucus carota immobilised in alginate beads approximately 3.5mm diameter   a. Dissolved using EDTA and phosphate.

Days of culture	OUR (μmol/bead/h)			Apparent $K_{0.5}$ (μM)		
	3	5	11	3	5	11
Whole bead	0.12	0.19	0.16	147	79	13
Quartered bead	0.14	0.20	–	44	38	–
Dissolved[a]	0.09	0.13	0.20	10	9	7

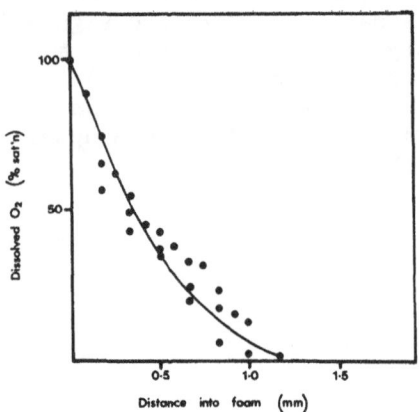

Figure 7.   Profile of oxygen concentration in an 8 x 8 x 8mm polyurethane foam cube fully loaded (approx. 80% viable volume) with cells of Cinchona pubescens grown in situ. Oxygen concentrations were determined with an oxygen micro-electrode; details to be published elsewhere.

CONCLUSIONS
Introducing plant cells into foam particles in a
controlled fashion to form an immobilised culture with very few free
cells present it now possible by seeding with a low density of cells
which become attached and grow to a high density. Such cells are
surrounded by a mucilagenous layer which, as well as probably playing
a role in acting to adhere the cells to the support, may affect the
micro-environment in which the cells exist. Gradients, of the nature
of that shown for oxygen, create a graded series of sub-micro-
environments within the particle. How this situation affects cellular
metabolism and relates to the observed effects on product synthesis
when cells are immobilised, where enhanced (Brodelius 1983; Lindsay &
Yeoman 1985) or decreased (Dainty et al. 1985; Berlin 1985) production
can occur, remains to be established. A better understanding of the
effect of the microenvironment on cellular behaviour might enable us
to tailor the immobilised particle to enhance particular
characteristics, in particular secondary product biosynthesis and
secretion.

REFERENCES
Bar-Or, Y., Kessel, M. & Shilo, M. (1985). Modulation of cell surface
        hydrophobicity in the benthic cyanobacterium Phormidium
        J-1. Arch. Microbiol., 142, 21-7.
Berlin, J. (1985). The use of immobilised plant cells - an
        evaluation. IAPTC Newsletter No. 46, 8-14.
Brodelius, P. (1982). Immobilised plant cells. In Immobilised Cells
        and Organelles (Mattiasson, Bo., ed.), 1, 27-55, CRC
        Press, Boca Raton, Fla.
Dainty, A.L., Goulding, K.H., Robinson, P.K., Simphins, I. & Trevan,
        M.D. (1985). Effect of immobilisation on plant cell
        physiology -real or imaginary? Trends Biotechnol., 3,
        59-60.
Fattom, A. & Shilo, M. (1984). Phormidium bioflocculant: production
        and activity. Arch. Microbiol., 139, 421-26.
Lindsey, K., Yeoman, M.M., Black, G.M. & Mavituna, F. (1983). A
        novel method for the immobilisation and culture of plant
        cells. FEBS Lett., 155, 143-9.
Lindsey, K. & Yeoman, M.M. (1985). Immobilised plant cells. In:
        Plant cell culture technology. Botanical Monograph series
        (Yeoman, M.M., ed.) Blackwell, Oxford.
Nilsson, I., Ohlson, S., Haggstrom, L., Molin, N. & Mosbach, K.
        (1980). Denitrification of water using Pseudomonas
        denitrificans cells. Appl. Microbiol. Biotechnol., 10,
        261-74.
Parr, A.J., Smith, J.I., Robins, R.J. & Rhodes, M.J.C. (1984).
        Apparent free space and cell volume estimation: a non-
        destructive method for assessing the growth and membrane
        integrity/viability of immobilised plant cells. Plant Cell
        Rep., 3, 161-4.

Read, N.D., Porter, R. & Beckett, A. (1983). A comparison of preparative techniques for the examination of the external morphology of fungal material with the scanning electron microscope. Can. J. Bot., 61, 2059-78.

Rhodes, M.J.C., Robins, R.J., Turner, R.J. & Smith, J.I. (1985). Mucilagenous film production by plant cells immobilised in a polyurethane or nylon matrix. Can. J. Bot., 63, 2357-63.

Rhodes, M.J.C., Smith, J.I. & Robins, R.J. (1986). Factors affecting the immobilisation of plant cells on reticulated polyurethane foam particles. Appl. Microbiol. Biotechnol., (submitted).

Robins, R.J., Hall, D.O., Shi, D.-J., Turner, R.J. & Rhodes, M.J.C. (1986). Mucilage acts to adhere cyanobacteria and cultured plant cells to biological and inert surfaces. FEMS Microbiol. Lett., 34, 155-60.

Robins, R.J. & Rhodes, M.J.C. (1986). Polymeric adsorbents stimulate anthraquinone production by cell suspension cultures of Cinchona ledgeriana. This vol., 35-40.

Scragg, A.H., Allan, E.J., Bond, P.A. and Smart, N.J. (1986). Rheological properties of plant cell suspension cultures. This volume, 24.

Tramper, J., Luyben, K.Ch.A.M. & Tweel, W.J.J. van den (1983). Kinetic aspects of glucose oxidation by Gluconobacter oxydans cells immobilised in calcium alginate. Appl. Microbiol. Biotechnol., 17, 13-8.

Wittler, R., Baumgartl, H., Schugerl, K. & Lubbers, D.W. (1984). Oxygen transfer in Penicillium chrysogenum pellets. Proc. 3rd. Eur. Congr. Biotech., 1, 513-20.

23. PRODUCT RELEASE FROM PLANT CELLS GROWN IN CULTURE

A.J. Parr, R.J. Robins and M.J.C. Rhodes

Plant Cell Culture Group,
AFRC Institute of Food Research,
Norwich Laboratory,
Colney Lane,
Norwich, NR4 7UA, U.K.

INTRODUCTION
    Plant cell cultures have great potential for the
biotechnological production of useful secondary products. One idea
which is currently attracting much attention is the use of immobilized
cells, produced either by embedding within a gel or by entrapment
within a reticulate matrix. Such systems are likely to operate most
effectively when the desired product is released into the growth
medium, where it can be harvested continuously without having to
destroy the cells. In cell cultures two situations arise: either (i)
cells have an endogenous secretory mechanism and naturally release the
desired product into the medium to some extent, or (ii) product is
accumulated and stored intracellularly. In exploiting such cultures
there is a need in the former instance to control and optimize the
release mechanisms, while in the latter there is a requirement to
alter membrane properties in order to effect product release. At
present relatively little is known about either aspect. In Cinchona
ledgeriana alkaloids are released to the medium but substantial
amounts are also accumulated intracellularly. We have therefore used
this species to study both natural and artificial modes of product
release. Further work on cell permeabilization has been done with
Beta vulgaris, in which the red pigment betanin is normally stored
within the vacuole.

Natural transport of alkaloids
    Kinetic analysis of the efflux of endogenously synthesized
quinoline alkaloids from C. ledgeriana cells revealed several release
phases. The major phase was slow (with a half-time of about 15
hours), and showed a distinct selectivity towards different alkaloids
(Figure 1). Uptake of exogenously added alkaloid was similarly
multiphasic, with the extent of uptake being increased by raising the
external pH (Figure 2). Over the short term, the uncoupler CCCP at a
concentration of 10μM had no effect on uptake.

In steady-state cultures there is a balance between uptake and release
processes. Alkaloid concentrations in the cell were always greater
than in the medium. The exact accumulation ratios (concentration in
cells/concentration in medium) however varied between different
alkaloids, and also depended on whether the alkaloids had been
synthesized endogenously, or had been added to the growth medium
(Table 1).

Figure 1. Efflux of endogenous alkaloids from
C.ledgeriana cells.

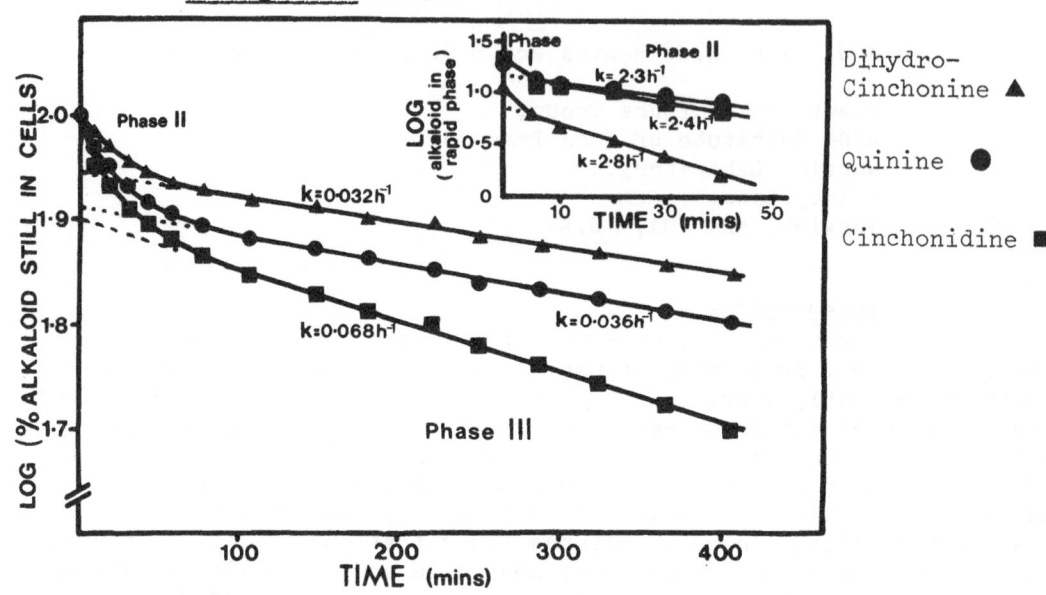

Figure 2. Uptake of exogenous cinchonine by cells of
C. ledgeriana.

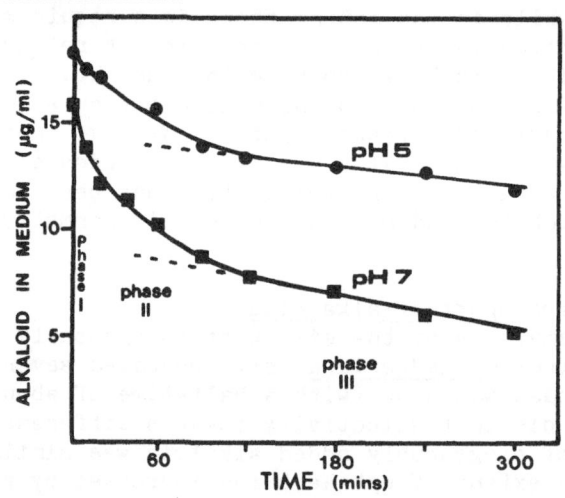

Table 1: Steady-state accumulation ratios for various
quinoline alkaloids in cultures of C. ledgeriana. (pH
4.7).

Alkaloid	Accumulation ratio for:	
	Endogenous alkaloids	Added alkaloids
Cinchonine	30	3.9
10.11-Dihydro-cinchonine	55	4.1
Cinchonidine	13	3.9

One published model for alkaloid transport (Renaudin, 1981) proposes
that uncharged alkaloid molecules are able to freely diffuse across
membranes, but protonated alkaloid cations cannot. There is thus a pH
dependent movement of alkaloids, with alkaloid distribution being
primarily determined by their accumulation in acidic compartments by
an 'ion-trapping' mechanism. Alkaloids might also bind to
intracellular material (possibly phenolics) and thus be accumulated to
a higher degree than predicted by 'ion-trapping' alone. In C.
ledgeriana some of the data can be explained by this model, but only
if binding is assumed to have a major role. This is required to
explain the discrimination between related alkaloids with similar pK
values. It would also explain the multiphasic kinetics observed,
although this could also reflect different membranes having different
permeabilities to alkaloid molecules. Despite this general agreement
with the model of Renaudin (1981), there are however a number of
unexplained results, noteably the different accumulation of endogenous
and exogenous alkaloids. It seems likely that additional processes
are operating. Endogenous alkaloids might conceivably be synthesised
in, or transported to, cell compartments not readily accessible to
externally added alkaloids. The results could also be explained by
cell specialization. Thus added alkaloids might be taken up by all
cells, via an 'ion-trapping' mechanism, but endogenous alkaloids might
be synthesized and/or stored in specialised cells, with a more
selective transport mechanism.

Although still incompletely understood, it is apparent that alkaloid
transport is quite complex, and is influenced by a variety of factors.
The potential therefore exists to manipulate the characteristics of
product release in a commercial system, for example by adjusting
intracellular or medium pH. It should also be possible to select for
strains showing enhanced efflux. While the precise details of such
manipulations are likely to be relevant only to the production of
alkaloids, the general concepts might usefully also be applied to
other systems - where, at present, product transport is often
extremely poorly understood.

Permeabilization of cells
The application of a wide variety of physical and chemical
treatments to cells has been found to produce release of
intracellularly stored material (Felix, 1982, see also Table 2). Not
all treatments are however equally effective, and the sensitivity of
the cells towards permeabilization may also vary during the growth
cycle (Table 2).

In a commercial system it is desirable to be able to reversibly
permeabilize cells, or otherwise permeabilize them without loss of
viability. It would then be possible to make extended use of the
available biomass. Unfortunately many organic solvents and detergents
are active only at a level which irreversibly disrupts cellular
integrity (Felix, 1982). In Beta vulgaris we have found that the

release of betanin which follows osmotic shock also seems to be
associated with cell damage, as seen under the microscope. Despite
these results, there is however a report in the literature (Brodelius
& Nilsson, 1983) that dimethylsulphoxide (DMSO) may reversibly
permeabilize cells of Catharanthus roseus so that alkaloids are
released while cell viability is maintained.

We have investigated the effect of DMSO on alkaloid release by C.
ledgeriana. Using the $3H_2O/^{14}C$-mannitol differential dilution method
it was shown that DMSO successfully permeabilized C. ledgeriana cells,
although somewhat higher levels were required than for C. roseus (Parr
et al. 1984). As expected the release of stored quinoline alkaloids
was also brought about (Parr et al. 1984). Following removal of the
DMSO we however found that membrane barriers were not restored (Figure
3), implying that permeabilized cells had been irreversibly damaged.
Those cells which had not been fully permeabilized by the levels of
DMSO used were found to remain intact. In C. ledgeriana it would
appear that DMSO is not a useful agent for inducing product release
while maintaining cell viability. This may well also be the case in
many other species. Thus secondary products are often stored within
the vacuole, and hence in order to induce product release both the
plasmalemma and tonoplast need to be permeabilized. The mixing of
cytoplasmic contents with phenolics and other potentially toxic
compounds present in the vacuole which would occur during
permeabilization might be expected to lead to permanent damage in many
cases.

Our experience with a variety of plant species and permeabilising
agents is that permeabilization is generally associated with permanent
damage to cells. Except in special cases it is thought unlikely that

Table 2:   The effect of various treatments on betanin release from
            cells of B. vulgaris.

Treatment	Percentage betanin released (cells in phase of rapid growth)	Time when cells are most sensitive
None	<1	-
1%(v/v) Toluene, 1 hour	86	Not tested
10%DMSO, 1 hour	26	NT
10%Ethanol, 1 hour	60	NT
0.5% Phenethyl alcohol, 1 hr	40	Early in growth cycle
5% Tween 20, 20 hr	25	NT
0.1% Triton X-100, 1 hr	50	All stages
0.1% Lysolecithin, 6 hr	5	NT
100ug/ml Nystatin, 7 hr	3	All stages
100ug/ml Poly-lysine, 24 hr	4	NT
0.8M Mannitol, continuous (24hr)	10	Late in growth cycle
0.8M Mannitol (2hr) followed by return to isotonic medium	23	Late in growth cycle.

Figure 3   Changes in membrane permeability following
treatment of C. ledgeriana cells with a one hour pulse of
either 10% or 20%(v/v) DMSO.

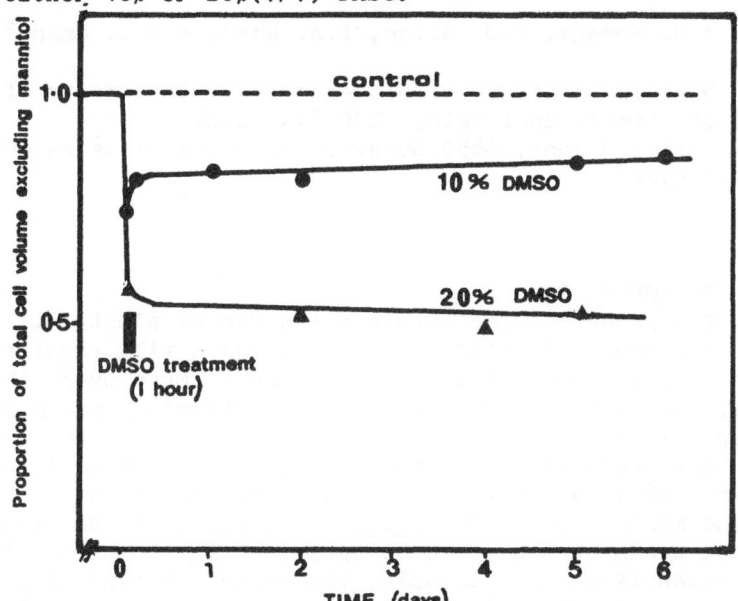

permeabilization methods will prove useful in inducing release of
stored products, unless further advances can be made. They may
however be more appropriate in the field of biotransformation, where
full metabolic integrity of the cells may not be necessary.

CONCLUSIONS
        Studies of both natural and artificially induced product
release from plant cells are still in their infancy. At present it
would appear that natural release mechanisms will be the most easily
adapted to commercial systems. Thus many substances, including
pharmaceutically important alkaloids, are naturally released by cells,
and the release processes are capable of being manipulated. In
contrast, the release of products by cell permeabilization appears
difficult to achieve without permanently damaging the culture.

REFERENCES
Brodelius, P. & Nilsson, K. (1983). Permeabilization of immobilised
        plant cells, resulting in release of intracellularly
        stored products with preserved cell viability. Eur. J.
        Appl. Microbiol. Biotechnol., 17, 275-80.
Felix, H. (1982). Permeabilized cells. Anal. Biochem., 120, 211-234.
Parr, A.J., Robins, R.J. & Rhodes, M.J.C. (1984). Permeabilization
        of Cinchona ledgeriana cells by dimethylsulphoxide.
        Effects on alkaloid release and long-term membrane
        integrity. Plant Cell Reports, 3, 262-5.
Renaudin, J-P. (1981). Uptake and accumulation of an indole
        alkaloid, [^{14}C]-tabernanthine, by cell suspension cultures
        of Catharanthus roseus (L.) G. Don and Acer pseudoplatanus
        L. Plant Sci. Lett., 22, 59-69.

24. RHEOLOGICAL PROPERTIES OF PLANT CELL SUSPENSION CULTURES

A.H. Scragg, E.J. Allan, P.A. Bond, & N.J. Smart[*]

Wolfson Institute of Biotechnology, University of
Sheffield, Sheffield, S10 2TN, U.K.
[*]Allelix Inc., 6850 Goreway Drive, Mississauga, Ontario,
Canada.

INTRODUCTION
Any process which involves the use of plant cell
suspensions for the production of fine chemicals will require mass
cultivation in volumes of at least 500 litres. More over the cell
density will vary during the growth cycle, attaining maximum biomass
levels of approximately 20 gram per litre. Plant cell suspensions are
viscous at high concentrations (Kato et al. 1978; Wagner & Vogelman,
1977; Tanaka, 1982), and this consequently affects mass transfer, heat
transfer, and mixing (Anderson et al. 1982; Blanch & Bhavaraju, 1976)
so that these parameters have to be optimised in order to obtain best
growth and production. For example, the effect of viscosity on a
transport process is shown in Table 1, where as the apparent viscosity
increases the volumetric oxygen transfer coefficient (K_{LA}) decreases.
These factors, the majority of which have been ascertained from
microbial cultures, imply that the rheological behaviour of plant cell
suspensions is of great importance in the development of a process
system. Despite this little information is available concerning the
rheological nature or viscosities developed by plant cell suspensions.
Wagner & Vogelman (1977) noted that the properties of suspension
cultures of Morinda citrifolia and Catharanthus roseus were
complicated showing shear thinning and thixotrophic behaviour. Tanaka
(1981) demonstrated that cell suspensions of Cudriania tricuspidata
and Vinca roseus were non-Newtonian in behaviour and obeyed the power
law.

In addition to their complicated rheological nature plant cells are
sensitive to hydrodynamic stress as a consequence of their large size,
extensive vacuole and rigid cell wall. In this article we have used

Table 1: Effect of suspensions on the volumetric oxygen
transfer coefficient (Kla). Data taken from Anderson et
al. 1982.

Suspended solids	Apparent viscosity (m Pa s^{-1}) at shear rate of 96 sec^{-1}	Kla (hr^{-1})	
		Agitation with Rushton turbine (5oo rpm)	Sparged with gas (0.123m sec^{-1})
O	O	5o5	313
15.8	86	176	117
20.8	193	1o8	79
25.8	393	56	43

cultures of C. roseus to investigate the rheological nature of plant
cell suspensions, the possible viscosities attainable at high biomass
levels and the effects of shear on viability.

Rheology

Rheology is the study of flow and the deformation of
matter by flow. The viscosity (η) of a liquid or gas is defined as
the ratio of shear stress (τ) to shear rate ($\dot{\gamma}$). If a liquid is
confined between two coaxial cylinders and one is rotated, a force
will be required to stop the other turning. With large diameter
cylinders the curvature can be disregarded and represented by two
plates moving in parallel (Fig. 1). At any point assuming a laminar
flow the tangential force divided by the area on which it acts is
defined as the shear stress.

$$\frac{\text{Force}}{\text{Area}} \quad = \quad \text{shear stress} \quad \text{(mPa, milliPascal)} \quad\quad (1)$$

The shear stress causes the liquid to flow having a maximum flow
velocity at the top plate which drops across the gap (Y) to zero at
the lower plate. If the drop in velocity is assumed to be linear then
the shear rate ($\dot{\gamma}$) is given by.

$$\frac{\text{velocity}}{Y} \quad = \quad \text{shear rate} \quad (\text{sec}^{-1}) \quad\quad\quad (2)$$

The shear stress is proportional to the shear rate thus;

$$\tau \; = \; \dot{\gamma}.\eta \quad\quad \text{being the viscosity (mPa sec}^{-1}) \quad (3)$$

$$\text{or} \quad\quad \eta \; = \; \tau/\dot{\gamma}$$

This relationship holds for Newtonian fluids such as water and sugar
solutions where the viscosity is constant at all shear rates. Other
fluids not obeying this formula are known as non-Newtonian. The
non-Newtonian fluids exhibit complex behaviour and can be classified
into three main groups (Figure 2).

Figure 1.   Flow between two parallel plates.

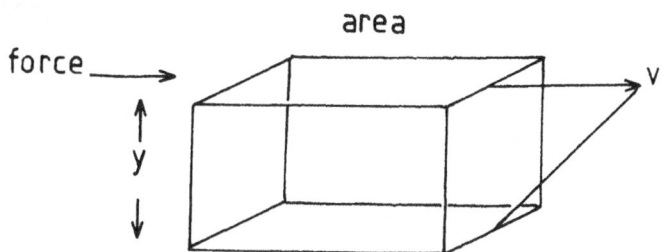

The first group includes fluids where viscosity varies with shear rate
but does not alter with time as shown in Figure 3. Pseudoplastic
fluids show a decrease in apparent viscosity with increasing shear
rate, whereas the dilatant fluids exhibit an increase. This is shown
in a log-log plot of apparent viscosity against shear rate (Fig. 3b).

Figure 2.   Classification of non-Newtonian fluids.

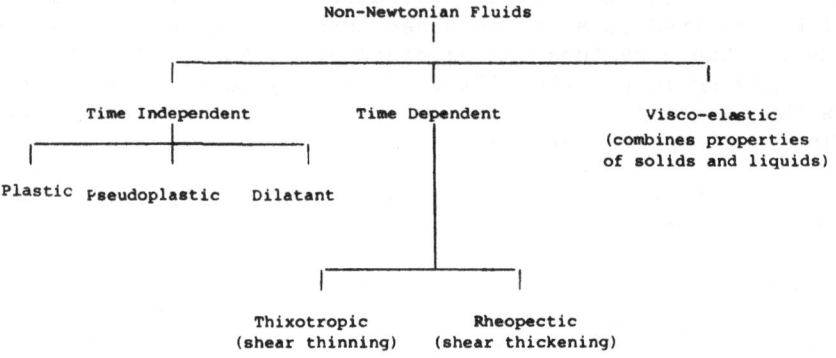

Figure 3.   Flow characteristics of non-Newtonian time-
independant fluids, as (A) shear rate against shear stress
or (B) log real or apparent viscosity against log shear
rate.

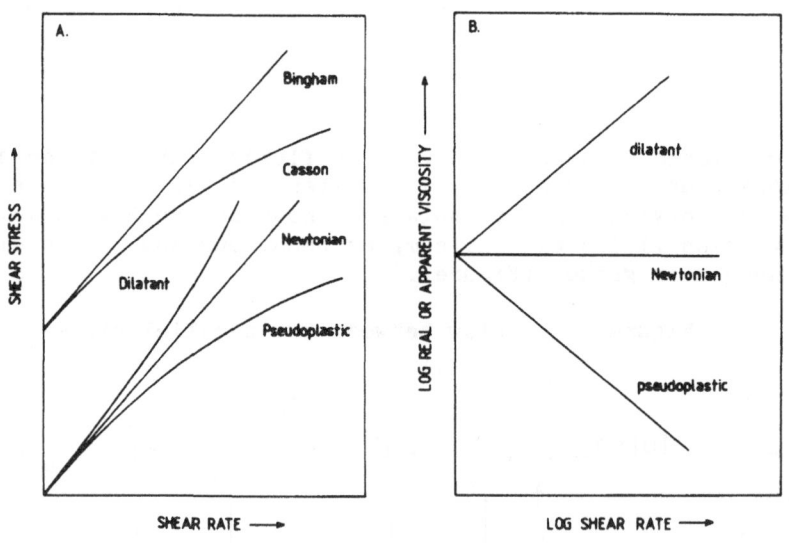

Some liquids will not flow until a minimum shear stress, known as the
yield stress, is applied. These are the plastic fluids represented by
Bingham or Casson fluids. Time-dependent fluids exhibit either an
increase (rheopectic) or decrease (thixotropic) in viscosity at a
fixed shear rate. If the liquid is allowed to stand after shearing
the viscosity may return to its original value i.e. reversible
thixotropy, or the internal structure may have been irreversibly
broken down and the viscosity will be different from that before
shear. When exposed to varying shear rate reversible rheopectic and
thixotropic fluids exhibit a hysterisis curve (Fig. 4). The third
group are the visco-elastic fluids which combine the properties of
solids and fluids.

Measurement of Viscosity
Viscosity as defined applies to conditions of laminar
flow, therefore complex flow patterns employed in some viscometers
make them unsuitable for precise measurement. In our study we have
used a Contraves Rheomat 115 fitted with a double gap rotating
cylinder (Figure 5). The bob, rotated at uniform speeds is centrally
positioned in a concentric space of 1.75 mm containing a known volume
(20 ml) of suspension culture. This particular configuration is
designed for measuring fluids of low viscosities.

Rotational viscometers are widely employed as they are both reliable
and versatile and are suitable for a rigorous measurement of
viscosity, as they approximate laminar flow over a wide range of
angular velocities. However, rotational viscometers are not

Figure 4.   Flow characteristics of non-Newtonian time-
dependent fluids.

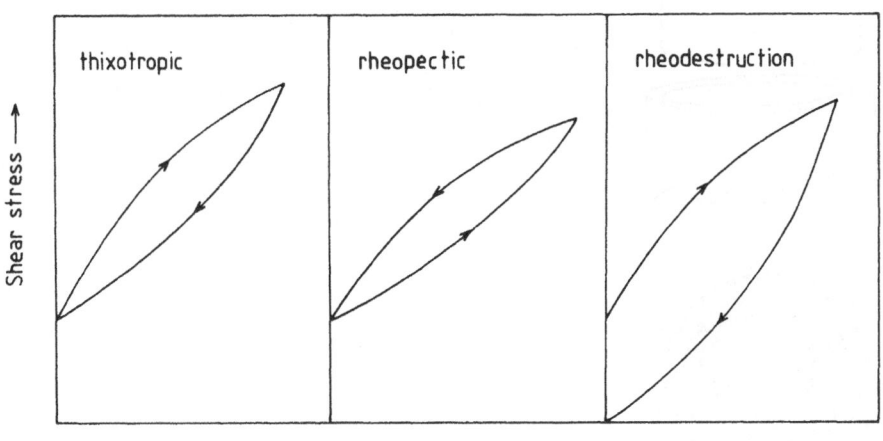

altogether suitable for particulate suspensions as they suffer from
phase separation in the region of the bob, gravity settling of
particles causing a non-homogeneous suspension, and destruction of
particles can also occur (Charles, 1978; Roels _et al_. 1974). To
overcome such difficulties Bongenaar et al. (1973) and Roels _et al_.
(1974) used a turbine impeller to reduce settling and phase
separation but the useful range of shear rates was limited. With
these constraints in mind we have used, in our initial studies the cup
and bob system which allows a comparison with the data of Kato _et al_.
(1978), Tanaka (1982) and Wagner & Vogelmann (1977). A helical ribbon
impeller system has been used to study the rheological properties of
Absidia corymbifera and this offers a relatively wide range of shear
rates and good mixing (Kim _et al_. 1983). Rapp _et al_. (1984) have used
an anchor impeller system to study the rheology of _Cellulomonas uda_
growing on newspaper and Vogelman _et al_. (1978) have used an anchor
impeller for _C. roseus_ suspensions.

Rheological Nature of Plant Cell Suspensions
Unless stated, all rheological properties have been
determined using cells, which have been retained on an 80 um nylon
mesh and resuspended in fresh culture media to a concentration of
100 g.l^{-1} wt weight. Before investigating the nature of _C. roseus_
cell suspensions the contribution supplied by the medium M3X (Smart _et
al_. 1982) before and after growth was determined.

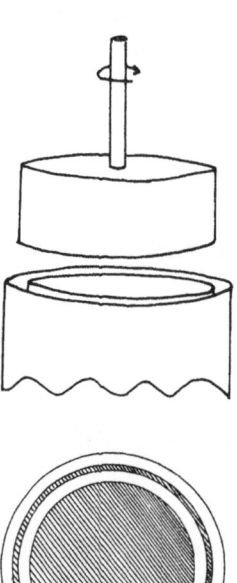

Figure 5.   Diagram of the double-gap
cylinder of the Contraves Rheomat 115
showing the cup and rotating bob.

0·1mm   2·225cm   2·4cm

Figure 6 shows the shear rate plotted against shear stress for M3X
medium before and after growth (14 days), and compares than with the
Newtonian fluid water. It is clear that both M3X media are Newtonian
in nature with the rise in apparent viscosity value above shear rates
of 900 sec^{-1} being due to the development of turbulent flow in the
rheometer.

During our studies we have used cell suspensions of C. roseus which
when examined consisted of cell aggregates of 100 to 1,000 μm in
diameter.   We are dealing, therefore with a suspension of aggregates
somewhat similar to pelleted mould cultures rather than individual
cells.   Kato et al. (1978) observed cell aggregates in Nicotiana
tabacum cultures of 100-800 μm with the mean size of 200-250 μm
changing to 350-400 μm during batch culture.   Tanaka (1982) has
estimated the aggregate size distributions in cultures of Cudriania
tricuspidata, Vinca roseus, and Agrostemina githago. The C. roseus
culture had a narrow distribution range of 44-297 μm, whereas C.
tricuspidata and A. githago had wider distributions of 44-2,000 μm. In
our experience the C. roseus suspension cultures used in this study
can be regarded as 'fine' as although large aggregates can be seen
these represent only a very small proportion, the mean size being
171-250 μm.

The size distribution of the cell aggregates can be altered by
hydrodynamic stress such as increasing the agitator or shaker speed

Figure 6.   The effect of varying the shear rate ($\dot{\gamma}$) on
the apparent viscosity (η) of water (O); M3X medium
(●); M3X medium after 14 day culture, all cells removed
by filtration (△).   These measurements were carried out
using the double-gap cylinder.

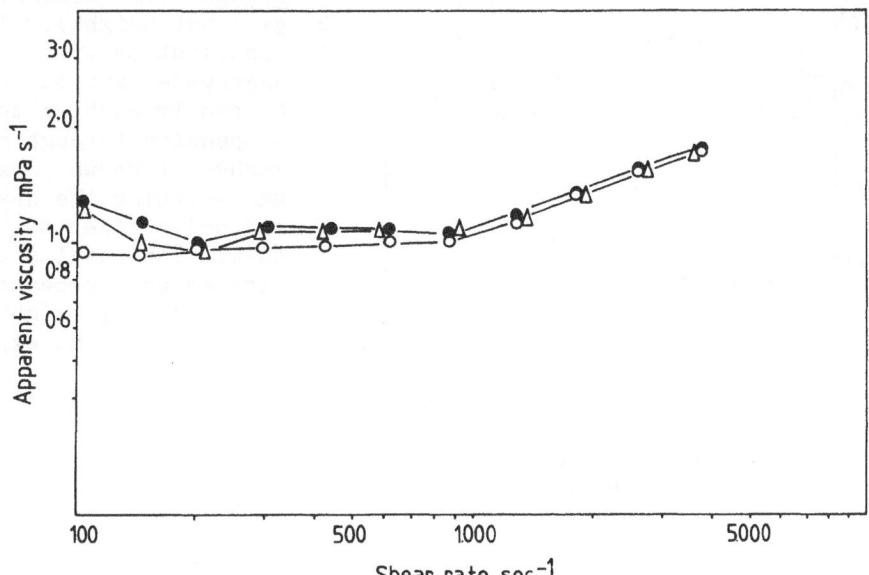

(Henshaw _et al_. 1966; Rajasekhar _et al_. 1971), although Tanaka (1981)
found that suspensions of _C. tricuspidata_ were unaffected by shaker
speed.

A suspension of _C. roseus_ at a concentration of 100 g.l^{-1} (wet weight)
was exposed to a constant shear rate of 297 sec^{-1} (Fig. 7).   The
apparent viscosity decreased rapidly with time.   This behaviour is
non-Newtonian showing time-dependent shear thinning. Analysis of the
cell size distribution before and after exposure to shear thinning was
undertaken by gently washing the cell aggregates through nylon meshes
of known pore size and taking weight measurements after drying at
60°C for 48 hours (Fig. 7). ·The results show a significant decrease
in the large cell aggregates and increase in smaller aggregates upon
treatment with no loss in total cell dry weight.   Whether this
alteration in aggregate distribution is due to the shear field itself
or that the larger aggregates are physically broken because they are
similar in size to the gap in the viscometer is unclear.   Large
particles have been shown to impair accurate measurement of viscosity
in cup and bob viscometers (Roels _et al_. 1974).   The torque readings
obtained with plant cell suspensions at low shear rates of 24.1 to
107.1 sec^{-1} gave considerable oscillations in both amplitude and
frequency which would indicate that the larger cell aggregates were
affecting the operation of the measuring system.

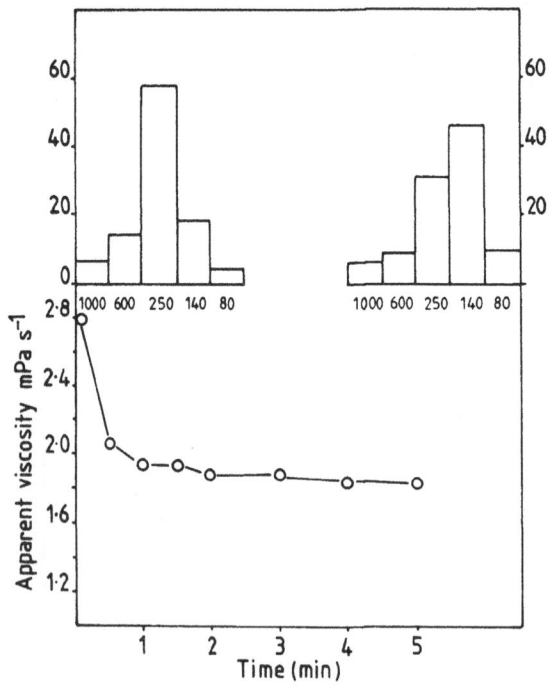

Figure 7.   The change in
apparent viscosity with
time during exposure to a
constant shear rate of 297
sec^{-1} (setting 8) of _C.
roseus_ cell suspension (100
gl^{-1} wet weight).   The size
distribution of the cell
aggregates was determined
by gently washing the cell
suspension through nylon
meshes of known pore size,
and weighing the dry mesh
and cells after 48 hours at
60°C.   The results are
plotted and percentages of
the total dry weight before
and after 5 mins exposure
to shear.

Exposure of C.roseus to two consecutive step-wise increases and decreases in shear rate resulted in non-Newtonian behaviour (Fig. 8). A yield stress was apparent which could be due to the presence of large aggregates although it was still present after one cycle. The non-Newtonian nature can be seen when the results are plotted as a log-log plot of apparent viscosity against shear rate (Fig. 9). The results show Newtonian behaviour at low shear rates with a pseudoplastic response at above shear rates of 200 sec^{-1}. The results shown in Figure 8 also indicate thixotrophy in the plant cell suspensions. The step wise increase and decrease in shear rate has a similar effect on the cell aggregate size distribution as constant shear conditions. Therefore, some of the features observed in Figures 8 and 9 may result from the method of analysis rather than the property of the suspension itself. Above shear rates of 900 sec^{-1} flow is no longer laminar, as apparent the trend is towards Newtonian behaviour (Charles, 1978) and the system can not be used as an indication of viscosity.

The effect of aggregate size on the torque reading obtained using the double-gap system was tested by separating a C. roseus suspension into two size groups retained on 250 and 80 μm mesh filters. These aggregates were resuspended at a concentration of 100 g.l^{-1} wet weight and exposed to a step-wise increase and decrease in shear rate (Fig. 10). At lower shear rates the 250 μm aggregates show the hysteresis curve characteristic of thixotrophic fluids whereas the 80 μm aggregates essentially showed Newtonian behaviour. These results confirm that many of the characteristics observed are due to the

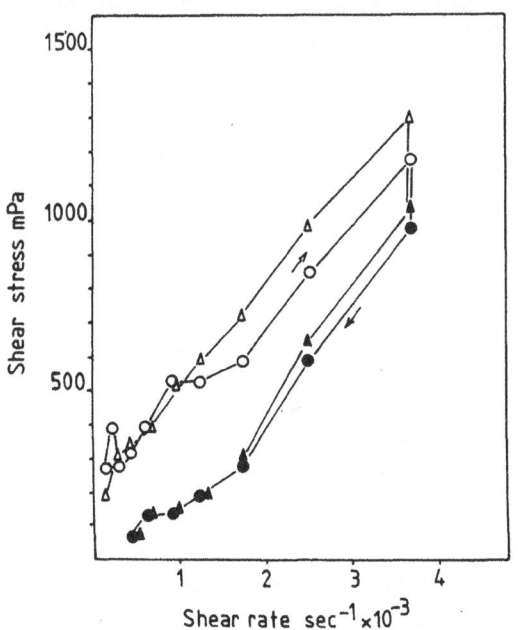

Figure 8.   The changes in shear stress for a C. roseus cell suspension in a double step-wise increase and decrease of shear rate. The cell suspension was made up at 100 gl^{-1} fresh weight and the stress readings were taken with one minute exposure at each shear rate in the double-gap rheometer. First run, increasing (○), decreasing (●); second run, increasing (△), decreasing (▲).

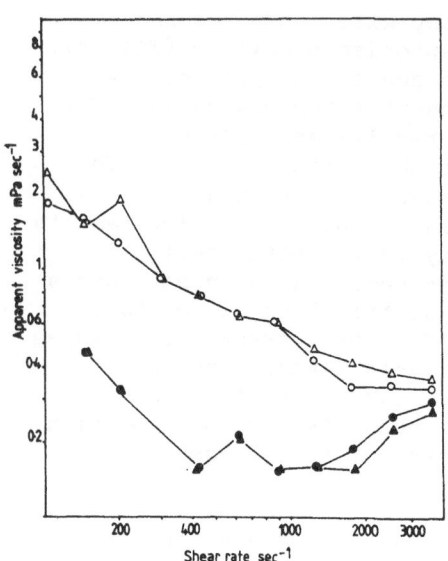

Figure 9.   The changes in
apparent viscosity for a C. roseus
cell suspension in a double step-
wise increase and decrease of
shear rate.   The results are those
obtained in Fig. 8.   First run,
increasing (○), decreasing
(●); second run, increasing
(△), decreasing (▲).

Figure 10.   The changes in apparent viscosity for C.
roseus aggregate suspensions in a step-wise increase and
decrease of shear rate.   The measurements were carried
using the double-gap rheometer.   The cell aggregates were
prepared by filtration of a 14 day old suspension using
nylon filter meshes of known size.   The separated
aggregates were resuspended in fresh M3X medium to a
concentration of 100 gl^{-1} wet weight before treatment.
Cell aggregates 250-600µm, increasing (○), decreasing
(●); cell aggregates 80-250µm increasing (△),
decreasing (▲).

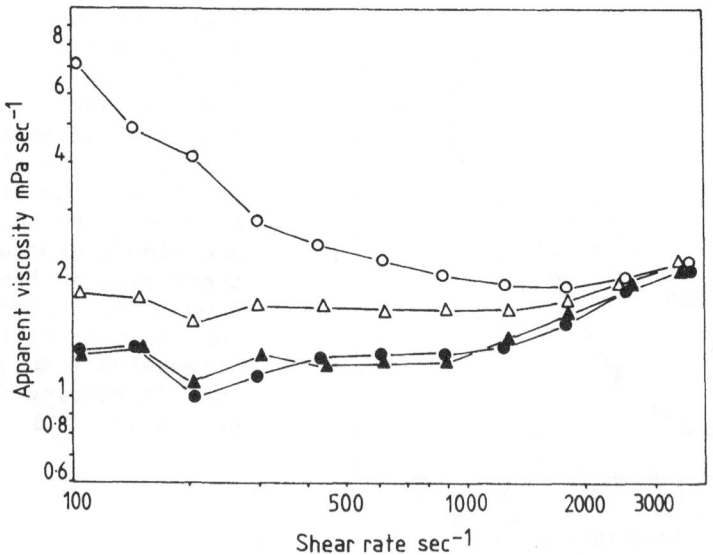

measurement system.   This can only be resolved by using another
system.

We have constructed an anchor impeller system which can be used with
the Contraves torque measurement system.   The rotor (Fig. 11) was
similar to that described by Rapp et al. (1984) and was used in a 250
ml. beaker containing 200 ml. cells.   A C. roseus suspension culture
exposed to a fixed rotational speed of 66 sec[-1] in this anchor system
showed no shear thinning or aggregate destruction (Fig. 12). The gap
between the rotor and beaker was 1.5 cm which is considerably larger
than the aggregate size.   The anchor impeller system was only
effective over a limited range of shear rates as no torque reading was
obtained below 207 sec[-1] and turbulence appeared to start at 1783
sec[-1].   A step-wise increase and decrease in shear rate using the
anchor impeller (Figure 13) with C. roseus suspensions indicates
pseudoplastic behaviour with no evidence of yield stress.

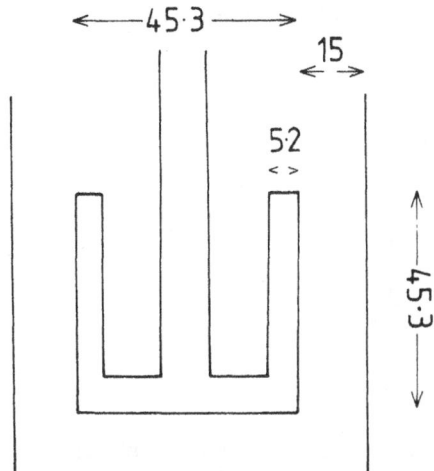

Figure 11.   Diagram and
dimensions of the anchor
impeller as fitted to the
Contraves Rheomat system. This
was operated in a 250ml beaker
with a sample volume of 200ml
and a gap between the beaker and
rotor of 1.5cm.

Viscosity of Plant Cell Suspensions
        The non-Newtonian nature of plant cell suspensions means
the apparent viscosity encountered will depend on the shear rates
involved.   Figure 14 shows the apparent viscosity at various shear
rates of C. roseus cells removed at various times from 100 l. (80 l.
working volume) air-lift bioreactor.   The culture starts off
essentially Newtonian but appears pseudoplastic by day 5 (mid-log).

The possible shear rates that plant cell suspensions may be exposed to
when grown in a stirred-tank are in the order of 20-200 sec[-1]
calculated from the relationship of shear rate to impeller speed
(Metzner & Otto 1957)

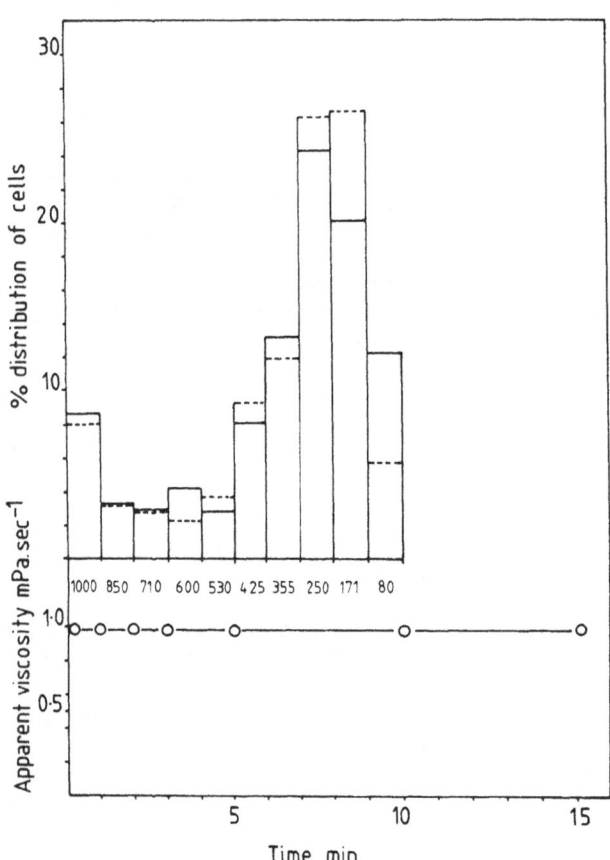

Figure 12.   Change in apparent viscosity of a C. roseus suspension with time using the anchor impeller at a shear rate of 66.4 sec^{-1}.   The C. roseus suspension was made up in fresh M3X medium to 100 gl^{-1} wet weight and 200ml were treated.   The size distribution of the cells was determined before (——) and after 15 minutes treatment (----).

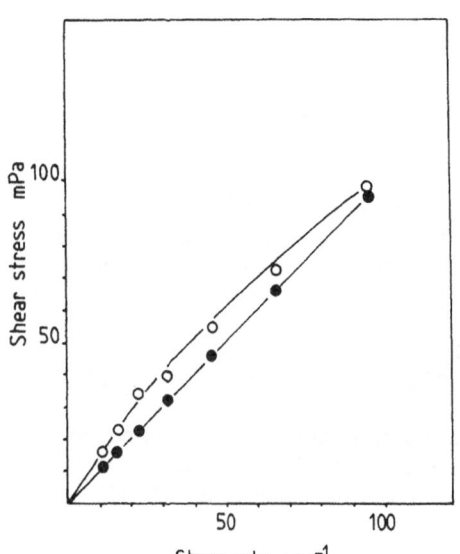

Figure 13.   The changes in shear stress with increases in shear rate using the anchor impeller.   The shear rate was calculated from the equation $\dot{Y} - kN$ where n is the impeller speed and k a constant which for this rotor was 15 (Rapp _et al_. 1984).   A C. roseus cell suspension at 100 gl^{-1} wet weight (o); water (●).

i.e $\dot{Y} = k N.$ when N is impeller speed   (4)
                    k is a constant = 10

When grown in an air-lift bioreactor the shear rate has been related
to the superficial gas velocity (V_S) (Van Suijdam 1980)

$$\dot{Y} = 50 \ Vs \qquad\qquad\qquad (5)$$

The Vs for the 100 l. air-lift bioreactor used in this investigation
would be in the range of 50-110 sec^{-1}. Therefore, a value of about
200 sec^{-1} would give an indication of the shear rates found during
the growth of plant cell cultures. The apparent viscosities at this
shear rate are below 10 mPa. sec^{-1} using the double-gap system and 1-2
mpa sec^{-1} with the anchor impeller. The values for plant cell
suspensions are therefore considerably lower than those found with
mycelial cultures and are unlikely to significantly affect their
culture.

The Effect of Shear on Viability
        It has been stated that plant cells in suspension are
sensitive to shear (Mandels; 1972). This observation has arisen from
the inability of workers to grow plant cells in stirred-tank
bioreactors other than at low agitation speeds (Wilson 1978; Fowler
1982; Dalton 1978; Smart & Fowler 1984). However, no quantitative
data is available on the shear rates which can affect growth. In our
initial studies using the Contraves rheomat as an accurate measure of
shear rate we attempted to study the effect of various rates on cell
viability using indicators such as fluorescein diacetate (FDA) or
oxygen uptake. These techniques are limiting in that although they
demonstrate the presence of an intact cell membrane they cannot
directly show whether the cell is viable.

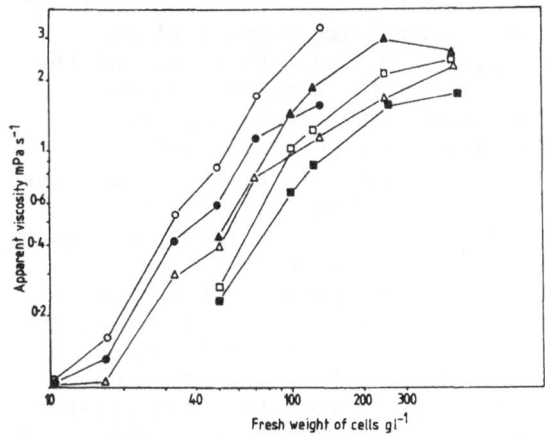

Figure 14. The effect of
cell density on the
apparent viscosity of plant
cell suspensions. The
apparent viscosity was
estimated using the double-
gap system at various shear
rates using cells grown in
either shake flasks or 80 l
air-lift bioreactor. Shake
flask cultures; 609 sec^{-1}
(△); 1245 sec^{-1} (■); 2250
sec^{-1} (□): bioreactor
cultures; 609 sec^{-1} (○);
1245 sec^{-1} (●); 3650 sec^{-1}
(△).

We therefore decided that the best measure of viability was the cells'
ability to grow and divide after inoculation into fresh medium at
concentrations above the critical inoculation density. In order to
carry out such experiments sterile conditions precluded the use of the
double-gap system. Thus a small stirred-tank (3 l.) was used for
these experiments. Table 2 shows the effect of subjecting a C.
roseus suspension to 1,000 rpm for up to 5 hours. This represents an
average shear rate of 166 sec^{-1}, and a maximum of 833 sec^{-1}. Samples
taken at intervals of 1 hour were inoculated into fresh medium and
their growth followed over 13 days. Although there was a loss in dry
weight of 29% over the 5 hours at high shear indicating some cell
damage the remaining cells showed no loss of viability . The C.
roseus cell line has subsequently been shown to be capable of growing
in a stirred tank bioreactor.

Discussion
Very little data is available concerning the rheological
nature of plant cell suspensions, and the possible viscosities
produced at high cell densities. This information is required if
large volumes of plant cells are to be grown without problems of
oxygen transfer and mixing.

The C. roseus cell suspension used in our studies consisted of
aggregates of 80 to 1,000 μm diameter with a mean value of 171-250 μm.
Therefore, this cell suspension would be somewhat similar to a
pelleted mould rather than a dispersed mycelial culture (Anderson
et al. 1982).

When exposed to constant shear, suspensions of C. roseus showed a
rapid decrease in apparent viscosity with time indicating
time-dependent behaviour. When similar cell suspensions were
subjected to step-wise increase and decrease in shear rates, a yield
stress was detected, the apparent viscosity decreased with increasing
shear rates and a hysterisis loop was formed. These features indicate
non-Newtonian behaviour with the yield stress suggesting a plastic
suspension of the Bingham or Casson type, with shear-thinning and the
hysteresis loop suggesting a thixotropic fluid. Both the shear-
thinning and increase and decrease of shear rate treatments were
accompanied by a decrease in the size distribution of the cell
aggregates. It was not initially clear whether this reduction was due
to the shear field or by attrition due to the large size of the
aggregates relative to the rheometer gap. The results shown in Figure
12 using aggregates of various sizes indicate that many of these
features may be a consequence of the method of measurement as the
smaller aggregates failed to exhibit a yield stress. In addition, at
low shear rates the torque values showed considerable oscillation in
amplitude and frequency which could be due to either interference or
settling of the larger aggregates. The oscillations were not observed
when small aggregates (80 μm) were examined. Replacement of the
double-gap cup and bob system with an anchor impeller gave no evidence
for time-dependent shear-thinning or a yield stress with C. roseus
cell suspensions. The cell suspensions did however exhibit non-

Newtonian pseudoplastic behaviour.   It can therefore be concluded that the shear thinning and yield stress are results of the method of measurement.

Van Suijdam (1980) investigated the rheology of _Penicillium chrysogenum_ pellet suspensions in a shear rate range below 20 sec^{-1} and recommended the use of the Casson model.   In contrast, Wittler _et al_. (1983) working with a pellet suspension of _P. chrysogenum_ at shear rates of 10 to 150 sec^{-1} found the suspension to be pseudoplastic fitting the Ostwald-de Waele model.

$$\tau = \gamma^n k \tag{6}$$

At high pellet concentrations and low shear rates the apparent viscosity is affected by the relative pellet volume.   In terms of plant cells this would be similar to a packed cell volume measurement. Cell suspensions of _M. citrifolia_ and _C. roseus_ exhibited characteristics of a plastic Bingham fluid showing a yield stress and linear slope (Wagner & Vogelman, 1977).   One _M. citrifolia_ culture did however show thixotropic behaviour.   In contrast, Tanaka (1981) using suspensions of _C. tricuspidata_ and _C. roseus_, found that these were non-Newtonian, (pseudoplastic) and obeyed the power law.   The previously observed yield stress and thixotrophic nature of _C. roseus_ suspensions appear to be features of the method of measurement and when the anchor impeller was used the plant cell suspensions were non-Newtonian at cell concentration above 100 gl^{-1}, exhibiting pseudoplastic characteristics of the Ostwald type.   A similar observation has been made for pelleted suspensions of _P. chrysogenum_ (Wittler _et al_. 1983).

The apparent viscosity at shear rates of 200 sec^{-1} of _C. roseus_ suspensions was only in the region of 1-8 mPa. sec^{-1} at concentrations of up to 300 gl^{-1} (wet weight) using the double-gap rheometer and 1-2 mPa sec^{-1} for the anchor impeller.   These values are considerably lower than those observed for both mycelial or pelleted mould cultures (Rapp _et al_. 1984; Anderson _et al_. 1982) where values of up to 1,000

Table 2:   The effect of shear on the viability of _C. roseus_ suspension cultures.

Time of exposure to shear (hr)	Dry and fresh weights of cultures after 9 days incubation	
	Dry wt.mg.ml^{-1}	Fresh wt.mg.ml^{-1}
Expt. 1.		
0	7.7 ± 0.2	171.9 ± 10
1	8.5 ± 2.8	164.7 ± 56
2	8.2 ± 1.9	230.7 ± 81
3	7.7 ± 1.2	198.4 ± 35
4	9.8 ± 1.5	206.1 ± 33
5	10.3 ± 1.0	233.0 ± 35
Expt. 2.	Dry and fresh weights after 14 days incubation	
0	7.2 ± 0.2	195.5 ± 6.9
2.5	7.9 ± 0.5	222.3 ± 2.1
5	7.7 ± 0.1	214.4 ± 6.1

mPa. sec[-1] can be found. It would therefore appear that the problems
of achieving good mixing in plant cell culture bioreactors is more
affected by the high settling rate of the cell aggregates rather than
viscosity.

It is generally accepted that plant cells in suspension are sensitive
to shear due to their large size, extensive vacuole and rigid cell
wall. It has been observed that plant cell suspensions cannot be
grown in stirred-tank bioreactors or require low agitation speeds
(Kato _et al_. 1978; Dalton, 1978; Fowler, 1982; Smart & Fowler, 1984;
Wilson, 1978). We have observed that cell suspensions became more
robust the longer they are in culture. The present C. roseus cell
line has been in culture for 5 years and although it was initially
sensitive to shear, it is now a more resistant culture. It is perhaps
no surprise that this culture was resistant to shear rates of up to
166 sec[-1] (average) and 800 (maximum). Although dry weight was lost
over the 5 hour treatment (29%) no evidence for loss of viability was
observed nor was the lag phase of growth extended in subsequent
culture. The shear resistance of this line has been further confirmed
by its successful growth in a stirred-tank reactor (Leckie, Allan,
Scragg & Cliffe, 1986). In contrast, suspension cultures of other
plant species have shown a loss of viability after 2.5 hours at shear
rates of 73 sec[-1].

In conclusion the rheological nature of plant cell suspensions
requires further study of different cell lines and using impellers of
different design before the non-Newtonian pseudoplastic flow
characteristics can be regarded as general. The viscosity of cells at
high density would seem to be less of a problem than the high rates of
settling. The shear sensitivity study will be extended to other cell
lines and species before any generalities can be made but it is clear
that some cell lines develop shear-resistance and can grow in stirred
tank reactors.

ACKNOWLEDGEMENTS
We gratefully acknowledge financial support from the
S.E.R.C. Biotechnology Directorate.

REFERENCES
Anderson, C., Le Grys, G.A. & Solomons, G.L. (1982). Concepts in the
    design of large-scale fermenters for viscous culture
    broths. Biochemical Engineering, Feb. 43-49.
Blanch, H.W. & Bhavaraju, S.M. (1976). Non-Newtonian fermentation
    broths : rheology and mass transfer. Biotech. Bioeng. 18,
    745-770.
Bongenaar, J.J.T.M., Kussen, M.W.F., Metz, B & Meijboom, F.W. (1973).
    A method of characterising the rheological properties of
    viscous fermentation broths. Biotec. Bioeng., 15, 201-
    206.
Charles, M. (1978). Technical aspects of the rheological properties
    of microbial cultures. Advances Biochem. Eng. 8, 1-62.

Dalton, C.C. (1978) The culture of plant cells in fermenters. Helio-
        synthesis et Aquaculture seminar de Martiques, CNRS, 1-
        11.
Fowler, M.W. (1982) The large scale cultivation of plant cells.
        Progress in Industrial Microbiology, 17, 209-229.
Henshaw, G.G., Jha, K.K., Mehta, A.R., Shakeshaft, D.J. & Street, H.E.
        (1966). Studies on the growth in culture of plant cells.
        I. Growth patterns in batch propagated suspension
        cultures. J.Expt.Bot., 17, 362-377.
Kato, A., Kawazoe, S. & Soh, Y. (1978). Viscosity of the broth of
        tobacco cells in suspension culture. J. Ferment.
        Technol., 56, 224-228.
Kim, J.H., Lebault, J.M. & Reuss, M. (1983). Comparative study on
        rheological properties of mycelial broth in filamentous
        and pelleted forms. Eur. J. Appl. Microbial. Biotechnol.
        18, 11-16.
Leckie, F., Allan, E.J., Cliffe, K.R. & Scragg, A.H. (1986). The
        effect of scale-up on growth and secondary metabolite
        accumulation of Catharanthus roseus. Poster presented at
        Symp. Biotech & Bioeng., Cambridge.
Mandels, M. (1972) The culture of plant cells. Advances in
        Biochemical Engineering, 2, 201-215.
Metzner, A.B. & Otto, R.E. (1957). Agitation of non-Newtonian fluids.
        A.I.Ch.E. Journal, 3, no. 1, 3-10.
Rajasekhar, E.W., Edwards, M., Wilson, S.B. & Street, H.E. (1971).
        Studies on the growth in culture of plant cells. XI. The
        influence of shaking rate on the growth of suspension
        cultures. J.Expt.Bot., 22, 107-117.
Rapp, P., Reng, H., Hempel, D-C. & Wagner, F. (1984). Cellulose
        degradation and monitoring of viscosity decrease in
        cultures of Cellulomonas uda grown on printed newspaper.
        Biotec. Bioeng. 26, 1167-1175.
Roels, J.A., Van Den Berg, J. & Voncken, R.M. (1974). The rheology of
        mycelial broth. Biotech. Bioeng. 16, 181-208.
Smart, N.J. & Fowler, M.W. (1984). An airlift column bioreactor
        suitable for large-scale cultivation of plant cell
        suspensions. J.Exp.Bot., 35, 531-537.
Smart, N.J., Morris, P. & Fowler, M.W. (1982). Alkaloid production by
        cells of Catharanthus roseus grown in airlift fermenter
        systems. In Proc. 5th Intl. Cong. Plant Tissue and Cell
        Culture, ed. A. Fujiwara pp. 397 Tokyo : Marazen Co.
        Ltd.
Tanaka, H. (1981). Technological problems in cultivation of plant
        cells at high density. Biotec. Bioeng. 23, 1203-1218.
Tanaka, H. (1982). Oxygen transfer in broths of plant cells at high
        density. Biotec. Bioeng. 24, 425-442.
Van Suijdam, J.C. (1980) Mycelial pellet suspensions biotechnological
        aspects. PhD thesis, Delft University of Technology.
Vogelmann, H., Bischof, A., Pape, D. & Wagner, F. (1978). Some
        aspects on mass cultivation. In Production of Natural
        Compounds by Cell Culture Methods, ed. A.W. Alfermann & E.
        Reinhard. pp. 130-146. Munich : Gesellschaft fur Strahlen
        und Umweltforschung mbH.

Wagner, F. & Vogelmann, H. (1977). Cultivation of plant tissue
        culture in bioreactors and formation of secondary
        metabolites. In Plant Tissue Culture and its
        Biotechnological Applications, ed. W. Barz, E. Reinhard, &
        M.H. Zenk. pp. 130-146. New York: Springer-Verlag.
Wilson, G. (1978). Growth and product formation in large scale and
        continuous culture systems. In Frontiers of Plant Tissue
        Culture, ed. T.A. Thorpe pp. 169-177. Calgary :
        University of Calgary.
Wittler, R., Matthes, R. & Schugerl, K. (1983). Rheology of
        Penicillium chrysogenum pellet suspensions. Eur. J. Appl.
        Microbial. Biotechnol., 18, 17-23.

DESCRIPTION OF A SYSTEM FOR THE CONTINUOUS CULTIVATION OF PLANT CELLS

P. Vienne
Givauden A.G.,   Switzerland

I.W. Marison
Wolfson Institute of Biotechnology,
The University,   Sheffield.   S10 2TN   U.K.

INTRODUCTION

A considerable amount of literature has been published concerning the growth of plant cells in liquid suspension culture as batch cultures (Fowler, 1971; Nash & Davies, 1972; Street, 1973; Wagner & Vogelmann, 1977; Wilson, 1980). Relatively few reports have been published concerning the growth of plant cell suspensions in continuous culture. Of these most of the work has centred on the use of Acer and Galium cells (King, 1976; Wilson, 1980; Wilson & Marron, 1978).

This lack of research into the growth and behaviour of plant cells under continuous culture conditions is surprising in view of the defined controlled growth conditions which can be achieved. It allows the study of the effects of various physical and nutrient parameters on plant cell growth, product formation and composition under conditions where the growth rate is maintained constant. The difficulties in maintaining viable, contaminant-free, disperse cell cultures over long periods of time and the availability of suitable culture vessels may be responsible for the lack of published data on continuous culture of plant cells.

It is well known that plant cells in batch culture undergo a characteristic pattern of growth and cell division termed the growth cycle (King et al., 1973; Gould et al., 1974; Wilson, 1980; Wilson & Marron, 1978). Such a growth cycle is not an inherent property of the cells but reflects the effect of the environment on cell growth and metabolism in a closed environment. Thus plant cells undergo a continually changing pattern of biosynthetic activity and cell composition during the batch culture 'cycle' which is probably a reflection of the continually changing physical and chemical composition of the environment (Wilson, 1980). Furthermore, since large inoculum densities are required to initiate batch cultures (usually of the order of 20% of the culture volume), the number of generations occurring before stationary phase is reached is between 2 and 3. Thus data concerning nutrient accumulation after inoculation and cell composition have limited value in such non-representative populations of cells, where no state of balanced growth is achieved.

With continuous (chemostat) culture the growth rate can be maintained indefinitely with no change in the environment other than the nature and concentration of the growth-limiting substrate. The effects of carbon, nitrogen, phosphate or some other nutrient-limitation on cell composition and metabolism can be examined whilst cells are growing at the same rate. These effects can further be studied by maintaining a specific nutrient limitation and varying the growth rate as desired simply by varying the rate of supply of fresh nutrient to the constant volume chemostat.

Such data would (i) aid the development of media suitable for both growth and product formation (ii) provide information concerning the role of key substrates in the metabolism and control of metabolic activity and (iii) enable a study of the effects of specific physical parameters such as pH, redox, mixing, shear and gassing regimes under defined environmental conditions.

Thus, photoautotrophic growth of spinach (Dalton, 1980) and asparagus (Peel, 1982) in chemostat cultures has been studied as a function of light intensity, and the variation in the biomass yield constant of Acer cells (Dougall et al., 1982) as a function of growth rate (= dilution rate) examined.

Sahai and Schuler (1982) found that mechanical, hydraulic and gassing systems seriously affected the growth of Nicotiana tabacum cells in chemostat culture and proposed a two-stage chemostat for the production of phenolics (Sahai & Schuler, 1984). The major difficulties were encountered in the design of the chemostat vessel, the difficulties associated with mixing and cell aggregation, and the problems associated with contamination of the cultures. The latter problem was overcome to some degree by the addition of antibiotics to the medium although adverse effects on cell growth in the presence of antibiotics have been reported (Walts & King, 1973; Umiel & Goldner, 1979).

In order to examine these problems and the stability of chemostat cultures of plant cells over extended periods the authors developed a 3.5 litre air-lift bioreactor previously used for batch culture in which cell suspensions of Catharanthus roseus were grown under glucose-limited conditions. C.roseus was chosen due to the ability to obtain fine, homogenous cell suspensions which enabled steady-state growth to be achieved. The design of the chemostat system is shown diagrammatically in Figure 1.

The chemostat involved a main cylindrical glass vessel ('riser') connected to an external loop ('downer'). Filter sterilised air entered the vessel via a sparger situated in the base of the riser. De-gassing of the medium occurred at the liquid/air interface at the top of the riser and downer. This gassing/de-gassing system allowed for efficient mixing, mass-transfer and short medium circulation times. The temperature of the medium was controlled at 25°C using a thermostatic circulator supplying water through a jacket surrounding the downer at a rate of 10 dm^3 min^{-1}. The pH and dissolved oxygen concentration of the medium was monitored continuously although no

effort was made to control these parameters since the buffering
capacity of the cells resulted in minimal changes in pH (5.8-6.5).   An
air flow rate of 0.5 dm^3 min^{-1} ensured that the dissolved oxygen
tension remained above 50% saturation.

The cultures of C.roseus were subject to severe foaming problems,
consequently anti-foam (Dow corning 1510, 0.05% w/v) was added to the
growth medium. In addition a further antifoam reservoir was connected
directly to the chemostat via a peristaltic pump controlled by an
interval timer. The timer was adjusted to inject 3 ml. of antifoam
into the riser every 16 hours.

A constant volume of medium in the chemostat (3 dm^3) was obtained
using an electrical contact system. Fresh medium was pumped into the
riser at a predetermined rate. As the culture volume rose, contact
was made between two electrodes placed in the downer, resulting in the
activation of the medium outlet pump. This system allowed for careful
control of the culture volume over a wide range of dilution rates.

Cultures were initiated using 0.8 dm^3 of a seven day suspension
culture. Approximately 4 days after inoculation, when the culture
fresh weight had reached 80-100 g dm^{-3}, the medium inlet pump was
switched on to provide a dilution rate of 0.10 day^{-1}.

During the initial 4 hours after inoculation a rapid decrease in the
medium phosphate concentration was observed with a corresponding rise
in the intracellular phosphate concentration (Figure 2).

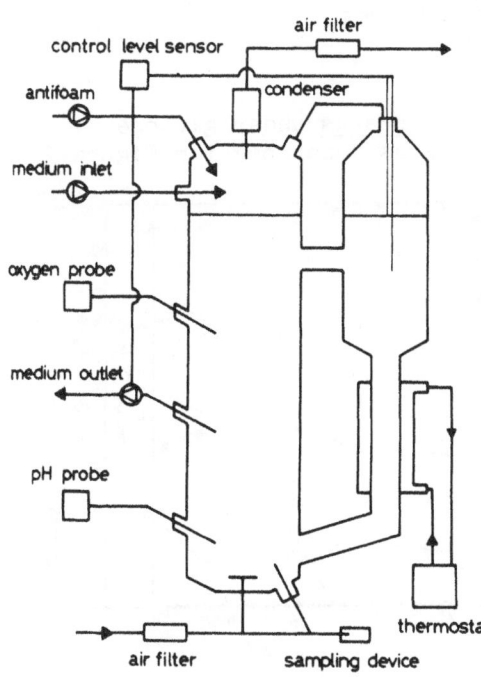

Figure 1.   Diagrammatic
representation of the 3.5 dm^3
air-lift, chemostat bioreactor.

After 7 hours the intracellular phosphate concentration fell slowly to reach a steady state value (2.5 μM [g fresh weight]$^{-1}$) once continuous culture conditions had been initiated [D=0.25 day^{-1}] (Figure 3).

Figure 2.   A glucose-limited chemostat culture of C. roseus line CIICA-7. The results represent the initial 24 hour period of batch growth prior to initiation of the medium inlet pump.
Medium : Gamborg's B5 containing glucose (10g dm^{-3}), 2,4-D (10^{-3} g dm^{-3}) and kinetin (10^{-4} g dm^{-3}). Symbols: FW, fresh weight, g dm^{-3} (■); DW, dry weight, g dm^{-3} (□); Pi, phosphate concentration in culture filtrate, mM (△); Pi intracellular phosphate concentration, μM [g fresh weight]$^{-1}$ (▲).

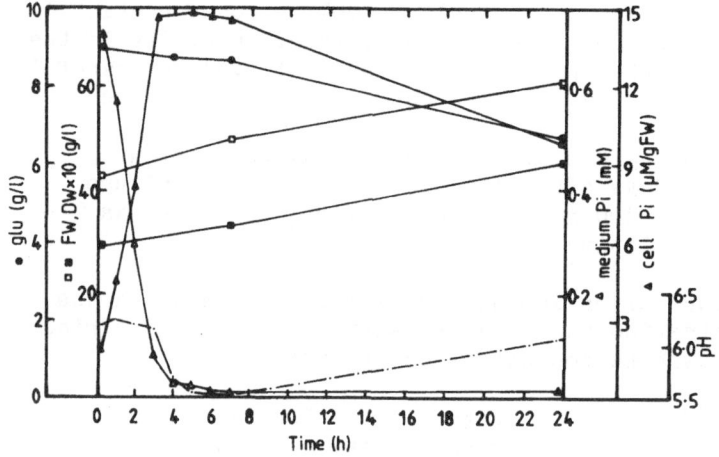

Figure 3.   65-day chemostat culture of C. roseus under glucose limitation. Medium composition and symbols as described for Figure 2.

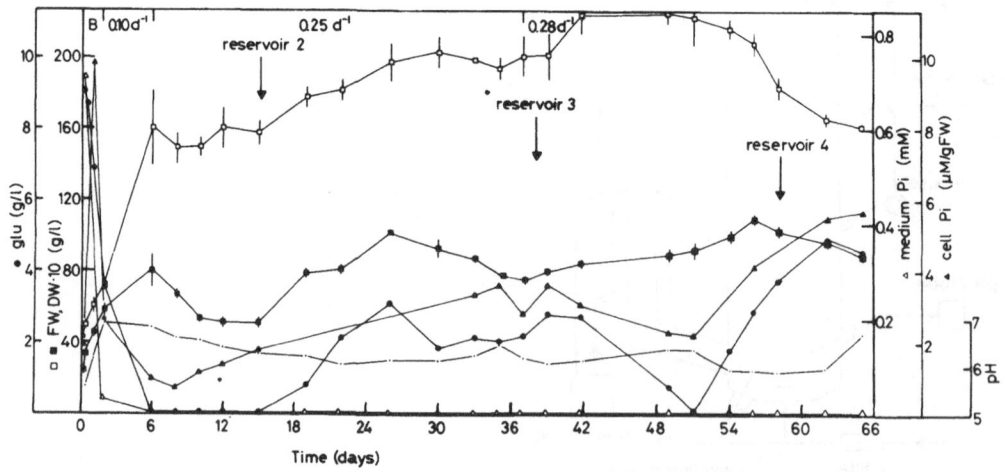

The intracellular phosphate concentration did not vary significantly upon increasing the dilution rate to 0.28 day^{-1} whilst the concentration in the medium was below the limits of detection.

This accumulation of phosphate by the cells is in agreement with the results of Wilson and Marron (1978) working with <u>Galium</u> cells under phosphate-limitation, and Sahai and Schuler (1984) with <u>Nicotiana tabacum</u> cells under glucose-limitation.

Over the 65 day period fluctuations in cell fresh weight, dry weight and intracellular phosphate concentration were observed (Figure 3) which suggests that a true steady state of balanced growth and cell composition was not achieved until after approximately 60 days. It appears that these fluctuations were partly correlated with a continuous change in cell size distribution (Figure 4). Thus in a manner similar to that reported for Galium (Wilson & Marron, 1978) and Nicotiana (Sahai & Schuler, 1984), the culture initially exhibited a mean cell size of 355 m^{-6} after inoculation which fell progressively with time, at dilution rates of 0.25 and 0.28 day^{-1}, to 171-250 m^{-6}.

The major reasons for the fluctuations in cell composition appear to be due to changing the medium reservoir (Figure 3), which indicates that some unidentified component of the medium is labile over the time-scale examined.

An important result of this work is that assuming a mean residence time of 3-4 days most of the cells in the chemostat were not subjected to large fluctuations in the internal or external environment particularly with regard to the size of the phosphate pool

Figure 4.   Cell size distribution with respect to dry weight for a 65-day glucose-limited chemostat culture of <u>C. roseus</u>.

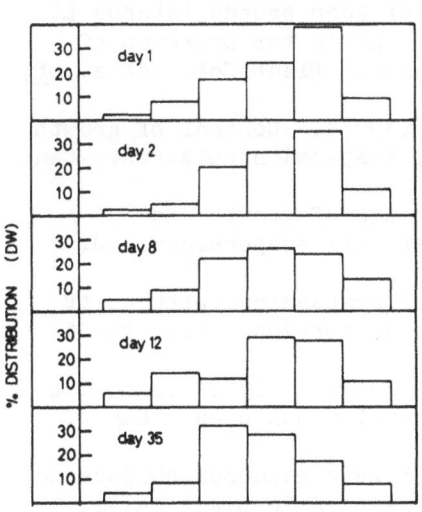
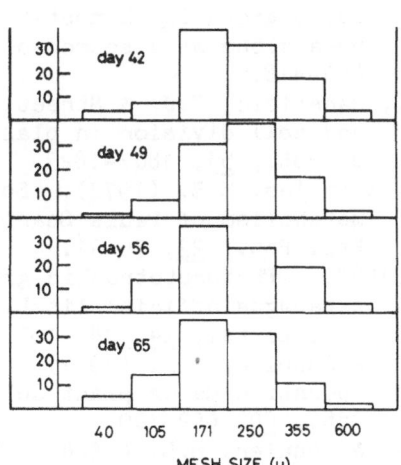

(Figure 3). However, in batch culture (Figure 2) these fluctuations were much greater making interpretation of the batch culture data difficult. This effect must be examined in more detail under conditions of phosphate-limitation.

CONCLUSIONS
The air-lift bioreactor described was capable of operating over periods of 65 days free from problems of contamination and enabled steady-state, glucose-limited continuous cultivation of C.roseus under defined environmental conditions, in contrast to those obtained in batch culture. Stability of medium components and the selection of small cell size need to be examined in more detail before longer term experiments over a wider range of dilution rates is attempted. However, the results indicate that continuous cultivation of plant cells is feasible and offers significant advantage over batch cultivation techniques.

REFERENCES
Dalton, C.C. (1980). Photoautotrophy of spinach cells in continuous culture : Photosynthetic development and sustained photoautotrophic growth. J. Exp. Biol., 31, 791-804.

Dougall, D.K., La Brake, S. & Whitten, G.H. (1982). Plant cells grown in chemostats. Proc. 5th Intl. Cong. Plant Tissue & Cell Culture, 51-54.

Fowler, M.W. (1971). Studies on the growth in culture of plant cells. XIV. Carbohydrate oxidation during the growth of Acer pseudoplatanus L. cells in suspension culture. J.Exp.Bot., 22, 715-724

Gould, A.R., Bayliss, M.W. & Street, H.E. (1974). Studies on the growth in culture of plant cells. XVII. Analysis of the cell cycle of asynchronously dividing Acer psuedoplatanus L. cells in suspension cultures. J.Exp.Bot., 25, 468-478.

King, P.J. (1976). Growth characteristics of Acer psuedoplatanus L. cells grown in chemostat conditions in the presence of urea alone as a source of nitrogen. Plant Sci. Letts., 6, 409-418.

King, P.J., Mansfield, K.J. & Street, H.E. (1973). Control of growth and cell division in plant cell suspension cultures. Can. J. Bot., 51, 1807-1823.

Nash, D.T. & Davies, M.E. (1972). Some aspects of growth and metabolism of Pauls Scarlet rose cell suspensions. J. Exp. Bot., 23, 75-91.

Peel, E. (1982). Photoautotrophic growth of suspension cultures of Asparagus officinialis L. cells in turbidostats. Plant Sci. Letts., 24, 147-155.

Sahai, O.P. & Schuler, M.L. (1982). On the nonideality of chemostat operation using plant cell suspension cultures. Can J. Bot., 60, 692-700.

Sahai, O.P. & Schuler, M.L. (1984). Multistage continuous culture to examine secondary metabolite formation in plant cells. Phenolics from Nicotiana tabacum. Biotechnol. Bioeng., 26, 27-36.

Street, H.E. (1973). Cell (suspension) cultures - techniques. In
        Plant Tissue and Cell Culture, ed. H.E. Street, pp. 59-99.
        Oxford : Blackwell Scientific Publications.
Umiel, N. & Goldner, R. (1976). Effects of streptomycin on diploid
        tobacco callus cultures and isolation of resistant
        mutants. Protoplasma, 89, 83-89.
Wagner, F. & Vogelmann, H. (1977). Cultivation of plant tissue
        cultures in bioreactors and formation of secondary
        metabolites. In Plant Tissue Culture and its
        Biotechnological Applications, ed. W. Barz, E. Reinhard &
        M.H. Zenk, pp. 245. Berlin : Springer-Verlag.
Watts, J.W. & King, J.M. (1973). The use of antibiotics in the
        culture of non-sterile plant protoplasts. Planta, 113,
        271-277.
Wilson, G. (1980). Continuous culture of plant cells using the
        chemostat principle. Adv. Biochem. Eng., 16, 1-16.
Wilson, G. & Marron, (1978). Growth and anthraquinone biosynthesis by
        Galium mollugo L. cells in batch and continuous culture.
        J. Exp. Bot., 29, 837-851.

26. THE ECONOMICS OF MASS CELL CULTURE

A. H. Scragg

Wolfson Institute of Biotechnology,
University of Sheffield
Sheffield   S10 2TN.   U.K.

INTRODUCTION
Plant cell culture as a technique has been with us for some time and it was in the early 1960's that it was proposed as an alternative to normal agriculture production for some phytochemicals. Plants have been for a long time a source of a wide range of chemicals including pharmaceutical, insecticides, flavours, fragrances, and colours. Even now with the advances in microbial produced pharmaceuticals some 25% are extracted from plants (Farnsworth & Morris, 1976). The advantages of a plant cell culture process for the production of fine chemicals has been outlined recently (Fowler, 1983; Scragg & Fowler, 1985). Despite these advantages plant cell culture will never be able to complete directly with microbial systems because of their intrinsic low productivity due to the slow growth of plant cells. As a consequence, at present, potential plant cell products must remain in the high price, low volume category. Even in this area the production of compounds by plant cell culture must have an advantage in cost or availability to be industrially exploited. Despite considerable effects only one commercial process has so far been announced, the production of shikonin from Lithospermum erythrorhizon by the Mitsui Co. Ltd. (Curtin, 1983).

Over the years there have been a number of excellent and comprehensive reviews of the compounds detected in plant cell cultures (Barz & Ellis, 1981; Dougall, 1979; Fowler, 1983; Berlin, 1984). The lists which have been produced are impressive in their range but unfortunately from the commercial side only a few of these compounds command sufficient a price to be potentially commercial. The list is shown in Table 1 which also lists possible world or U.S. markets.

Costs of a Plant Cell Culture Process
The inclusion of the compounds in the list of potential commercial targets (Table 1) is based on plant derived products of the high value category or products where demand has exceeded supply due to various problems eg. diosgenin. There is only a limited amount of information concerning the costs of products from plant cell cultures. Goldstein et al. (1980) proposed values of between 17-555 $.Kg depending on the scale. Fowler (1983) and Smart (1984) suggest values of between 250 and 1,500 £.Kg. Recently Sahai & Knuth (1985) gave values of 228-5900 $.Kg depending again on the yield of secondary product. The figures suggest that the inclusion of spearmint in Table 1 is perhaps premature. In addition, the other flavours and

fragrances are complex mixtures which, with present technology would be
extremely difficult to match in plant cell culture and therefore
should be also eliminated at this stage. Shikonin and perhaps
berberine are of a sufficient price to be included in Table 1 but are
only used in Japan and are of little value in the European or US
markets. Mitsui Co. Ltd., are beginning to market shikonin as the dye
in a bio-lipstick but the volume and price of this market is unknown.
This leaves only a limited number of compounds which can be as
potential plant cell culture products with quinine being of only
marginal value.

Scale of Production
No new process will command in its initial stages the
whole market share, in fact on of the advantages proposed for plant
cell processes would be the ability to reduce variations in plantation
production. Therefore, a realistic figure for the initial market
penetration would be in the order of 10% of the US or World market. In
the case of products such as serpentine, codeine and digoxin this
would be in the region of 300 Kg per year. This figure is 10 fold
less than that suggested by Goldstein et al. (1980) and Sahai & Knuth
(1985), but it is five times the shikonin production claimed by the
Mitsui Co. Ltd. (Curtin, 1983). The overall annual production has a
considerable influence on the total cost of any product in both
capital and production costs. The effects can best be seen in Table
2. In this table Goldstein et al. (1980) have costs of 551 $.Kg at
10,000 p.a. which drop to 106 $.Kg for 1,000,000 Kg. p.a. Sahai &
Knuth (1985) base their estimates on 20,000 Kg per annum. These
figures do appear far in excess of the 300 Kg p.a. suggested or the 50
Kg of shikonin produced at present. This would suggest, at least in
the initial stages, a plant cell culture process will have an annual
production and use vessels far smaller than was at first thought.
This in time will affect the overall costings.

Productivity
Given a target of some 300 Kg per year what features of
plant cell culture will affect the final costings. One of the more
important factors in the economics of any process is productivity,
that is yield of product, per litre, per day. In a normal batch
culture productivity is a consequence of the final yield of product
and the time required to reach that figure. The time require for a
batch fermentation can be divided into: the time require to turn
around the vessel, wash it out etc., the time required to fill with
media and sterilise, the lag in growth, and the growth time itself.
This can be represented.

$$t_{total} = \frac{1}{\mu_m} . \ln \frac{X}{X_o} + t_T + t_S + t_L$$

μ_m = growth rate
X = biomass final
X_o = biomass initial
t_T = turnaround time
T_S = time for sterilisation
t_L = lag time

With doubling times of days plant cell cultures will have very long
run times compared with microbial systems especially if the desired
product is formed in stationary phase. The effect of run time on
overall production is shown in Figure 1.   The following assumptions
have been made; the yield of product is 1% of the dry weight, the
biomass yield is 20 $g.l^{-1}$ dry weight and the bioreactor volume is
1,000 l. The normal run time would perhaps be 20 days for C. roseus
which would give some 15 runs p.a. The 300 day year is proposed as
the other days will be required for maintenance, contaminated runs
etc.   Thus to produce 300 Kg p.a. under these conditions would require
a 100,000 l bioreactor, a reduction of run time by 50% would only
decrease the volume to 50,000 l. The dramatic effect of changes in
yield upon the cost of plant cell products can be seen in Table 2, and
is best illustrated by Figure 2. Here the same figures have been used
and it can be seen the dramatic effect small changes in yield can have
on the bioreactor volume required to produce 300 Kg per year.   In this
area there are reasons to be encouraged as yields of secondary
products in plant cell cultures have recently reached values
equivalent to microbial systems. Ulbrich et al. (1985) report of
figure of 21.4% for rosmarinic acid from Coleus blumei and

Table 1: Possible costs and requirements of plant cell
culture products. Data taken from Curtin (1983), and
Sahai and Knuth (1985), and McHale (1986).

Compound	Use	Price $Kg	World Annual Requirement (Kg)	
ajmalicine (serpentine)	circulatory diseases	1,500	3,000	
codeine	sedative	650	7,700	(US)
digoxin	heart stimulant	3,000	6,000	(US)
diosgenin	steroids	674	200,000	
jasmine	fragrance	5,000	100	
quinine (quinidine)	antimalarial, bittering agent (heart stimulant)	100	300-500,000	
rose oil	fragrance	3,300	4,000	
shikonin	dye, antibacterial	4,500	150	JAPAN
spearmint	flavour	30	3,000,000	
vincristine, vinblastine	anti-leukemic	5,000,000	5	

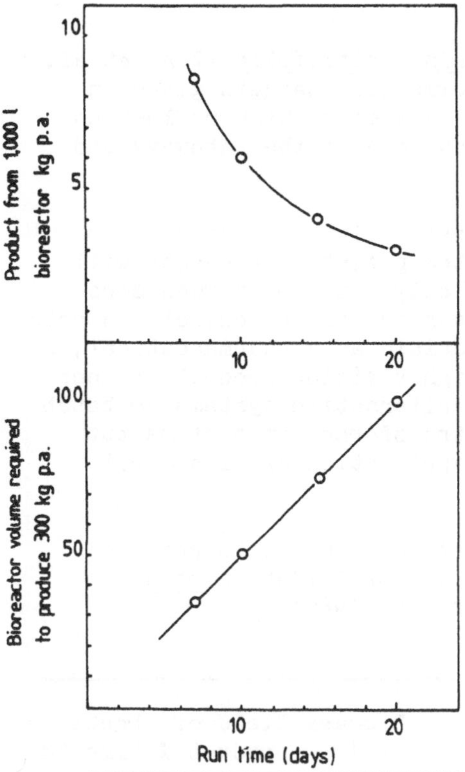

Figure 1.   The effect of run
time on bioreactor volume and
the production from a 1,000 l
bioreactor.

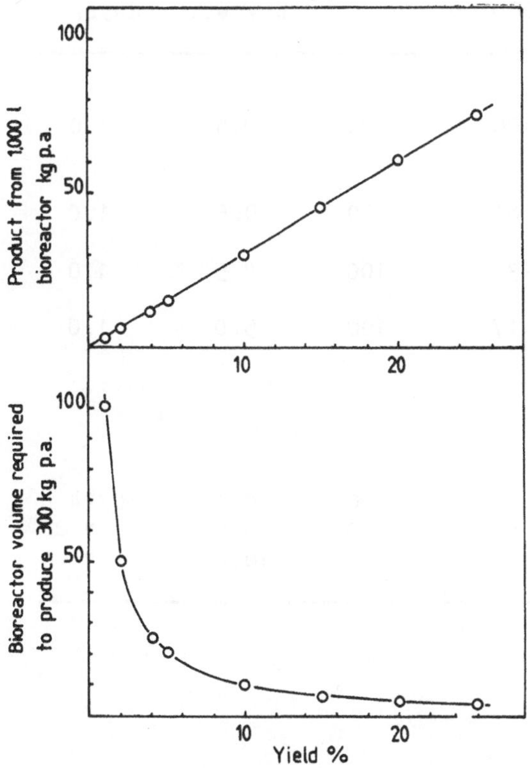

Figure 2.   The effect of
yield on bioreactor volume and
the production from a 1,000 l
bioreactor.

18.0% anthraquinone has been found in <u>Morinda citrifolia</u> (Zenk et al.
1975). However, many of the potential commercial targets given in
Table 1 have resisted attempts at improving yields which reflect no
lack of effort but rather the lack of knowledge of the pathways and
controls of these systems.

In conclusion it is clear that if plant cell culture is to develop
into commercial processes the yields of the potential products will
have to be increased. This increase can only come about when more
information is available about the pathways and their control. A note
of encouragement is possible use of rosmarinic as a pharmaceutical, a
case of looking for an application for high yielding products rather
than the reverse. I have not considered alternative systems to batch
culture such as immobilised cells or biotransformation systems but
these do all seem very promising in the application of plant cell
culture.

Table 2: The possible cost of phytochemicals produced
by plant cell culture. (1) Data of Goldstein <u>et al</u>.
(1980), (2) Data of Sahai & Knuth (1985).

Production per annum Kg	Price per Kg $			Biomass gl^{-1}	Yield of product % dry wt.	Doubling time Td days
	10,000	100,000	1,000,000			
(1)	551	241	106	20	0.5	1.0
90% reuse of biomass	323	142	61	20	0.5	1.0
	165	72	31	100	0.5	1.0
	91	40	17	100	5.0	1.0
(2)						
Production per annum Kg	20,000					
	5,900			20	0.1	60/24 d
	1,045			20	1.0	"
	228			20	10.0	"

REFERENCES

Barz, W. & Ellis, B.E. (1981). Potential of plant cell cultures for pharmaceutical production. Ber. Deutsch. Bot. Ges. 94, 1-26.

Berlin, J. (1984). Plant cell cultures - a future source of natural products. Endeavour 8, 5-8.

Curtin, M.E. (1983). Harvesting profitable products from plant tissue culture. Biotechnology 1. 649-657.

Dougall, D.K. (1979). Factors affecting the yields of secondary products in plant tissue culture. In: Plant Cell and Tissue Culture - Principles and Applications. Ed. W.E. Sharp. pp.727-743. Columbia, Ohio State University Press.

Farnsworth, R.N. & Morris, P.M. (1976). Higher plants - the sleeping giant of drug development. Amer. J. Pharm. 148, 46-52.

Fowler, M.W. (1983). Commercial applications and economic aspects of mass plant cell culture. In: Plant Biotechnology ed. S.H. Mantell & H. Smith pp.3-37. Cambridge: Cambridge University Press.

Goldstein, W.E., Ingle, M.B. & Lasure, L. (1980). Product cost analysis. In: Plant Tissue Culture as a Source of Biochemicals. ed. E.J. Staba. pp.191-234. Boca Raton, Florida: CRC Press.

McHale, D. (1986). The Cinchona Tree. Biologist 33, 45-53.

Sahai, O. & Knuth, M. (1985). Commercialising plant tissue culture processes: Economics, problems and prospects. Biotechnology Progress 1, 1-9.

Scragg, A.H. & Fowler, M.W. (1985). The mass cultivation of plant cells. In: Cell Culture and Somatic Cell Genetics of Plants. ed. I.K. Vasil. pp.103-128. New York: Academic Press.

Smart, N.J. (1984). Plant cell technology as a route to natural products. Lab. Practice. Jan, 11-17.

Ulbrich, B., Wiesner, W. & Arens, H. (1985). Large-scale production of rosmarinic acid from plant cell cultures of Coleus blumei Benth. In: Primary and Secondary Metabolism of Plant Cell Cultures. eds. K.-H. Neumann, W. Barz & E. Reinhard. pp.293-303. Berlin: Springer-Verlag.

Zenk, M.H., El-Shagi, H. & Schulte, U. (1975). Anthraquinone production by cell suspension cultures of Morinda citrifolia. Plant Med. Suppl. 79, 101-136.

SECTION 5

SELECTION, STABILITY AND VARIATION IN
SECONDARY METABOLISM

27. SOMACLONAL VARIATION

S.H. Mantell

Unit for Advanced Propagation Systems
Department of Horticulture
Wye College (London University)
Wye, Ashford, Kent. TN25 5AH U.K.

INTRODUCTION
The term "clone" has been used in traditional horticulture
to describe a collection of plants derived by asexual propagation
methods from a single mother stock plant. The resulting clonal line
is therefore expected to consist of individuals identical in all
respects with the exception of an occasional `sport', which usually
originates from somatic mutation within shoot meristem initials. The
frequency with which somatic mutants or `sports' arise, depends on the
species and the conditions under which plants are being grown, but as
a general rule, normal rates of mutation are expected to be expressed
at frequencies of only 10^{-6} - 10^{-8}. This apparently low incidence of
somatic abnormality at the whole plant level means that clones can be
maintained provided off-types are recognised and separated from the
clonal line. Similar requirements for stable cloning extend to the
micropropagation of plants. Increasing experiences with in vitro
culture indicate that stable cloning can only be achieved adequately
when axillary or apical meristem culture systems are used (Krikorian,
1982; Hussey, 1983; Mantell, 1986). Instability in the phenotypes of
plants cloned by tissue culture techniques becomes more apparent
(frequencies of off types may be as high as 50%) when plants are
regenerated from dedifferentiated tissue like callus or from
adventitious meristems on tissue explants.

MERISTEMS AND SOMATIC MUTATION
The majority of angiosperms contain structured stochastic
meristems, each of which consists of multiple impermanent initials - a
meristeme d'attente - such that meristems do not assume an abberant
phenotype until the majority of these initials contains somatic
mutations in their genomes. Positive diplontic selection also plays a
determinative role in maintaining the basic diploid number in the germ
line of plants. According to recent studies by Klekowski et al.
(1986), apical meristems may influence mutation frequency either by
reducing the mutation rate or by enhancing diplontic selection.
Isolation of plant tissues in vitro through cell or protoplast culture
into discrete cellular units separated from the physiological
gradients and controlling influences of apical meristems, frequently
leads to increased incidence of abberant forms of regenerated plant.
The term somaclonal variation was coined by Larkin & Scowcroft (1981)

to describe this type of variation. Somaclonal variation presents
some obvious advantages to researchers in the field of crop improvement, but to propagators it emphasises the need for tighter
restrictions as to the use of certain plant tissue culture methods for
true-to-type clonal propagation. The genetic basis of somaclonal
variation is not fully understood but there is an increasing amount of
information coming to hand on the nature of some of the genotypes and
phenotypes present in somaclones. The purpose of this chapter is to
review some of this information as a background to the subsequent
specialist sections in this part of the book.

EXPRESSION OF SOMACLONAL VARIATION AT THE WHOLE PLANT AND THE CELL CULTURE LEVELS

The large scale (mass) propagation of horticultural crops
by tissue culture methods and the subsequent assessments of the field
performances of the somaclones so produced has permitted an evaluation
of the range of phentoypic variability in plants. Swartz & Lindstrom
(1986) have recently collated results of a survey of soft fruit stocks
propagated in the USA and other countries through tissue culture.
Their results indicate that axillary shoot culture systems in
strawberry, raspberry and blueberry provide a reliable form of cloning
with off types only occurring at a frequency between 0.01 and 0.05%.

TABLE 1:   Examples of somaclonal variant phenotypes of
crop plants expressed at the whole plant level.

Species	Phenotypes altered in somaclones
Barley	Plant height, tillering, fertility
Lettuce	Leaf weight, length, width, flatness and colour, bud number
Maize	Pollen fertility
Oats	Plant height, heading date, leaf striping, twin culms, aion morphology, heteromorphic bivalents, ring chromosomes
Onion	Bulb size and shape, clove number, aerial bulbil germination
Pelargonium	Leaf shape, size and form, flower morphology, plant height fasciation, pubescence, anthoujanin pigmentation, essential oil composition
Rape	Flowering time, glucosinolate content, growth habitat

However, in the cases where shoot proliferation was inadvertantly of
both axillary and adventitious origin, frequencies of off types
increased to the 1-10% level for genetic and phenotypic abnormality
and as high as 50% for breakdown of periclinal chineras of raspberry
and blackberry. The `clonal' regenerants obtained from
protoplast-derived calluses or cell suspension and callus cultures have bee
assessed for several important crop species such as potato (Shephard,
et al., 1980), sugarcane (Heinz, D.J. et al., 1977), tomato (Evans et
al., 1985), sorghum, maize oats, barley and others. Examples of the
types of somaclonal variant phenotypes obtained at the whole plant
level is summarised in Table 1.

At the cell culture level, there are numerous cases of somaclonal
variability, particularly with respect to the expression of phenotypes
related to secondary metabolite accumulation. For instance, one
example of the wide range of variability exhibited by small cell
aggregates and single cells (clones) derived from a single callus
suspension culture of tobacco is shown in Figure 1. To demonstrate
that such variability was not due solely to the heterogeneity of cell
development states associated with batch cultures, clones from the
extremes of the nicotine accumulation range were isolated and assessed
for nicotine accumulation levels after two consecutive callus culture

Figure 1: Scatter diagram of the fresh weight and
nicotine content of 'single cell' clones of N. tabacum CV
NC 2512. Source Unpublished data of D. Pearson.

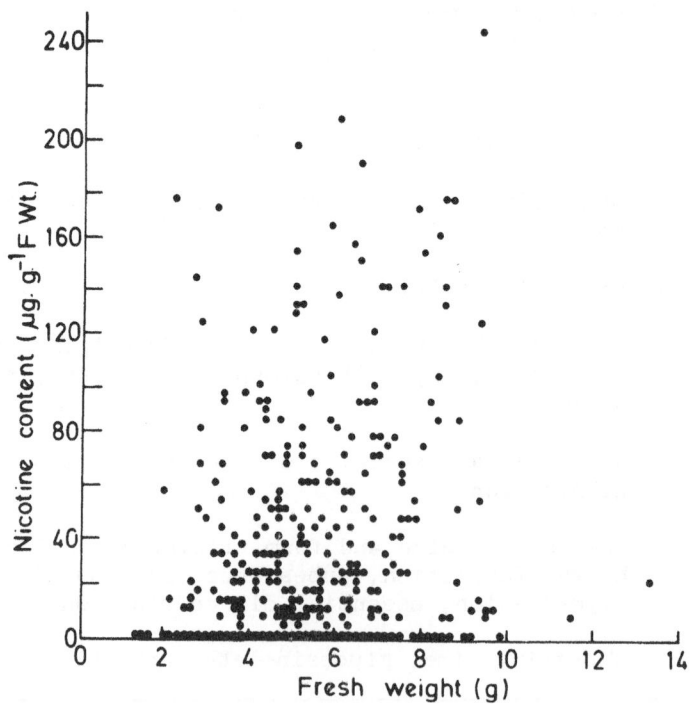

generations on solid media for 40 days each. The 'high' producing
lines retained significantly higher nicotine levels than the 'low '
producing lines (Table 2), demonstrating that the original differences
in nicotine accumulation levels were probably the expression of some
form of somaclonal variation. Variation of this type is common in
plant cell culture and as such poses a severe problem for the
maintenance of high product yields over prolonged sub-cultures in
liquid or immobilised cell culture. These particular problems are
dealt with in the subsequent chapters in this section. This chapter
deals with the possible origin of these different manifestations of
somaclonal variation. At the outset it should be made clear that
since so little is known about the underlying dynamic mechanisms
involved in the expression of somaclonal phenotypes, then somaclonal
variation in any one instance might be the consequence of not one but
a number of random or non-random events. Over the last five years, it
has become clear that several genomic and somatic events can take
place before and during in vitro culture that could lead to expression
of variable phenotypes. These are now reviewed.

GENETIC BASES OF SOMACLONAL VARIATION
        Several types of genetic changes are known to occur in
tissue cultured plant cells especially in those which pass through a
callus stage of development. These include gross karyotypic changes,
chromosomal re-arrangements, somatic crossing-over with sister
chomatid exchange, activation of transposable sites along certain
regions of nuclear DNA and re-arrangements of nucleotide sequences in
both nuclear and organellar DNA. It is now apparent that many of
these changes can take place in the mother plant itself as part of
normal differentiation processes.

a) Genetic variation within somatic tissues in vivo
        In approximately 90% of angiosperms, at least part of the
somatic cell population undergoes chromosome endoreduplication

TABLE 2:   Fresh weight and nicotine accumulation in
'high' and 'low' nicotine producing lines of tobacco.
Source Unpublished data of D. Pearson.

Culture generation	Clones	Fresh weight (g) Mean ± S.E.	Nicotine content (μg g^{-1}FW) Mean ± S.E.
1	'High'	4.9 ± 0.3	56.6 ± 6.2
	'Low'	5.0 ± 0.6	16.7 ± 5.7
2	'High'	6.2 ± 0.2	44.8 ± 5.1
	'Low'	6.1 ± 0.8	10.1 ± 2.5
3	Unselected	5.3 ± 0.1	33.9 ± 2.3

concomitant with differentiation (D'Amato, 1985) and what is more
significant is that the degree of polysomaty may vary from tissue to
tissue so that in any given group of cells, the level of ploidy may
vary considerably (Chriqui & Bercetche, 1985). For example a variable
number of cells may not endoreduplicate (remaining with 2C or 4C
nuclear DNA components), while others will undergo a different number
of endoreduplication cycles (8C, 16C, 32C, 64C etc.). Meristematic
tissues (e.g. pericycle, procambium and cambium) remain diploid and
when not dividing generally rest in 2C. Moreover, a condition known
as aneusomaty in which aneuploid chromosome complements arise by the
occasional organization of two or more separate spindles in an
uninucleate cells has been observed frequently in certain genera such
as Poa, Orobanche, Hymenocallis and Saccharum (particularly clone
H50-7209) and in tetraploid Ribes clones (for review see D'Amato,
1985). Increasing observations on plants at the cytogenetic level
have indicated that aneusomaty once established within a meristem can
persist in a remarkably stable manner. Other more subtle genetic
changes (gene mutations) which are only detectable after careful
genetic analysis are probably significant routes of somaclonal
variation. It is known that under some physiological stress
conditions, for instance single deficiencies of anions (phosphorous,
nitrogen, sulphur) and cations (calcium and magnesium), spontaneous
mutations arise in root tip tissues in vivo. Both chromosomal and gene
mutations have been reported under such conditions in Antirrhinum
majus, Oenothera and Tradescantia (Steffensen, 1961; D'Amato, 1970).
What the direct or indirect mechanisms for these mutagenic effects are
is not known but it was suggested some time ago that since many plants
metabolites like amines, amides, aldehydes, alkaloids, phenols,
quinones, coumarins and the degradation products of nucleic acids
themselves are mutagenic agents (D'Amato & Hoffman-Ostenhof, 1956),
serious consideration should be given to elucidating the effects of
physiological stress metabolites on automutagenesis within somatic
tissues both in vivo and in vitro. Other effects of stress are
discussed further below.

Spontaneous gene mutations are suspected to occur at surprisingly high
frequencies. For instance, Lorz & Scowcroft (1983) in an elegant
study of somaclonal variation in which mesophyll protoplasts of the
Su/su mutant of tobacco were isolated and the expression of its aurea
phenotype used as a marker of mutation at the Su locus, estimated that
as many as 25% of the observed variants obtained had arisen from pre-
existing genetic variation in planta. Another proposed route of
genetic change known to operate in somatic cells is by transposition
of genetic elements from one part of the genome to another.
Transposable genetic elements are implicated in somatic mutations
which induce spotting (or colour breaking) phenotypes in somatic
tissues of some plants. Somatic cross-overs between chromosomes have
been shown to operate between more than one gene locus in tobacco and
three other crop species (Evans & Paddock, 1976).

Many forms of gross and subtle genetic changes therefore take place in
somatic tissues in planta which could influence the genotypes of

plants regenerated from protoplasts, cells and tissues in vitro.

GENETIC VARIATION INDUCED AS A CONSEQUENCE OF AN IN VITRO CULTURE

Evidence is accumulating which demonstrates that processes occurring during the initial stages of culture may be important determinants of somaclonal variation. For example, Barbier & Dulieu (1983) concluded that the majority of somatic mutations occurring at the yg^+/yg and a^+/a, loci in tobacco in regenerants of mesophyll-derived protocalluses took place during the first and not the subsequent culture generations. Also results of a study of changes in the karyotypes of potato protoclones by Sree Ramulu et al. (1984) have shown that endoreduplication, polyploidisation and aneuploidy are most common during the initial stages of protoplast growth. The association of cell division with the onset of cell wall regeneration in protoplasts means that the initial mitotic divisions in which the cytoskeletal structure of the cells is far from that pertaining in cells held in a matrix of tissue. The precise activity of the mitotic spindle must therefore be studied in greater detail during initial stages of protoplast culture to determine whether many of the observed changes in karyotype are in fact due to abnormal spindle formation associated with altered cytoskeletal properties of cells in isolation from one another.

The possible implications of physiological stress during in vitro cultures establishment on the levels of spontaneous mutations caused by over production of certain mutagenic metabolites in individual cells is a distinct possibility and should be borne in mind. For instance, recently it has been found that significant changes take place in endogenous polyamine levels during excision trauma and the first few hours of explant placement on culture media (Mantell et al. 1986). Further studies are underway to assess the importance of this stress-related response on somaclonal variability in the tissue culture products of woody solonaceous species particularly because polyamines are known to be important regulators of gene conformity and genetic expression (see Smith, 1985 for review).

Some of the many gross karyotypic changes which have been observed in cell and tissue cultures and their regenerants have been thought to be due to prolonged exposure of cells to growth regulators, particularly the artificial auxins like 2,4-dichlorophenoxyacetic acid and naphthalene acetic acid (eg Bayliss, 1980). However, recent evidence from studies in which garlic root tips were exposed to increasing amounts of auxin up to levels which were well in excess of those generally used for in vitro culture work, showed that these treatments did not produce significant increases in gross karyotypic changes to mitotic configurations compared to controls in which basal MS medium was employed as an incubation medium (Dolezel & Novaks, 1984).

Clearly more definite studies at the molecular level now need to be carried out to examine short-term changes which occur during the establishment and dedifferentiation of somatic tissues in vitro. One

of the main findings of recent studies of this type has been the
presence of abnormal DNA restriction patterns in the mitochrondial
genomes of a range of calliclones, protoclones and cell suspension
culture materials. Chourey et al. (1986) report that specific site
hybridisation probing of restriction endonuclense digests of mito-
chrondrial DNA in genomes of cell suspensions and protoclones of maize
and sorghum, revealed that repeated DNA regions of the mitochondrial
genome were specifically associated with a high capacity for loss
and/or rearrangement. Similar high levels of molecular variability
were reported previously by Kemble & Shephard (1984) in restriction
fragment analyses of mitochondrial DNA isolated from potato protoclones. More recently, molecular analysis of genomic DNA of potato
protoclones by Landsmann & Uhrig (1985), using a probe consisting of
the central part of 25SrDNA to search for DNA re-arrangements in
particular sections of the nuclear DNA fractions in Southern-blot
procedures, has revealed that two out of twelve clones contained
altered DNA configurations. Such alterations to these ribosomal genes
were stable after one cycle of conventional tuber propagation.
Ribosomal RNA genes are multicopy DNA sequences known to be subject to
amplification processes in insects and animals.

This is of particular significance to the discussion above on
physiological stress, when it is considered that deamplification of
rDNA is known to occur under stress conditions in flax cell suspension
cultures (Cullis & Charlton, 1981). Therefore stresses associated
with the in vitro culture process itself clearly need urgent
consideration and should be taken into account when determining the
possible causes of somaclonal variation.

EPIGENETIC BASES OF SOMACLONAL VARIATION
Some of the alterations in phenotypes observed in tissue
culture regenerants may be the result of changes in gene expression,
(ie. modifications in the transcription, translation and/or post-
translational processes). Epigenetic changes are manifested as
developmental modifications which are transient and potentially
reversible in nature. For example, cytokinin habituation - the
altered ability of wild type cells to divide in the absence of an
exogenous supply of cytokinin and yet cells retain their totipotency -
is reversible and is the result of epigenetic changes rather than
permanent alterations to the cell genome. Also, the reversibility of
cytokinin habituation in tobacco cells is far more rapid than that
expected if the phenotype was caused by somatic mutation.
Habituation, therefore, results in changes in the expression of the
genes which are normally silent (Meins, 1983).

There is a wide range of temporal changes in phenotypes of plants
regenerated from tissue culture. For example, in strawberry there are
characteristics which are only evident for four months to one year
after transplantation from in vitro to in vivo conditions. In the
field, a proportion of micropropagated plants show increased levels of
runner production, lateness in flowering, reduced fruit size and
increased susceptibility to diseases, pests and herbicides (Swartz &

Lindstrom, 1986). Some of these temporary changes in phenotype are
similar to those found in juvenile seedling materials. Indeed the use
of prolonged subculture in the presence of cytokinin results in
habituation which confers juvenile phenotypes expression. Loss of
secondary metabolic expression in cell cultures has been attributed to
epigenetic changes of this type but this is equivocal. For example,
inability of Solanum aviculare cell suspension in cultures to produce
steroidal sapogenins after several subcultures was attributed to the
prolonged exposure of cell cultures to cytokinin (Mantell & Smith,
1983).

The changes in gene expression which occur in strawberry tissue
cultures is currently under study. Examination of four different
isoenzyme banding patterns by means of starch gel electrophoresis
during shoot multiplication of the cultivar 'Tioga' in vitro has
revealed that over three initial subcultures of 40 days there is a
corresponding increase in multiple forms to three out of four of the
marker systems (Table 3). There is an increased level of gene
expression associated more with the first two culture generations than
with the presence of growth regulators in the multiplication medium.
Clearly significant changes take place during the initial phases of in
vitro culture and work is now in progress to determine whether
permanent genetic changes through re-arrangements or mutation are

TABLE 3: Isozyme bands present in starch gels of
buffered leaf extracts of micropropagated strawberry
plantlets and mother plants of cv. Tioga. * MS salts
+ 1 mg.l^{-1} IBA + 1 mg.l^{-1} kinetin + 0.1 mg.l^{-1} GA. ** MS
salts + no growth regulators. Source Unpublished data
of A. Ebida and the author

Isoenzyme marker	Mother plant	* Multiplication cultures Subculture No.			** Maintenance cultures Subculture No.		
		1	2	3	1	2	3
Esterase	4	13	5	5	9	5	4
Acid phosphatase	5	12	8	10	15	8	6
Leucinpamino peptidase	9	9	7	6	9	6	9
Phosphoglucose isomerase	3	4	2	2	3	3	2
Total	21	38	22	23	36	22	21

taking place at this stage, or whether the observed variations are
merely due to developmental changes in gene expression.

ONTOGENIC BASES OF SOMACLONAL VARIATION
The majority of angiosperms contain meristems which are
multilayered and structured with an outer tunica layer overlaying one
or two tissue layers referred to as the corpus. Some plants consist
of tissue mixtures of more than one genotype as in chimeras. When
adventitious buds arise from cell initials in either or a combination
of tissue layers, then the products of tissue culture vary as a
result. Some somaclonal variants arise by this route since easily
distinguished variegated perclinal chimeras can be shown to breakdown
following regeneration of plants via adventitious shoots on either
callus or even directly from explants (Skirivin, 1978). Not only
variegated chimeras breakdown under these conditions. Thornless
blackberry and loganberry in which L1 is thornless and L2/L3 is
thorny, provide ideal model systems for studying chimera breakdown in
vitro (eg. Rosati et al. 1986). The question which should now be
asked is whether or not chimerism is in fact more common in nature
than previously supposed. The discussion above concerning the high
levels of polyploidy and aneuploidy in somatic tissues on the mother
plant are relevant when it is considered that chimeral shoots have
been known to breakdown and produce variants at levels of 50% or more.

CONCLUSIONS
There are many biological routes through which variation
arises when plants are passed through tissue culture. They range from
changes which occur to the gross karyotype by natural processes of
endoreduplication occurring in the mother plant prior to explantation,
to more subtle changes which occur to organellar and nuclear DNA
particularly in the early stages of protoplast culture, to changes to
the gross karyotype of cells and tissues exposed to prolonged in vitro
culture, to changes in the manner in which shoots are initiated from
different histogenic layers and to temporal and reversible changes
which occur in gene expression. Recent findings indicate that the in
vitro culture process itself causes modifications to all these
factors. It is anticipated that physiological stresses which are
imposed upon protoplasts, cells and tissues during the initial
establishment phases of culture will be shown to have a significant
bearing on the induction of somaclonal variation.

REFERENCES
Barbier M. & Dulieu H. (1983). Early occurrence of genetic variants
            in protoplast cultures. Plant Sci. Lett., 29, 201-6.
Bayliss M.W. (1980). Chromosomal variation in plant tissues in
            culture. Int. Rev. Chytol. (Suppl.) 11A, 113-44.
Chourey P.S., Lloyd R.E., Sharpe D.Z. & Isola N.R. (1986). Molecular
            analysis of hypervariability in the mitochondrial genome
            of tissue cultured cells of maize and sorghum. In: The
            Chondriome -mitochondrial and chloroplast genomes of
            plants. eds S.H. Mantell, G.P. Chapman & P.F.S. Street.
            London, Longman Press Ltd. (in press).

Cullis C.A. & Charlton L. (1981). The induction of ribosomal DNA
     changes in flax. Plant Sci. Lett. 20, 213-7.
D'Amato F. (1970). Other cases of mutations. In: Manual on Mutation
     Breeding. Chapter 4. International Atomic Energy
     Association, Vienna, 1970.
D'Amato F. (1985). Cytogenetics of plant cell and tissue cultures
     and their regenerates. CRC Critical Reviews 3, 73-112.
Dolezel J. & Novak F.J. (1984). Cytogenetic effect of plant tissue
     culture medium with certain growth substance on Allium
     sativum L. meristem root tips. Biologia Plantarum 26,
     293.
Evans D.A. & Paddock E.F. (1976). Comparison of somatic crossing
     over frequency in Nicotiana tabacum and three other crop
     species. Can. J. Genet. Cytol. 18, 57.
Evans D.A., Sharp W.R. & Medina-Filho H.P. (1985). Somaclonal and
     gametoclonal variation. Amer. J. Bot. 71, 759-74.
Heinz D.J., Krishamurthi M., Nickell L.G. & Maretzki A. (1977). Cell,
     tissue and organ culture in sugarcane. In: Plant Cell
     Tissue and Organ Culture. eds. J. Reinert & Y.P.S.
     Bajaj., pp.3-17. Berlin, Springer-Verlag.
Hussey G. (1983). In vitro propagation of horticultural and
     agricultural crop. In: Plant biotechnology, eds S.H.
     Mantell and H.Smith pp.111-138, Cambridge, Cambridge
     University Press.
Kemble R.J. & Shepard J.F. (1984). Cytoplasmic DNA variation in a
     potato protoclonal population. Theor. Apple Genet. 69,
     211-6.
Klekowski E.J. Jr., Mohr H., Kazarinova-Fukshansky N. (1986).
     Mutations, apical meristems and developmental selection.
     In: Genetics, Molecules and Evolution, ed J.P. Gustafson.
     New York, Plenum (in press).
Krikorian A.D. (1982). Cloning higher plants from aseptically
     cultured tissues and cells. Biol. Rev. 57, 151-218.
Landsmann, J. & Uhrig, H. (1985). Somaclonal variation in Solanum
     tuberosum detected at the molecular level. Theor. Appl.
     Genet. 71, 500-5.
Lorz H. & Scowcroft W.R. (1983). Variability among plants and their
     progeny regenerated from protoplasts of Su/su heterozy-
     gotes of Nicotiana tabacum. Theor. Appl. Genet. 66, 67-
     75.
Mantell S.H. & Smith H. (1983). Cultural factors that influence
     secondary metabolic accumulations in plant cell and tissue
     cultures. In: Plant biotechnology eds S.H. Mantell & H.
     Smith. pp.75-108. Cambridge, Cambridge University
     Press.
Mantell S.H., Russell C. & Knights M. (1986). Relationship between
     endogenous titres of polyamines and excision trauma in
     epidermal explants of the woody solanaceous plant Solanum
     aviculare. (in preparation).
Mantell S.H. (1986). Problems of stability in mass propagation. In:
     Micropropagation in Horticulture - practice and commercial
     problems. Proc. Inst. of Hortic. Symp. Sutton Bonnington,
     March 1986. (in press).

Meins F., Jr. (1983). Heritable variation in plant cell culture.
          Annu. Rev. Plant Physiol. 34, 327-46.
Rosati P., Gaggioli D. & Giunchi L. (1986). Genetic stability of
          micropropagated 'Loganberry' plants. J. Hort. Sci. 61,
          33-41.
Shepard J.F., Bidney D. & Shahin E. (1980). Potato protoplasts in
          crop improvement. Science, 208, 17-24.
Skirvin R.M. (1978). Natural and induced variation in tissue
          culture. Euphytica 27, 241-66.
Smith T.A. (1985). Polyamines. Annu. Rev. Pl. Physiol. 36, 117-43.
Steffensen D.M. (1961). Chromosome structure with special reference
          to the role of metal ions. Int. Rev. Cytol. 12, 163.
Swartz H.J. & Lindstrom J.T. (1986). Small fruit and grape tissue
          culture from 1980 to 1985: commercialization of the
          technique. Proc. Conf. Tissue Culture as a Plant
          Production System for Horticultural Crops. USDA,SEA,NER.
          Agricultural Research Report September 1985 (in press).

28. <u>GENETIC BASIS OF INSTABILITY IN CELL CULTURES</u>

A. Stafford

Wolfson Institute of Biotechnology,
University of Sheffield,
Sheffield,   S10 2TN. U.K.

INTRODUCTION

While culture variation can be advantageous in providing
the basis for effective selection programmes for higher - yielding
cell lines, continuing phenotype instability is obviously undesirable.
The phenotypic variation often observed in plant cell cultures can be
conveniently divided into two categories. In the first place there is
short term variation, which is specifically associated with the
establishment of new cultures. Anyone familiar with this process will
be aware of the morphological changes which can occur with progressive
subcultures of recently initiated callus. In the long term,
variability in product yield over successive subcultures is
potentially a problem and has often been observed. Tabata & Hiraoka
(1976) observed wide variation in nicotine yield among single-cell
clones derived from <u>Nicotiana rustica</u> callus tissue. Ogino <u>et al</u>.
(1978) isolated single-cell clones from <u>N. tabacum</u> cell suspensions
and noted fluctuations in their nicotine contents in the absence of
selection over 14 months. Similar observations were made by Deus-
Neumann & Zenk (1984) on the accumulation of serpentine in selected
cell clones of <u>Catharanthus roseus</u>. In these situations, the culture
may be growing under apparently the same environmental conditions for
many years but still exhibit yield variability. One form of long-term
variation which has particular relevance to the success of any
commercial process is exemplified by a study of Zenk (1978) on
ajmalicine production by <u>C. roseus</u> cell cultures. In this case the
level of alkaloid accumulation fell gradually and quite consistently
over a period of many months.

However, phenotypic instability is not necessarily the rule. The
accumulation of isoquinoline alkaloids in <u>Coptis japonica</u> (Fukui <u>et
al</u>., 1982) and rosmarinic acid in <u>Coleus blumei</u> cell cultures (Zenk <u>et
al</u>., 1977) were found to be stable. Other examples are quoted in
reviews (e.g. Yamada & Hashimoto, 1984).

MECHANISMS UNDERLYING CULTURE VARIATION

The mechanisms underlying culture instability are poorly
understood. Genetic factors have been invoked, though the evidence
mainly derives from numerous investigations of recognisable karyotype
abnormalities in cultured cells. Attempts have been made to link
gross genomic changes to alterations in secondary metabolite yield.
Deus & Zenk (1982) found that a low alkaloid-producing cell line of <u>C.</u>

roseus possessed a greater range of ploidy levels than a high alkaloid
cell line. Directed ploidy increases have been brought about in
cultures by colchecine treatments. In a recent paper, Becker &
Chavadej (1985) reported a substantial increase in valepotriate
production in colchecine-treated Valeriana wallichii cell lines.
However, while the increased product level was stable, the ploidy
level of the cultures fell to their original level upon the removal of
colchecine.

While gross differences in karyotype may at least partially explain
the differences observed between secondary product yields of stable
cell lines, they do not provide an explanation for the long-term
culture instability recorded in so many cases. A significant feature
in this respect is the fact that not all cultures of the same species
exhibit yield variability (Deus-Neumann & Zenk, 1984). Furthermore,
it is possible to derive both stable and unstable sub-clones from the
same source culture (Ellis, 1982; Ohta & Yatazawa, 1982). These
reports indicate that while variation in the culture environment and
subculture regime may contribute to yield variation (Morris, 1986) not
all cultures respond to these stimuli in the same manner.

The culture conditions imposed upon new explant tissues are likely to
exert considerable trauma on the surviving cells. For instance, the
inclusion of cytotoxic components such as 2,4-dichlorophenoxyacetic
acid (2,4-D) in tissue culture media is likely to introduce a
selection pressure for those cells most able to respond positively and
divide. The newly initiated culture may derive from several explant
cells at different developmental stages, alternative states of
differentiation or even diverse genotypes. After several subcultures
under the same conditions, undefined selection pressures inherent in
the culture process should have achieved some degree of equilibrium,
such that those cell types most capable of continued division under
the prevailing conditions attain dominance. If this situation is
accepted then cases of instability of product yield as characterised
by a gradual loss of synthetic potential may be explained quite
simply. The appearance and proliferation of a low-producing mutant
cell type capable of higher rates of cell division than the dominant
form would result in higher growth rates but decreasing product
accumulation over successive subcultures. Similarly an unconscious
alteration in the culture regime may introduce a selection pressure
favouring the emergence of a formerly insignificant cell type. On the
other hand, if the cells of an explant which have been induced to
divide are homogeneous and genetically stable, it should in theory be
possible to produce a phenotypically stable cell line.

INTRINSIC GENETIC INSTABILITY
          The production of unstable cell lines which display no
consistent trend towards the gradual loss or acquisition of particular
properties is most easily explained by proposing that the cells of
certain explant tissues are themselves intrinsically unstable. One
mechanism which could produce this condition has received considerable
attention recently but was first noted in the 1940's, when McClintock

proposed an explanation for long-term instability of kernel phenotype
in Zea mays (McClintock, 1984). She postulated the existence of
"controlling elements", genetic factors which were capable of moving
around the genome and modifying or completely suppressing gene
expression. Since this pioneering work was done, the presence of
several controlling elements, for example the so-called Ds, Ac, and Mu
elements has been demonstrated at the molecular level in maize
(Freeling, 1984). Recent work has also shown that the flower pigment
variation in Antirrhinum majus observed by Harrison & Carpenter (1979)
is produced by the insertion and excision of a controlling element,
Tam 1 (Bonas et al., 1984). Both A. majus and Z. mays have been
involved in rigorous breeding programmes and the inheritance of flower
and kernel colouration respectively has been subjected to critical
genetic analysis. It is quite likely that genomes of other plant
species which have not received such attention are subject to the
control of similar genetic elements. Because of the imprecise nature
of controlling element excision and the consequent generation of
sequence diversity at the site of excision, it has been suggested that
these insertions may make an important contribution to the genome
heterogeneity necessary for evolution (Schwarz-Sommer et al., 1985).

A number of factors have been found to affect the rate at which
controlling elements excise from their site in the genome, thereby
allowing reversion to the wild-type phenotype. McClintock found that
chromosome breakage could activate controlling elements in maize
plants, and predicted that environmental shocks could also activate
these events. The injection of maize plants with barley stripe mosaic
virus induces insertional mutations (Freeling, 1984). Temperature may
have a marked effect upon excision rates; in A. majus these are
dramatically increased at lower growth temperatures, resulting in
enhanced reversion to wild-type petal pigmentation (Harrison &
Carpenter, 1979). The presence of modifying genes, and the
developmental stage of the plant may influence excision. Finally,
there is some evidence that the culture process itself might affect
the mobility of genetic elements. In Drosophila, copia-type mobile
elements increase in number in cultures by a factor of ten or more
(Rubin, 1983). Similarly, at a recent symposium, Evola et al. (1984)
reported work with maize which indicated an activation of controlling
elements during tissue culture, resulting in a tenfold increase in the
mutation rate at one locus.

The consequences of controlling element activation are reversion to
"wild-type" phenotype, and because of the demonstrated imprecise
nature of excision, mutations in the formerly repressed gene may
result. It is obvious that if the explant tissue to be used for
setting up a new culture consists of cells in which controlling
elements are present in the genome, then there is a possibility of
producing excisions by the culture process. Especially if the element
were capable of repeated insertion and excision, the resulting culture
would be subject to apparently spontaneous phenotypic changes. Other
as yet unknown mechanisms which produce genetic instability in plant
cells may also exist.

Whatever the mechanism may be, there is considerable circumstantial evidence which points to an intrinsic predisposition to phenotypic and perhaps genetic instability in some cell cultures. The indications are clear, in that a range of varieties and tissues should be used for the purpose of setting up new cultures. These cultures, of potentially diverse genetic backgrounds, should ultimately be screened not only for yield level, but also for yield stability over consecutive subcultures.

REFERENCES

Becker, H. & Chavadej, S. (1985). Valepotriate production of normal and colchecine-treated cell suspension cultures of _Valeriana wallichii_. J. Nat. Prod. **48**, no. 1, 17-21.

Bonas, U., Sommer, H., Harrison, B.H. & Saedler, H. (1984). The transposable element Tam 1 of _Antirrhinum majus_ is 17 kb long. Mol. Gen. Genet. **194**, 138-43.

Deus-Neumann, B. & Zenk, M.H. (1982). Exploitation of plant cells for the production of natural compounds. Biotechnol. Bioeng. **24**, 1965-74.

Deus-Neumann, B. & Zenk, M.H. (1984). Instability of indole alkaloid production in _Catharanthus roseus_ cell suspension cultures. Planta media **50**, 427-43.

Ellis, B. (1982). Cell-to-cell variability in secondary metabolite production within cultured plant cell populations. _In_: Plant Tissue Culture 1982, Proc. 5th Intl. Cong. Plant Tissue & Cell Culture, ed. Fujiwara, A., pp.395-6. Tokyo; Abe Photo Printing.

Evola, S.V., Burr, F.A., & Burr, B. (1984). The nature of tissue culture induced mutations in maize. _In_: Abstracts to the Eleventh Aharon Katzir-Katchalsky Conference on Plant Molecular Biology, January 8-13, 1984: Jerusalem.

Freeling, M. (1984). Plant transposable elements and insertion sequences. Ann. Rev. Plant Physiol. **35**, 277-98.

Fukui, H., Nakagawa, K., Tsuda, S., & Tabata, M. (1982). Production of isoquinoline alkaloids by cell suspension cultures of _Coptis japanica_. _In_: Plant Tissue Culture 1982, Proc. 5th Intl. Cong. Plant Tissue & Cell Culture, ed. Fujiwara, A. pp.313-4. Tokyo: Abe Photo Printing.

Harrison, B.J. & Carpenter, R. (1979). Resurgence of genetic instability in _Antirrhinum majus_. Mutation Research. **63**, 47-66.

McClintock, B. (1984). The significance of responses to the genome to challenge. Science **226**, 792-801.

Morris, P. (1985). Long-term stability of alkaloid production in cell cultures of _Catharanthus roseus_. _In_: this volume.

Ogino, T., Hiraoka, N. & Tabata, M. (1978). Selection of high nicotine-producing cell lines of tobacco callus by single-cell cloning. Phytochem. **17**, 1907-10.

Ohta, S. & Yatazawa, M. (1982). Selection and stable preservation of high nicotine producing tobacco cell lines through repeated transfer under defined culture conditions. _In_: Plant Tissue Culture 1982, Proc. 5th Intl. Cong. Plant

Tissue and Cell Culture, ed. Fujiwara, A. pp.321-2. Tokyo: Abe Photo Printing.

Rubin, G.M. (1983). Dispersed repetitive DNAs in Drosophila. In: Mobile Genetic Elements, ed. Shapiro, J.A. pp.329-62.   New York, London: Academic Press.

Schwarz-Sommer, Z., Gierl, A., Cuypers, H., Peterson, P.A. & Saedler, H. (1985).   Plant transposable elements generate the DNA sequence diversity needed in evolution.   EMBO J. 4, No. 3., 591-7.

Tabata, M. & Hiraoka, N. (1976).   Variation of alkaloid production in Nicotiana rustica callus cultures.   Physiol. Plant. 38, 19-23.

Yamada, Y. & Hashimoto, T. (1984).   Secondary products in tissue culture.   In: Applications of Genetic Engineering to Crop Improvement. ed. Collins, G.B. & Petolino, J.G.   pp.561-604. Dordrecht/Boston/Lancaster: Martinus Nijhoff/Dr. W. Junk.

Zenk, M.H., El-Shagi, H. & Ulbrich, B. (1977).   Production of rosmarinic acid by cell-suspension cultures of Coleus blumei. Naturwissenschaften 64, 585-6.

Zenk, M.H.   (1978).   The impact of plant cell culture on industry. In: Frontiers of Plant Tissue Culture, Proc. 4th Intl. Cong. Plant Tissue and Cell Culture, ed. Thorpe, T.A., Alberta.

29.   A RAPID SCREENING TECHNIQUE FOR THE SELECTION OF HIGH
      YIELDING CAPSAICIN CELL LINES OF CAPSICUM FRUTESCENS MILL.

M.M. Aitken & M.M. Yeoman

Department of Botany,
University of Edinburgh,
King's Buildings, Mayfield Road,
Edinburgh. U.K.

INTRODUCTION
    It has been shown that tissue cultures of Capsicum
frutescens can accumulate capsaicin, the hot element of chilli pepper,
under certain conditions (Yeoman et al., 1980). The amount of
capsaicin accumulated is considerably enhanced in immobilised cell
cultures (Lindsey and Yeoman, 1984). However, there are considerable
variations between cultures. In order to further investigate this
variation and isolate high yielding clones, a large number of cell
lines were isolated from a single culture. Subsequently a study was
carried out in which preliminary analysis showed that different cell
lines accumulated different amounts of capsaicin (Holden et al.,
1986).

In order to discover which of the large number of cell lines
synthesised and accumulated the most capsaicin, a screening technique
was devised in which an artificial antigen was chemically synthesised
by combining capsaicin with Human Serum Albumin. This antigen was
subsequently used to raise antisera in mice, rabbits and sheep and the
purified antisera used as a basis for a selection procedure in
conjunction with an Enzyme-Linked Immuno Sorbent Assay (ELISA).

METHODS AND RESULTS
    Screening Techniques Used in Cell Line Selection.   At
first all of the samples were analysed by ELISA and HPLC to establish
specificity and reproducibility of the ELISA. The ELISA is a highly
sensitive immunological assay which can detect capsaicin below
nanogram levels. The HPLC technique can detect not only capsaicin but
also a wide range of soluble plant phenolics (Holden et al., 1986) and
is the method of choice for detailed analysis of the products of plant
metabolism.

An ELISA for the Detection of Capsaicin.   Antibodies to capsaicin
were raised in mice, rabbits and sheep by inoculation with a
capsaicin-protein conjugate. The conjugate was prepared by linking
capsaicin to p-Amino Benzoic Acid (PABA) by the method of Gross et
al., (1968), with the ommission of the benzene/methanol final
extraction which is inappropriate for capsaicin. This product was
linked to Human Serum Albumin using the carbodiimide technique of
Langone et al., (1973). The final product was capsaicin-p-Amino
Benzoic Acid-Human Serum Albumin (capsaicin-PABA-HSA). This was

purified by precipitation with saturated ammonium sulphate and
extensive dialysis against 0.1M carbonate/bicarbonate buffer.

For immunisation purposes the conjugate was mixed with Alhydrogel
adjuvant as described by Herbert, (1973), and injected subcutaneously
over a three month period. Serum was tested for sensitivity to
capsaicin using either a Bentonite Flocculation test as described by
Bozioevech (1963), or by a Ring Precipitin test as described by Garvey
et al. (1977). The immunoglobulin fraction was isolated from the
whole serum by the method of Clarke and Adams (1977). The
immunoglobulin fraction was further purified by passing it through an
affinity column which was prepared by conjugating capsaicin to
epoxy-sepharose (Sigma, London) using the technique described by
Vretblad (1976). Antibody specific to capsaicin was bound to the
sepharose and subsequently eluted with 0.1M glycine/HCl buffer.

Specificity of Antisera to Capsaicin.   Antisera to capsaicin was
raised in mice, rabbits and sheep. All three antisera reacted
strongly with capsaicin concentrations down to 1ng in the Bentonite
Flocculation test but antisera from mice and rabbits showed a number
of cross-reactions with some compounds which are structurally similar
to capsaicin. A range of substances was therefore tested to determine
the specificity of the antisera from different sources. As Table 1
shows the only antiserum which showed no detectable cross-reactions
was that raised in sheep and this was the antiserum used in all
subsequent assays.

From Table 1 it can also be seen that there was no detectable
cross-reaction with any of the constituents of SH medium used in the
the culture of chilli peppers, thus enabling assays to be carried out
on unextracted growth medium which greatly reduced the time necessary
to carry out an assay.

TABLE 1.   Cross-Reaction Tests of Antisera Raised to
Capsaicin.

| Test Substance | Reaction of Antisera | | | (+ =cross-reaction |
	Mice	Rabbits	Sheep	− =no cross-reaction)
Valine	−	−	−	
Vanillin	+	−	−	
Phenylalanine	−	−	−	
Caffeic Acid	+	+	−	
Ferulic Acid	+	+	−	
Isocapric Acid	+	+	−	
Gallic Acid	−	−	−	
p-OH-Benzoic Acid	−	−	−	
p-Coumaric Acid	−	−	−	
m-Coumaric Acid	−	−	−	
trans-Cinnamic Acid	−	−	−	
Protocatechuic Aldehyde	−	−	−	
Full SH Medium	−	−	−	
Control-Capsaicin	+	+	+	

The ELISA Method.   The ELISA method used was the Double-Antibody
Sandwich technique as described by Voller _et al_., (1979). A typical
standard curve for capsaicin concentration is shown (Fig. 1).

Advantages of ELISA as a First Step Screening Technique.   The main
advantages gained by the use of ELISA in a screening programme are
speed and sensitivity. Many cell lines can be assayed simultaneously
in 96-well microtitre plates and no prior extraction of the medium is
necessary. The sensitivity of the assay allows the screening of very
small amounts of tissue (c. 10 mg) which is a great advantage when
dealing with large numbers of cell lines as any unproductive cultures
can be discarded at an early stage in the clonal isolation programme.
The availability of a specific antiserum to capsaicin also enables the
use of immunofluorescence microscopy in the localisation of capsaicin
within the tissues and cells of callus or fruits.

Detailed Clonal Analysis – The Use of HPLC.   Although the ELISA has
proved extremely useful in the initial screening programme for high
capsaicin producing cell lines, it does not give information about the
other phenolic products in the biosynthetic pathway which may be
present in the cultures. For example, it was found that analysis by
HPLC provided a profile of the soluble phenolic products of chilli
pepper cultures and that this differed between cell lines. Therefore,
a detailed HPLC analysis of a number of cell lines was carried out in
order to determine the amount of variation in phenolic products which

Fig. 1.   Standard curve of capsaicin concentration as
measured by ELISA. (Mean values with standard error).

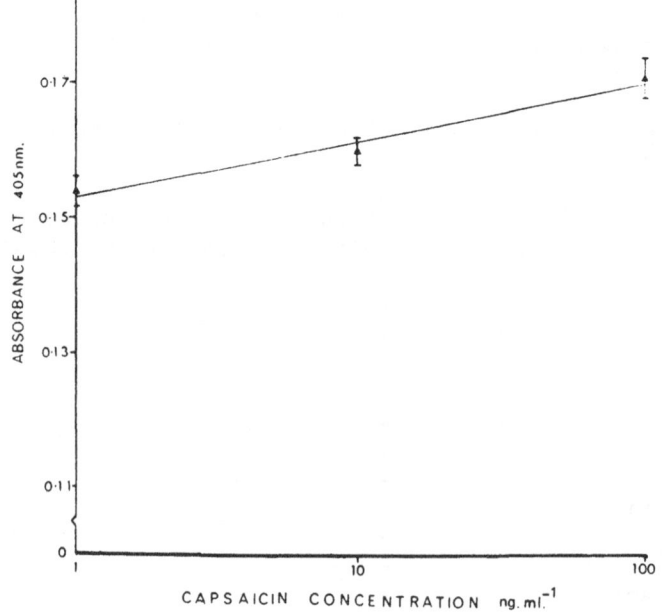

CAPSAICIN CONCENTRATION   ng.ml^{-1}

existed between the lines. It was discovered that each of the cell
lines studied showed a unique pattern of soluble phenolics which was
consistent between replicates (Fig. 2). A long term study of a small
number of these cell lines is now being undertaken in order to
determine the stability of this variation in clonal material in
culture.

CONCLUSIONS
Initial results have shown that a pool of variability
exists within cell suspension cultures of the chilli pepper. Whether
this variability can be exploited in the industrial production of
capsaicin depends on the stability of high yielding cell lines in
culture. More research in this area is essential if the production
potential of plant metabolites on a large scale is to be realised.

Fig. 2.   HPLC profiles of pepper cell lines showing the
amount of variability between cell lines with respect to
the accumulation of phenolics in the culture medium.

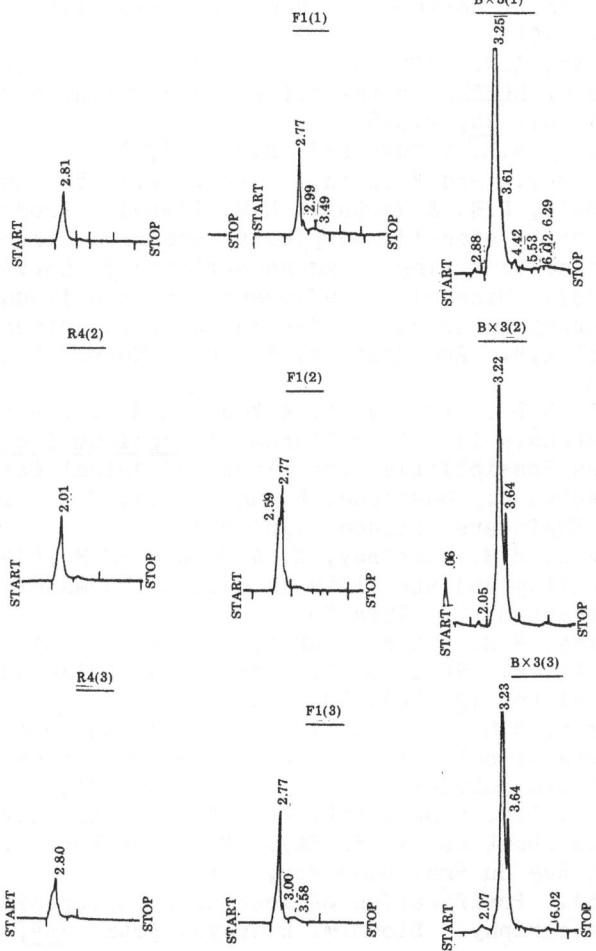

The use of modified culture techniques such as slow growth in an immobilised state may be necessary to maintain the culture stability. The use of cryopreservation techniques already established as a major means of culture storage (Withers, 1985) could also be used to maintain high yielding cultures in a stable condition.

The use of continuous cell line selection may be considered if instability in culture is a problem. The development of rapid screening techniques, such as the ELISA make this approach feasible.

The way forward in the field of plant metabolite production as an industrial process must therefore be in a combination of the techniques described above and not simply in the mass culture of unselected plant cells.

REFERENCES

Bozioevich, J., Scott, H.A. & Vincent, M.M. (1963). The Bentonite Flocculation Test for detection of plant viruses and titration of antibody. Proc. Soc. Exp. Biol. and Med. 114, 789-94.

Clarke, M.F. & Adams, A.N. (1977). Characteristics of the microplate method of ELISA for the detection of plant viruses. J. Gen. Virol. 34, 475-83.

Garvey, J.S., Cremer, N.E. & Sussdorf, D.H. (1977). Methods in Immunology. 3rd Edition, 275-79. W.A. Benjamin, Inc.

Gross, S.J., Campbell, D.H. & Weetall, H.H. (1968). Production of antisera to steroids coupled to proteins directly through the phenolic A ring. Immunochemistry 5, 55-65.

Herbert, W.J. (1973). Mineral-oil adjuvants and the immunisation of laboratory animals. In "Handbook of Experimental immunology". 2nd Edition, Ed. D.M. Weir. Blackwell, Oxford.

Holden, M.A., Hall, R.D., Lindsey, K. & Yeoman, M.M. (1986). Capsaicin biosynthesis in cell cultures of Capsicum frutescens. In: Process Possibilities for Plant and Animal Cell Cultures". Eds. Webb, C., Mavituna, F. and Faria, J.J. Institute of Chem. Engineers, London. (In Press).

Holden, P.R., Aitken, M.M., Lindsey, K. & Yeoman, M.M. (1986). Variability and stability in cell cultures of Capsicum frutescens Mill. This Volume 31.

Langone, J.J., Gjika, H.B. & Van Vunakis, H. (1973). Nicotine and its metabolites. Radioimmunoassay for nicotine and cotine. Biochemistry 12 (24), 5025-30.

Lindsey, K. & Yeoman, M.M. (1984). The viability and biosynthetic activity of cells of C. frutescens immobilised in reticulate polyurethane. J. Exp. Bot. 35, 1684-96.

Voller, A., Bidwell, D.E. & Bartlett, A. (1979). The enzyme-linked immunosorbent assay (ELISA). Dynatech Europe, Borough House, Rue du Pre, Guernsey, U.K.

Vretblad, P. (1976). Purification of lectins by biospecific affinity chromatography. Biochim. Biophys. Acta. 434, 169-76.

Withers, L. (1985). Cryopreservation of cultured cells and meristems.
        In: "Cell Culture and Somatic Cell Genetics of Plants".
        Ed. Indra. K. Vasil. Acad. Press, pp. 254-316.
Yeoman, M.M., Miedzybrodzka, M.B., Lindsey, K. & McLauchlan, W.R.
        (1980). In "Plant Cell Cultures: Results and
        Perspectives", 327-43. Eds. F. Sala, B. Parisi, R. Cella
        & O. Ciferri. Elsevier.

30.   SELECTION STUDIES ON CATHARANTHUS ROSEUS

R. C. Cresswell

Wolfson Institute of Biotechnology,
University of Sheffield,
Sheffield   S10 2TN.   U.K.

The Madagascan Periwinkle Catharanthus roseus produces a
number of alkaloids as secondary metabolites, some of which are known
to have therapeutic activity (Svoboda 1964).   In cell culture the main
alkaloids produced by C. roseus are ajmalicine (used for the treatment
of hypertension) and the related compound serpentine.   Research was
initiated on a low alkaloid-producing cell line in order to
investigate the feasibility of selecting higher yielding cell lines.
Two techniques have frequently been employed elsewhere in attempts to
select cell lines producing high yields of secondary metabolites.

One of these techniques employs the selection of variant cell lines
with resistance to antimetabolites such as amino acid analogues (Ranch
et al. 1983; Widholm 1974).   Often, such variant cells are found to be
resistant by virtue of their over-production of normal metabolites.
For example a p-fluorophenylalanine-resistant cell line of sycamore
was found to contain increased concentrations of phenylalanine,
tyrosine and phenolics (Gathercole & Street 1978), and
m-fluorophenylalanine-resistant cell lines of Nicotiana were found to
contain 3-8 times higher levels of cinnamoyl putrescines than normal
cells (Berlin et al. 1981).

C. roseus cell lines with resistance to the tryptophan analogue, 5-
methyl tryptophan, have previously been found to over produce
tryptophan (an intermediate in alkaloid synthesis) but not to over
produce alkaloids (Scott et al. 1979).   In this report the selection
of two 5-methyl tryptophan-resistant lines of C. roseus is described
and some of their characteristics are discussed.

Suspensions of C. roseus (L) G. Don. were maintained in M3 medium at
pH 5.8 by transferring 28ml cell suspension into 100ml every 7 days.
For experiments on the selection and growth of 5-methyl tryptophan-
resistant cell lines, solutions of 5-methyl tryptophan at
approximately pH 6.5 were filter sterilized and added to previously
autoclaved media.

Estimates of cell viability and alkaloid extractions were as described
elsewhere (Stafford et al. 1985).

When normal cells, sensitive to 5-methyl tryptophan, were inoculated
into growth media containing various concentration of 5-methyl

tryptophan there was a loss of viability (Fig. 1A) and inhibition of growth (Fig. 1B). The loss of viability and growth inhibition were greater with higher concentrations of 5-methyl tryptophan and total loss of apparent viability was achieved after 7 days with 60 and 80 mg.l^{-1} 5-methyl tryptophan. The cells in all the flasks turned brown in colour during this period and remained so for several weeks. After approximately 50 days of incubation there was a reappearance of cream coloured (normal) cell-aggregates and an increase in viability in two suspensions containing 40 mg.l^{-1} 5-methyl tryptophan (Fig. 2).

FIGURE 1.   Selection of 5-methyltryptophan-resistant C. roseus
Viabilities (A) and growth (B) of C. roseus cells in the presence of 0 (▲), 10 (▼), 20 (△), 40 (△), 60 (◇) and 80 (◆) mgl^{-1} 5-methyltryptophan. Symbols for the first 13 days show average figures from two cultures. After 13 days only the cells on 40 mgl^{-1} 5-methyltryptophan (ie. lines 40A (○) and 40B (□)) are considered. Lines 40A and 40B were subcultured into fresh medium with 40 (open symbols) and 80 mgl^{-1} (closed symbols) 5-methyltryptophan on day 68. Measurements of growth (B) were not made after 13 days.

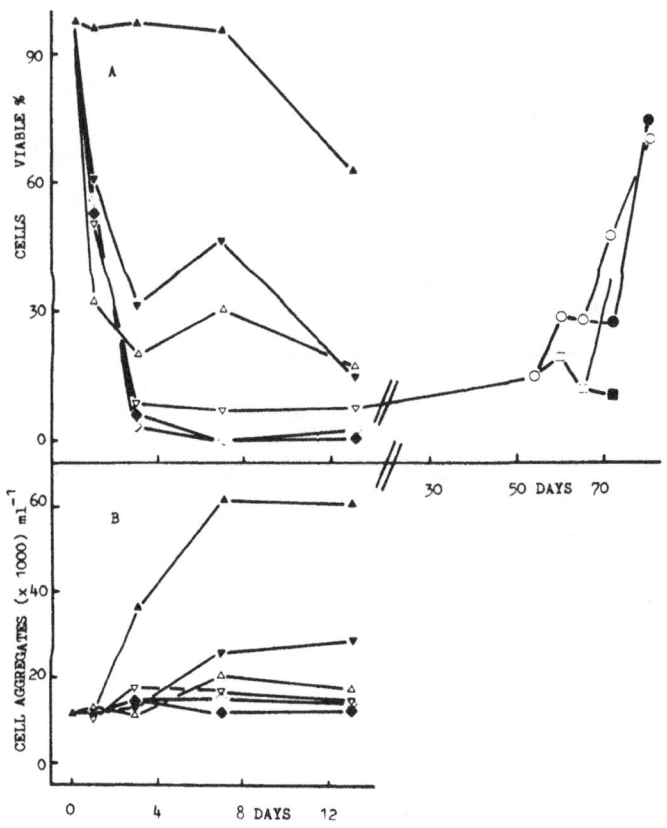

These suspensions were each then subcultured into fresh medium
containing 40 mg.l^{-1} 5-methyl tryptophan which resulted in increases
in cell viability (Fig. 1A) and growth. Two weeks later the two
growing cell suspensions (40A and 40B) were each used to inoculate
media containing 40 and 80 mg.l^{-1} 5-methyl tryptophan and further
increases in cell viabilities were observed. Eventually one of the
cell suspensions (40A) growing in medium containing 80 mg.l^{-1} 5-methyl
tryptophan was inoculated into medium containing 200 mg.l^{-1} 5-methyl
tryptophan and the cells grew.

FIGURE 2.   The effects of 5-methyltryptophan on the
growth and viability of normal and resistant cells
Resistant cells for this experiment were obtained from the
resistant line 40A and were growing previously on 80 and
200 mgl^{-1} 5-methyltryptophan. The graphs show growth (A)
and viability (B) of:- 1.   normal cells in the presence
of zero (●), 80 (○) and 200 mgl^{-1} (□) 5-methyltryptophan.
2.   resistant cells (previously growing on 80 mgl^{-1} 5-
methyltryptophan) in the presence of zero (▼) and 80 mgl^{-1}
(▲) 5-methyltryptophan. 3.   resistant cells (previously
growing on 200 mgl^{-1} 5-methyltryptophan) in the presence
of zero (▽) and 200 mgl^{-1} (△) 5-methyltryptophan. Symbols
show average figures from two cultures.

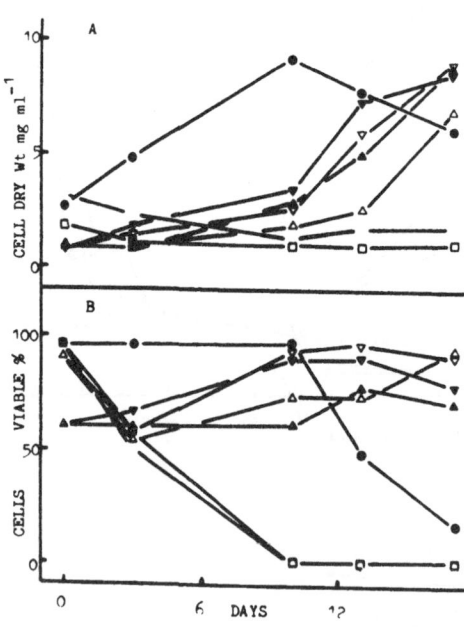

Figure 2 compares the effects of 5-methyl tryptophan on the growth and
viability of one of the resistant lines (40A) and normal cells.   In
the presence of 80 or 200 mg.l^{-1} 5-methyl tryptophan normal cells
failed to grow and lost viability within 10 days.   Resistant cells
however were able to grow in 80 or 200 mg.l^{-1} 5-methyl tryptophan
(Fig. 2A, 2B) and showed increased viability during incubation (Fig.
2B).

The resistant cell lines were maintained as suspensions and the
lengths of the growth periods were gradually extended from 7 to 14
days over many weeks.   At the end of each growth period during this
process cell samples were freeze-dried for alkaloid extraction and
analysis.   It was found however that the maximum accumulation of
alkaloids detected in the resistant cells was less than the maximum
alkaloid accumulation detected in the parent cell line (Table 1).

Table 1.   **Maximum alkaloid production by normal and
resistant cells**

Cell line	Serpentine produced mg g^{-1} dry wt cells	Ajmalicine produced mg g^{-1} dry wt cells
AV16 (parent line)	1.0	0.1
40B (resistant cells)	0.5	0
40A (resistant cells)	0.16	0

Table 2.   **Characteristics of manually selected cell-
aggregates**.   104 cell-aggregates were selected, of which
40 were non-fluorescent and the remainder showed varying
degrees of blue and green autofluorescence in ultraviolet
light.   9 of the selected cell-aggregates were
successfully raised to produce suspension cultures and
their characteristics at the time of selection are
recorded.

Cell line	Total number cells in cell-aggregate	Number of blue-autofluorescing cells in cell-aggregate	Number of green-autofluorescing in cell-aggregate
1D	30	12	3
6A	45	17	3
1C	41	8	12
4C	25	0	0
2D	19	7	0
6C(1)	40	20	2
6C(2)	37	0	0
24B	35	0	0
2B	43	14	4

Cell lines resistant to 5-methyl tryptophan have been selected with
<u>Datura innoxia</u> (Ranch <u>et al</u>. 1983), <u>Daucus carota</u> (Widholm 1974),
<u>Solanum tuberosum</u> (Carlson & Widholm 1978) and <u>Catharanthus roseus</u>
(Scott <u>et al</u>. 1979). Frequently such resistance has been accounted
for by the presence in resistant cells of an altered anthranilate
synthetase less sensitive to feedback inhibition by tryptophan. Such
cells were shown to accumulate levels of free tryptophan up to 48
times the level found in normal cells (see eg. Carlson & Widholm
1978). A carrot cell culture, however, was found to be resistant to
5-methyl tryptophan apparently by decreased uptake of the tryptophan
analogue (Widholm 1974). The modes of resistance to 5-methyl
tryptophan by the two cell lines selected in the present study are not
yet known, and it is possible that they contain mixtures of different
resistant cell types. Attempts to increase alkaloid production by
these cell lines would require characterization of the modes of
resistance and, possibly, further selection.

In the second selection technique frequently used in plant cell
culture, cells containing large quantities of a visible or readily
measured product (e.g. a pigment) have been selected and raised to
give high-producing cell lines. This approach has provided high
producing lines of e.g. nicotine-producing tobacco (Ogino <u>et al</u>. 1978)
and pigment-producing <u>E. millii</u> cells (Yamamoto & Mizuguchi 1982). The
alkaloids produced by <u>C. roseus</u> autofluoresce under ultraviolet light
and are therefore useful markers for similar selection studies.
Attempts were made to select small cell aggregates with high (or low)
autofluorescence in ultraviolet light at follows. A suspension of
cell aggregates (<530um) was spread over a thin (1mm) agar layer in a
petri dish mounted on an inverse microscope. Selected cell aggregates
were picked out by the capillary action of a 10ul capillary tube,
placed immediately onto conditioned agar and incubated in the light
for 20 days. After this period growing cell aggregates were
transferred to fresh agar medium and then sequentially into 10ml, 50ml
and 100ml liquid medium.

Table 3.   <u>Alkaloid production by manually-selected cell
lines</u>. Figures show the maximum observed values for
alkaloid production in 9 manually selected cell lines when
raised to produce suspension cultures of up to 100 ml.

Cell line	Serpentine produced mg g^{-1} dry wt.	Ajmalicine produced mg g^{-1} dry wt.
1D	1.86	1.76
6A	0.04	0
1C,4C,2D,6C(1),6C(2),24B	0	0
2B	0.02	0
AV16 (parent line)	1.0	0.1

Over 100 cell-aggregates of less than 45 cells each were selected
manually for high or low autofluorescence and their characteristics
noted. Many of the selected cell-aggregates failed to grow at the
early stages and only 9 were eventually raised to produce cell
suspensions of up to 100ml. (Table 2). When these cell lines were
analysed for alkaloid production only three were found to produce
alkaloids and only one (ID) produced more serpentine and ajmalicine
than the parent line (Table 3). Line ID was maintained on agar medium
and continued to produce at least 1.5mg alkaloids g^{-1} dry wt when
returned to suspension. The parent line however was maintained in
suspension and after one year produced only negligible quantities of
alkaloids.

Manual selection was restricted by the poor growth of the selected
cell-aggregates. Very small cell-aggregates (less than 45 cells each,
<200 μm diameter) were selected in an attempt to increase the homo-
geneity of the culture but many of the small cell-aggregates failed to
grow on conditioned medium. The failure of the cells to grow was most
likely due to a combination of 1) exposure to ultraviolet light during
selection, 2) the poor growth of plant cells at low cell density in
vitro and 3) the possibility that fluorescent cells were moribund.
Recently a report of an eight year selection study on C. roseus
demonstrated increased alkaloid yields after selection of larger (2mm)
fluorescent, growing, cell-aggregates (Deus-Neumann & Zenk 1984). The
increased alkaloid yields were however unstable and repeated selection
was necessary for their maintenance.

ACKNOWLEDGEMENT
          This work was supported by a grant from the Ministry of
Agriculture, Fisheries and Food.

REFERENCES
Berlin, J., Kukoschoke, K.G., & Knobloch, K.H. (1981). Selection of
          tobacco cell lines with high yields of cinnamoyl
          putrescines. Planta Medica. 42, 167-72.
Carlson, J.E. & Widholm, J.M. (1978). Separation of two forms of
          anthranilate synthetase from 5-methytryphophan susceptible
          and resistant cultured Solanum tuberosum cells. Physiol.
          Plant. 44, 251-55.
Deus-Neumann, B. & Zenk, M.H. (1984). Instability of indole alkaloid
          production in Catharanthus roseus cell suspension
          cultures. Planta Medica. 50, 427-31.
Gathercole, R.W.E. & Street, H.E. (1978). A p-fluorophenylalanine-
          resistant cell line of Sycamore with increased contents of
          phenylalanine, tyrosine and phenolics. Z. Pflanzenphysiol.
          89, 283-87.
Ogino, T., Hiraoka, N. & Tabata, M. (1978). Selection of high
          nicotine-producing cell lines of tobacco callus by single
          cell cloning. Phytochem. 17, 1907-10
Ranch, J.P., Rick, S., Brotherton, J.E. & Widholm, J.M. (1983).
          Expression of 5-methyltryptophan resistance in plants
          regenerated from resistant cell lines of Datura innoxia.
          Plant Physiol. 71, 136-40

Scott, A.I., Mizukami, H. & Lee Sin-Leung (1979). Characterization of
        a 5-methyltryptophan resistant strain of Catharanthus
        roseus cultured cells. Phytochem. 18, 795-98.
Stafford, A., Smith, L. & Fowler, M. W. (1985). Regulation of product
        synthesis in cell cultures of Catharanthus roseus (L) G.
        Don. Plant Cell Tiss. & Org. Culture 4 83-94.
Svoboda, G.H. (1964). The current status of Catharanthus roseus
        (Vinca rosea) research.   Llyodia 27, 275-76
Widholm, J.M. (1974). Cultured carrot cell mutants:
        5-methyltryptophan-resistant trait carried from cell to
        plant and back. Plant Sci. Lett. 3, 323-30
Yamamoto, Y. & Mizuguchi, R. (1982). Selection of a high and stable
        pigment-producing strain in cultured Euphorbia millii
        cells. Theor. Appl. Genet. 61 113-16

31. VARIABILITY AND STABILITY OF CELL CULTURES OF CAPSICUM
FRUTESCENS

P.R. Holden, M. Aitken, K. Lindsey, M.M. Yeoman.

Department of Botany,
University of Edinburgh,
The King's Buildings,
Mayfield Road,
Edinburgh. EH9 3JH   U.K.

INTRODUCTION
        Temporal and spatial variation in morphology and
metabolism within plant cell cultures creates difficulties when trying
to maintain cells in vitro for experimental purposes. Progress in the
development and exploitation of many advanced in vitro techniques is
often prevented by the inability to overcome problems arising in
apparently basic procedures such as maintenance of growth or
regeneration of plants from manipulated cells. It is a well known
fact that with increasing time in culture cells exhibit gross changes
in their karyotype as well as a reduction or loss of their
biosynthetic or regenerative potentials (Skirvin, 1978; Thomas et al,
1979).

Direct genetic changes are very common in cultured plant cells and
there are many well documented examples of increased ploidy levels
(Bennici et al, 1968; Renfroe & Berlyn, 1985) as well as structural
rearrangements of chromosomes (Mitra & Stewart, 1961; Bayliss, 1980).
Likewise the incidence of epigenetic change is very high although not
so readily recognised as chromosomal modifications (Meins & Binns,
1977; Meins, 1983). The reduction or loss in the capacity of the
cells to produce secondary metabolites while in culture is also widely
reported (Tabata & Hiraoka, 1976; Deus-Neumann & Zenk, 1984). This
instability has prevented progress in the commercial production of
useful compounds by mass cell culture (Scragg & Fowler, 1985).
Similarly the gradual loss in regenerative potential with time in
tissue culture has caused problems with genetic manipulation (Thomas
et al., 1979; Dale, 1983) and germplasm storage (Withers, 1983).

The variation expressed in plant cell cultures, though considered to
be detrimental to the stability and homogeneity of the cultures, can
be exploited as a source of genetic diversity. If cultured cells can
be induced to differentiate back to the whole plant then variation
which has arisen in culture can be manifested in the regenerated
plant. This phenomenon, termed somaclonal variation (Larkin &
Scowcroft, 1981) can give rise to regenerants with new and potentially
'useful' characteristics and therefore can be considered to be an
important technique in plant improvement (Heinz et al, 1977; Secor and
Shepherd, 1981). With these facts in mind we have set out to find
some of the causes and effects of variation within cultured cells and

regenerated plants of Capsicum frutescens Mill. Cv. annuum, the chilli
pepper.

Clonal lines of Capsicum frutescens were established in order to
examine interclonal variability and intraclonal stability with respect
to various aspects of growth, structure and metabolic activity,
namely, degree of friability, chlorophyll, protein, soluble phenolics,
capsaicin and DNA contents. Comparisons were also made between these
lines under different experimental regimes.

ISOLATION AND MAINTENANCE OF CLONES

Single cells were isolated from a single suspension
culture from which three hundred callus cultures (clones) were
established (Fig. 1). They were maintained, with a subculture every 4
weeks, on Schenk and Hildebrandt growth medium (Schenk & Hildebrandt,
1972) containing 0.5 mg l^{-1} 2,4-D, 2.0 mg l^{-1} pCPA, 0.1 mg l^{-1}
kinetin, and 30 g l^{-1} sucrose. Investigations of the clones were made
while maintained as callus, in suspension, as immobilised cells and as
regenerated plants (Fig. 1)

Procedure for isolation of single cells   A cell plating technique was
used to isolate single cells using friable callus from a single
suspension culture. Aggregates were sieved through a series of mesh
filters down to 100uM in size, then centrifuged and diluted with
conditioned medium to a volume of approx. 10^3 cells/ml^{-1}. They were
then mixed with an equal volume of fresh medium and 1% low temperature
agarose and plated out on a gridded Petri dish. The plated cells were
examined microscopically for single cells which were marked so that
their growth could be followed. When colonies were approx. 1mm in
diameter they were pricked out and placed on 50% conditioned medium
for one growth period and then transferred to normal SH medium. About
300 cell lines were established this way.

Figure 1.   Isolation and maintenance of clones of
Capsicum frutescens and their manipulation under differing
environmental conditions.

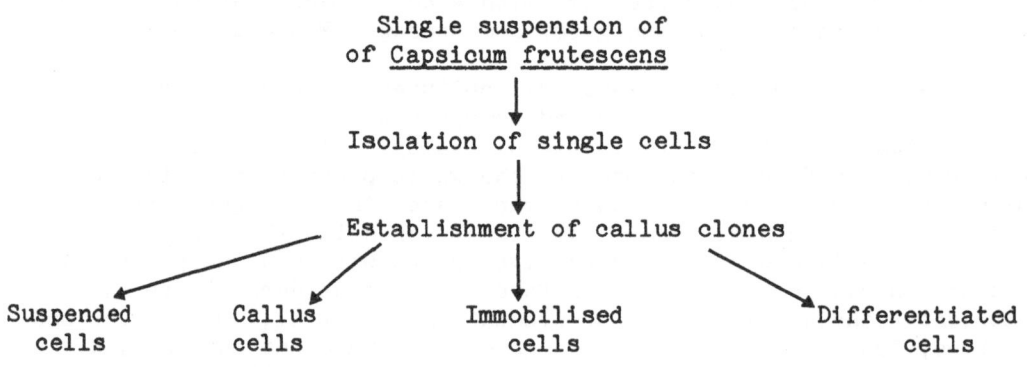

VARIABILITY BETWEEN CLONES

<u>Morphological differences</u> Morphological differences are apparent as shown by the variation in the growth rates and friability of the clones (Table 1). Such differences affect the ability of the clones to immobilise in reticulate polyurethane. It can be seen that clones which are friable and contain little pigment have considerably higher growth rates (Table 1).

<u>Metabolic differences</u> Variation between clones is apparent in the chlorophyll and capsaicin contents (Table 1). Chlorophyll contents vary considerably and the most friable cultures tend to accumulate the lowest levels of chlorophyll (Table 1).

<u>Table 1</u>.   <u>Interclonal Variability</u> Appearance and growth index, chlorophyll and capsaicin contents of selected clones of <u>Capsicum</u> <u>frutescens</u>.

Clone	Appearance	Growth Index	Chlorophyll/ Culture (μg)	Capsaicin/ Suspension culture (μg)
BV8	Green/Compact	1.35	15.0	1.08
N4	Yellow/friable	2.47	1.4	ND
CB6	Pale/Compact friable	3.85	2.4	0.028
B24	Yellow/Compact friable	2.21	1.8	ND
G3	Pale/Compact friable	0.57	2.9	3.56
CD5	Pale/Compact	0.08	2.8	0.1
AG2	Yellow/Green compact	1.36	12.91	1.15
Q5	Pale/friable	2.28	ND	ND
BQ1	Yellow/friable	0.06	ND	0.19
AN2	Pale/friable	1.32	ND	ND

ND = Not Detectable

<u>Table 2</u>.   <u>Interclonal Variability</u> Apperance, growth index and capsaicin content from immobilised cultures of selected clones of <u>Capsicum</u> <u>frutescens</u>.

Clone	Appearance	Growth Index	Capsaicin/Immobilised Culture (μg)
AN2	Pale/friable	1.32	6.15
AE2	Yellow, green/ compact, friable	2.29	ND
F2	Yellow, green/ compact, friable	0.627	3.237
Z1	Pale green/friable	0.198	9.62
Y3	Pale green/friable	1.02	ND

$$\text{Growth Index} = \frac{\text{Final Weight} - \text{Initial Weight}}{\text{Initial Weight}}$$

Capsaicin production has been determined in suspended cell cultures (Table 1) and, in a small number of cases, in immobilised cell cultures (Table 2). Much variation is noticeable and in general cultures derived from callus with characteristically low growth rates tend to accumulate the highest yields (TABLES 1,2). In one instance immobilisation has resulted in the production of capsaicin, although the suspended clone failed to produce any detectable levels (AN2) and a second clone (Y3) was further found to begin production of dihydro-capsaicin on immobilisation. Each of the suspended clones examined was found to have a distinct profile of chloroform soluble compounds (mostly soluble phenolics) which accumulate in the liquid medium suggesting that each cell line has a unique metabolic finger print (Aitken & Yeoman 1986).

INSTABILITY OF THE CLONES

The clones show instability over a period of time as it can be seen that in some of the selected clones growth and appearance have changed over the 20 month maintenance period (Table 3). Clones which were initially slow growing with green compact growth habits were found to be stable in appearance and growth rate (Table 3). The initially faster growing clones, however, developed a reduced growth rate and became more compact and hard in structure (Table 3). It has been found that the morphologically more stable clones retained the capacity to produce capsaicin whereas the unstable ones accumulated increased level of phenolics but not capsaicin. There is also, in at least one case, evidence for the development of auxin habituation. Apart from an ability to grow in the absence of exogeneous auxin, cultures characteristically have an increased growth rate, increased friability, and decreased pigmentation and capsaicin contents.

TABLE 3. INTERCLONAL VARIABILITY   Appearance and growth index of selected clones of Capsicum frutescens taken on two consecutive dates.

Clone	August 1983 Appearance	Growth Index	March 1985 Appearance	Growth Index
BV8	Green/compact	1.35	Hard, nodular	0.96
N4	Yellow/friable	2.47	Brown, compact	1.14
AE2	Yellow/friable compact	2.18	Yellow, friable	2.05
B24	Yellow/friable	2.21	Brown/compact	1.11
N3	Yellow/friable	1.09	Green/compact	0.54
CB6	Pale/compact friable	3.85	Pale, Brown/compact	1.94
AG2	Yellow, green/friable	1.36	Dark/compact	0.44

VARIABILITY WITHIN CLONES

A fast growing and friable clonal line (AE2) was isolated to examine the amount of variation that occurs within a clone. An increasing number of isolates of the clone were maintained under the three environmental regimes and analysis of growth and metabolic activity was made. It is apparent that there is variation in growth, protein and chlorophyll content within the clone (Table 4). It was noticeable that there were differences between the environmental regimes especially in the chlorophyll content of cells on agar and in suspension (Table 4). Work is being carried out to look at variation within immobilised cell cultures with regard to production of capsaicin and soluble phenolics.

Table 4   Intraclonal Variability Growth index, protein and chlorophyll contents of isolates of Clone AE2 of Capsicum frutescens on agar and in suspension media.

AE2 Repli- cates	CALLUS			SUSPENSION		
	Growth Index	Chlorophyll (µg/g FW)	Protein (mg/g FW)	Growth Index	Chlorophyll (µg/g FW)	Protein (mg/g FW)
1	3.97	0	2.56	4.49	0	1.55
2	7.86	16.9	2.4	7.83	12.55	1.6
3	8.94	33.38	2.58	9.39	7.14	1.08
4	8.35	4.31	1.35	6.41	12.49	1.5
5	13.45	10.68	1.34	6.75	10.58	1.88
6	8.71	40.87	2.38	7.77	10.55	1.6
7	14.85	4.84	1.92	8.39	11.04	1.73
8	10.70	14.52	1.56	8.22	-	1.2
9	14.63	15.62	2.15	8.34	-	1.6

Table 5   Stages of differentiation and regeneration associated with different ages and positions of explants from Capsicum frutescens

Explant (age)	Degree of Regeneration
Callus (18 months)	+
Callus (8 months)	+
Callus (4 months)	++
Leaf Disc (2 month plant)	++
Stem (2 month plant)	+++
Cotyledon/Hypocotyl	++++

Key: +   Compaction: slowing of growth
       ++   Differentiation: greening
       +++   Rooting or shooting
       ++++ Rooting and shooting

CLONE REGENERATION: VARIABILITY AMONG REGENERANTS
        Work is in progress to develop a technique to regenerate the clones. It has been found that the period of time that cells are left in the 'undifferentiated' state is significant for successful

regeneration. A gradient of organogenic potential seems to exist with time in culture (Table 5). Therefore attempts to develop the necessary balance of growth regulators for regeneration has been prevented by the unstable state of the callus clones although aided by the use of physiologically younger hypocotyl and cotyledon material (Table 5). Work is being carried out to look at the stage at which the callus clones 'lose' their capacity to regenerate.

CONCLUSIONS AND FUTURE PROSPECTS

Preliminary results indicate that variability in morphological and biochemical markers is apparent between clones derived from a single suspension culture and that instability in these markers occurs over successive subcultures, particularly in initially rapidly-growing clones. The reasons for the variability and instability are not openly apparent and require further investigation. One approach being developed is to analyse changes in karyotype by an estimation of cellular DNA contents using fluorescent cell sorting techniques. A comprehensive examination of the effects of the environment - changes in pH, light, temperature - will be made to create the most favourable and stable conditions for growth. Instability does not necessarily mean a loss but rather a change in developmental potential and it is through an understanding of the nature of this change which will help make growth in vitro more stable and the manipulations of cultured tissues more accessible.

REFERENCES

Aitken, M.M. & Yeoman, M.M. (1986). The use of ELISA in cell line selection in tissue cultures of Capsicum frutescens. This volume 29.

Bayliss, M.W. (1980). Chromosomal variation in plant tissue in culture. Int. Rev. Cytol. Supply. 11A, 113-144.

Bennici, A., Buiatti, M. & D'Amato, F. (1968). Nuclear conditions in haploid Pelargonium in vitro and in vivo. Chromosoma 24, 194-201.

Dale, P.J. (1983). Protoplast culture and plant regeneration of cereals and other recalcitrant crops. Experientia Suppl. 46, 31-41.

Deus-Neumann, B. & Zenk, M.H. (1984). Instability of indole alkaloid production in Catharanthus roseus cell suspension cultures. Planta Medica 50, 427-431.

Heinz, D.J., Krishnamurthi, M., Nickell, L.G. & Maretzki, A. (1977). Cell tissue and organ culture in sugarcane improvement. In Applied and Fundamental Aspects of Plant Cell, Tissue and Organ Culture, ed. J. Reinert and Y.P.S. Bajaj, pp. 3-17. Berlin and New York: Springer-Verlag.

Larkin, P.J. & Scowcroft, W.R. (1981). Somaclonal variation - a novel source of variability from cell cultures for plant improvement. Theor. Appl. Genet. 60, 197-214.

Lindsey, K. & Yeoman, M.M. (1985). Immobilised plant cell culture systems. In Primary and Secondary Metabolism of Plant Cell Cultures, ed. K.H. Neumann, W. Barz & E. Reinhard, pp. 304-315. Berlin Heidelberg: Springer-Verlag.

Meins, F. & Binns, A.N. (1977). Epigenetic variation of cultured
    somatic cells: Evidence for gradual changes in the
    requirement for factors promoting cell division. Proc.
    Natl. Acad. Sci. U.S.A. 74, 2928-2932.
Meins, F. (1983). Heritable variation in plant cell culture. Ann.
    Rev. Plant Physiol. 34, 327-346.
Mitra, J. & Steward, F.C. (1961). Growth induction in cultures of
    Haplopappus gracilis cells. IV. The behaviour of the
    nucleus. Am. J. Bot. 47, 358-68.
Renfroe, M.H. & Berlyn, G.P. (1985). Variation in nuclear DNA content
    in Pinus taeda L. Tissue cultures of diploid origin. J.
    Plant Physiol. 121, 131-39.
Scragg, A.H. & Fowler, M.W. (1985). The mass culture of plant cells.
    In Cell Culture and Somatic Cell Genetics of Plants, Vol.
    2, ed. I.K. Vasil, pp. 103-128. London: Academic Press.
Secor, G.A. & Shepard, J.F. (1981). Variability of protoplast-
    derived potato clones. Crop Sci. 21, 102-105.
Schenk, R.U. & Hildebrandt, A.C. (1972). Medium and techniques for
    induction and growth of monocotyledonous and
    dicotyledonous plant cell cultures. Can. J. Bot. 50,
    199-204.
Skirvin, R.M. (1978). Natural and induced variation in tissue
    culture. Euphytica 27, 241-266.
Tabata, M. & Hiraoka, N. (1976). Variation of alkaloid production in
    Nicotiana rustica callus cultures. Physiol. Plant. 38,
    19-23.
Thomas, E., King, P.J. & Potrykus, I. (1979). Improvement of crop
    plants via single cells in vitro - an assessment. Z.
    Pflanzenzüchtg. 82, 1-30.
Withers, L.A. (1983). Germplasm storage in plant biotechnology. In
    Plant Biotechnology, SEB Seminar Series, 18 eds. S.H.
    Mantell & H. Smith, pp. 187-218. London and New York:
    Cambridge Univ. Press.

32. SELECTION FOR RESISTANCE TO LATE BLIGHT IN CELERY

A. Donovan, H. Collin & S. Isaac.
Department of Botany,
University of Liverpool,
Liverpool   L69 3BX. U.K.

INTRODUCTION
The major disease of celery is late blight caused by the
Ascomycete Septoria apiicola. This disease results in black leaf and
stem lesions of up to 1 cm in diameter and ultimately in plant death.
Although some varieties show partial resistance to this disease none
are immune (Bohme, 1960). Little is known about the actual mechanism
of resistance but it may be related to the presence of the flavour
compounds, phthalides and terpenoids, since the more strongly
flavoured varieties are more resistant (Cochran, 1932; Sheridan, 1964
& Pink-pers.comm).

Despite a long period of breeding using conventional intact plant
selection techniques, it has not been possible to breed for resistance
in the self-blanching varieties. Evidence from a wide range of crops
suggests that useful variation can arise during a short tissue culture
stage (Larkin & Scowcroft, 1981). A proportion of regenerated plants
show a greater level of resistance to a range of diseases than the
parent plants (Chaleff, 1983).

The approach to selection for increased resistance to late blight in
celery was two fold. It was necessary, firstly to examine the
mechanism of resistance in intact celery plants in order to clarify
the role of greening and flavour production and secondly, to initiate
tissue cultures of a commercial variety to induce somaclonal
variation.

MECHANISM OF RESISTANCE TO LATE BLIGHT
Earlier evidence suggested a correlation between
resistance and concentration of flavour compounds in the essential oil
(phthalides and terpenoids) but no quantitative information was
available to support these claims. To examine this relationship,
resistance and composition of the essential oil were compared in two
contrasting varieties of celery, a mild flavoured self-blanching
variety, Celebrity, and a strongly flavoured green variety, Cutting.
Resistance was measured using a leaf inoculation test. Fully
developed leaves were removed from plants (20 weeks old) placed on wet
filter paper in Petri dishes and inoculated with a spore suspension (1
x 10^6. ml^{-1}) of S. apiicola. Development of lesions with time is
shown in Fig. 1. Results indicated that leaves of the strongly
flavoured Cutting variety were more resistant to infection than leaves

of the self-blanching mildly flavoured Celebrity. Chlorophyll content
of the celery from plants (20 weeks old), was determined by Arnon's
method (Arnon, 1949). Cutting contained 1.406 ± 0.027 and Celebrity
contained 0.718 ± 0.015 mg chlorophyll g.fresh weight^{-1}.

Flavour compounds were extracted from celery using the method
described by Collin & Watts (1983) and analysed by gas liquid chromatography. Traces obtained from extracts of Cutting and Celebrity are
shown in Fig. 2. These results showed a difference in concentration
in some components of the essential oil. The most obvious difference
was that β-pinene, γ-terpinine and β-eudesmol are at a significantly
higher concentration in Cutting. Limonene, β-pinene and γ-terpinine
were major components in both varieties. When tested at physiological
concentrations, the complete extract and also the separate components,
limonene, β-pinene and γ-terpinine were found to inhibit spore
germination and hyphal growth of S. apiicola. The specific inhibitory
effects of components of the essential oil on the growth of S.
apiicola requires further investigation.

Fig. 1. Leaf bioassay showing differences in resistance
to Septoria apiicola of celery varieties: Cutting (▲)
and Celebrity (■).

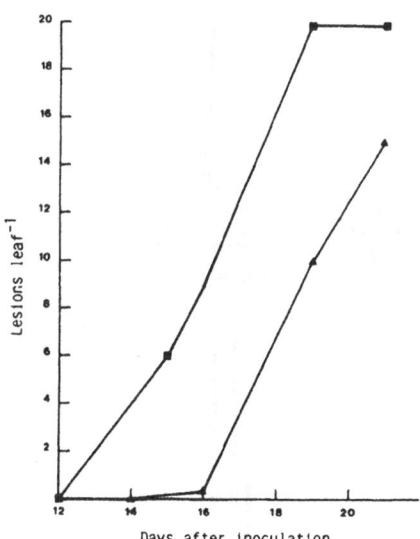

Days after inoculation

INITIATION OF SOMACLONAL VARIATION

Somaclonal variation can be generated by exposing parent material to a short period in tissue culture and subsequently regenerating the tissue culture back to plants. In order to generate such variation in celery, callus cultures were initiated from a total of 2,250 seedling petiole explants, each 1 cm long, with four explants taken per seedling. The explants were obtained from the widely used self-blanching variety Celebrity, which is also susceptible to late blight. These were incubated on a callus induction medium containing Murashige & Skoog medium, 3% sucrose, 0.5 mg.l^{-1} 2,4-dichlorophenoxy acetic acid (2,4-D) and 0.6 mg.l^{-1} kinetin (Williams & Collin, 1976). An assessment was made of the colour, size

Fig. 2. GC trace of extracts from celery varieties.
Cutting (a) and Celebrity (b).
3 µl samples injected on to a 9ft x 4mm internal diameter glass column. Packing: SP2100 (10%) on 60-100 mesh Chromasorb W. Temperature programme: 60-250°C at 5°C. min^{-1}, Flow rate: 30 cm nitrogen. min^{-1}. Preliminary identification: 1 = β pinene, 2 = limonene, 3 = γ-terpinene, 4 = caryophyllene, 5 = sesquiterpenes, 6 = 3 butyl phthalide, 7 = β-eudesomol, 8 = sedanenolide.

(a)

(b)

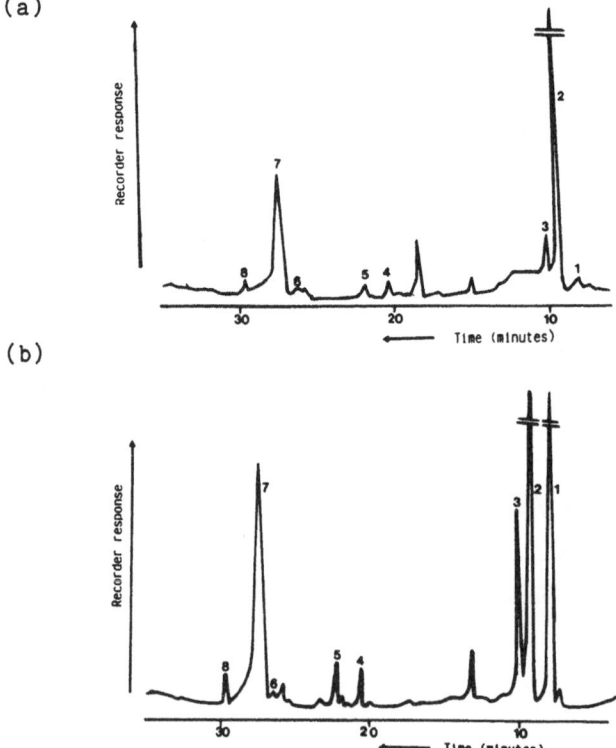

Fig. 3.   Variation of petiole explants in tissue cultures
of Celebrity: (a) size of callus, (b) differentiation of
callus, (c) plantlet formation.

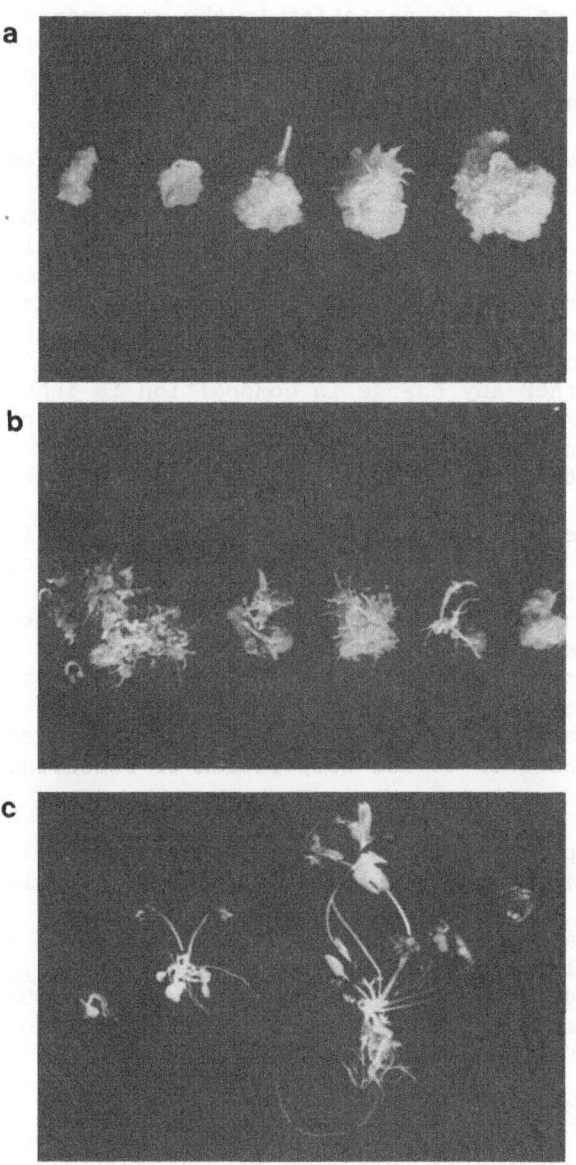

and differentiation of the callus from the same and from different
seedlings. After one subculture the explants and callus were
transferred to a medium with 2,4-D and kinetin omitted. Plantlets
have been regenerated from the callus cultures and the plants
eventually transferred to soil in individual pots in the greenhouse.
There was a wide variation in size of the callus (Fig. 3a) and
differentiation into roots and shoots (Fig. 3b) during the first
subculture. There was also a wide variation in the appearance of
shoots, roots and embryos and in the frequency of these structures on
each callus (Fig. 3c). Regenerated plantlets continued to show
variation in morphology after planting out.

The next step in the selection programme is to identify the resistant
individuals amongst the regenerated plants and their progeny. A rapid
screening method is a major advantage where large numbers of plants
are being screened. The leaf bioassay outlined earlier fulfills such
a role. Having identified resistant plants it is important to
establish the mechanism of resistance. GC analysis of selected
regenerated plants and progeny will show whether the resistant plants
have the same composition of phthalides and terpenoids as resistant
celery varieties and whether any change in composition is inherited.
By relating change in resistance to callus size, shape, colour and
degree of differentiation it may also be possible to identify
potentially resistant material very early in the tissue culture stage.

Differences in resistance to S. apiicola in commercial varieties have
now been correlated with greening and amount and composition of the
essential oil. The growth of S. apiicola has also been shown to be
sensitive to physiological concentrations of the major components of
the essential oil such as limonene, β-pinene and γ-terpinine. The
tissue cultures and regenerated plantlets demonstrate a wide variation
in morphology. Using the leaf bioassay and the GC analysis,
regenerated plants and particularly their progeny can now be screened
for resistance and this related to the composition of essential oils.

REFERENCES
Arnon, D.I. (1949). Copper enzymes in isolated chloroplasts
          polyphenoloxidase in Beta vulgaris. Plant Physiol. 24,
          1-15.
Bohme, H. (1960). The causes of the different resistances of two
          different celeraic varieties to leaf spot (Septoria
          apii-graveolentis Dorogin). Phytopath. Z. 37, 195-213.
Chaleff, R.S. (1983). Isolation of agronomically useful mutants from
          plant cell cultures. Science, 219, 676-82.
Cochran, L.C. (1932). A study of two Septoria leaf spots of celery.
          Phytopath. 22, 791-812.
Collin, H.A. & Watts, M. (1983). Flavour production in culture. In :
          Handbook of plant cell culture. Vol. 1. Ed. Evans, D.A.,
          Sharpe, W.R., Ammirato, P.V., & Yamada, Y. MacMillan
          Press.

Larkin, P.J. & Scowcroft, W.R. (1981). Somaclonal variation - A novel
        source of variability from cell cultures for plant
        improvement. T. Appl. Gen. 60, 197-214.
Sheridan, J.E. (1964). Septoria blight of celery. Ph.D. Thesis,
        Imperial College, London.
Williams, L. & Collin, H.A. (1976). Embryogenesis and plantlet
        formation in tissue cultures of celery. Ann. Bot. 40,
        325-32.

33.   EFFECTS OF MODIFICATION OF THE PRIMARY PRECURSOR LEVEL BY
<u>SELECTION AND FEEDING ON INDOLE ALKALOID ACCUMULATION IN</u>
<u>CATHARANTHUS ROSEUS</u>

A. Stafford & L. Smith

Wolfson Institute of Biotechnology,
University of Sheffield,
Sheffield S10 2TN. U.K.

INTRODUCTION
The effect of increasing intracellular levels of primary
precursors on subsequent indole alkaloid production in cell cultures
of <u>Catharanthus roseus</u> has previously been investigated in two ways;
in the first place, by supplying precursors to the medium, and
secondly by producing variant cell lines, resistant to normally
inhibitory doses of precursor analogues.

In <u>C. roseus</u> the primary precursors which have received most attention
in this respect are tryptophan, tryptamine and secologanin, the condensation product of the last two compounds being the first indole
alkaloid on the pathway leading to both monomeric and dimeric
alkaloids. Feeding of tryptophan to cell cultures has produced con-
flicting results. Zenk <u>et al</u>. (1977) found that while precursors of
tryptophan did not effect an increase in alkaloid yield, L-tryptophan
itself supplied at 500 mg l^{-1} medium produced a three-fold stimulation
of the serpentine level. The addition of loganin at 250 mg l^{-1} bought
about a two-fold increase in serpentine level whereas secologanin had
no effect. Doller (1978) found that the addition of tryptophan or
tryptamine brought about inhibition of both growth and serpentine
accumulation.

Attempts to produce precursor-analogue resistant variants have
resulted in cell lines producing 30 times the normal level of intracellular tryptophan (Schallenberg & Berlin, 1979; Scott <u>et al</u>., 1979);
but only one report has been made of selecting a cell line with
enhanced alkaloid levels using this method (Deus-Neumann & Zenk,
1984). These authors combined a selection for 5-methyl tryptophan
(5MT) resistance with a screen for high autofluorescent properties.

The work presented here summarises an attempt made to select high
serpentine-producing cell lines of <u>C. roseus</u> on the basis of 5MT
resistance, and results obtained from feeding these cell lines with
both tryptophan and the monoterpenoid precursor secologanin.
Serpentine was the major alkaloid produced in source and selected
cultures.

ANALYTICAL METHODS
Alkaloid analyses were performed using the method of P.
Morris as described in Stafford <u>et al</u>. (1985).

Tryptophan analyses were performed using a Locarte bench model amino-
acid analyser equipped with a Motorola M6800 datalogging integration
system.   Samples were loaded onto a 9 x 250mm column of sulphonated
polystyrene resin beads in 0.1N HCl.   Elution was with citrate buffer
0.2M pH 3.3/30 min; citrate buffer 0.2M pH 4.05/50 min; citrate-borate
buffer 0.2M pH 9.55.   Tryptophan eluted at 127 min in citrate-borate
buffer with a flow rate of 60 ml min^{-1}.   0.2M NaOH/20 min and citrate
buffer 0.2M pH 3.3/50min were used to regenerate the column. Detection
was made using ninhydrin reagent at 30 ml min^{-1} and the absorbance was
recorded at 570nm and 440nm.   In the feeding experiments described,
tryptophan was determined using the HPLC alkaloid method described
above.   Where possible, direct comparisons were made between amino-
acid analyser and HPLC values for tryptophan in the same samples, and
found to be similar.

SELECTION OF 5-METHYL TRYPTOPHAN-RESISTANT CELL LINES
Selection for analogue-resistance was made on 2
"production" media: (a) MZ (Zenk et al., 1977) and (b) M3 (Murashige &
Skoog containing naphthaleneacetic acid at 1 mg l^{-1} and kinetin at 0.1
mg l^{-1}; Fowler, 1983).

FIGURE 1.   Effect of 5-MT on growth of C. roseus cultured
on maintenance medium and production medium.

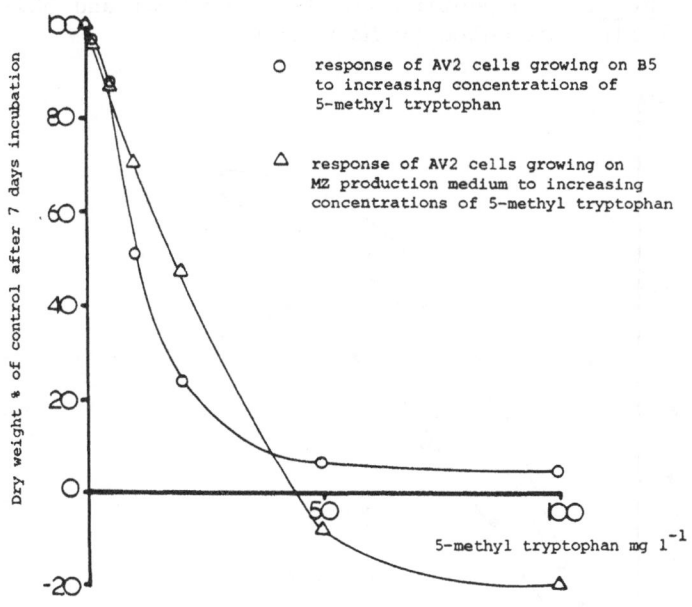

The effects of a range of 5MT concentrations on the growth of C. roseus were first determined (Figure 1). The growth response of one cell line designated AV2 was assessed on B5, a maintenance medium (Gamborg et al., 1968) supporting cell division but no alkaloid synthesis, and on MZ, as the possibility that different media could confer altered cellular resistance to the analogue could not be discounted. From the data obtained, 60 mg l^{-1} 5MT was selected as a standard inhibitory dose. Similar dose-response curves were obtained using other cell lines.

Table 1   Tryptophan levels in controls and in cell lines selected on MZ medium.

Day	tryptophan μg g^{-1} dry weight cells		
	Control	MV14/5MTR(2)	MV14/5MTR(3)
0	40.85	426.16 $\pm$ 269.17	616.09 $\pm$ 46.83
8	-	505.13 $\pm$ 133.38	243.72 $\pm$ 83.29
15	23.83 $\pm$ 7.98	603.16 $\pm$ 73.09	243.03 $\pm$ 68.49
43	33.70 $\pm$ 1.44	329.83 $\pm$ 79.42	-

FIGURE 2.   Serpentine yields in control and 5MT-resistant cell lines selected on MZ medium.

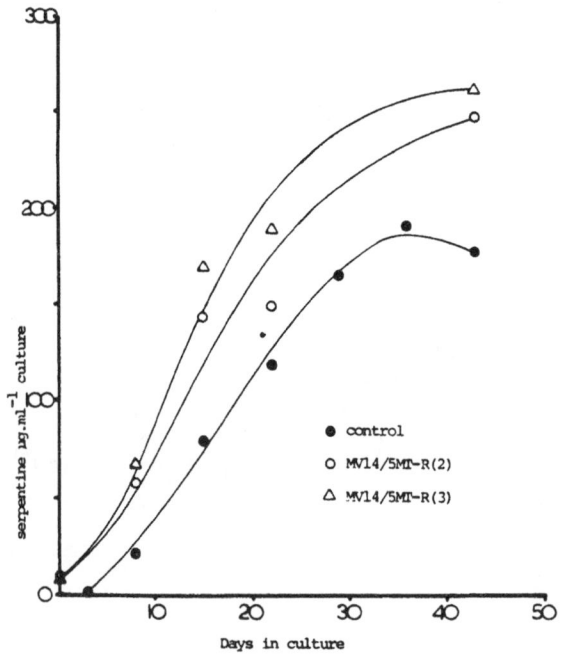

All selections were made in suspension cultures and no attempt was
made to derive single-cell clones. M3 medium is able to support both
growth and alkaloid production in C. roseus, so selection was performed simply by maintaining M3 suspensions in the presence of 5-
methyl tryptophan. MZ medium supports biomass increase and alkaloid
production but not cell division. Therefore, in order to undertake a
selection for analogue-resistance in this medium, a standard procedure
was adopted in which suspensions were cultured for 21 days on MZ and
analogue, then for two consecutive growth cycles on B5 medium without
analogue, to allow division of the selected resistant cells. Two
resistant cell lines were selected on MZ medium from a single source
culture, and designated MV14/5MTR(2) and MV14/5MTR(3). One resistant
cell line was selected on M3 medium, and designated MV57. Viabilities
and growth rates were stable in all selected cell lines four months
after initial exposure to 5-methyl tryptophan.

INTRACELLULAR TRYPTOPHAN AND SERPENTINE LEVELS IN CELL LINES RESISTANT TO 5MT

Intracellular tryptophan levels were enhanced
approximately 20 fold in MV14/5MTR(2) and 7 fold in MV14/5MTR(3) in 15
day old cultures grown on MZ + 5MT (Table 1). Growth rates and final

Table 2   Percentages of autofluorescing cells in MZ-grown cultures of control and 5-methyl tryptophan resistant C. roseus suspensions.

| Day after | | % blue autofluorescent cells | |
subculture	Control	MV14(2)5MT	MV14(3)5MT
0	none	c.50% +	c.50% +
15	50% ++	50% ++	50% ++
22	50% +++	50% +++	50% +++
43	1% +	90% +++	70-80% +++

Table 3   Growth and alkaloid productionon M3 medium of C. roseus cell lines AV11 MV57

Day after subculture	Dry weight mg ml^{-1} culture		Serpentine µg ml^{-1} culture		Autofluores-cence % under uv	
	AV11	MV57	AV11	MV57	AV11	MV57
0	2.23±0.20	1.97±0.09	6.53	0.46	30	5-10
14	10.36±1.84	9.9 ±0.16	65.40±8.49	1.86±0.33	40	1
29	5.79±0.20	4.49±0.20	21.80±1.16	0.59±0.26	40	1

biomass yields of both resistant lines were slightly depressed
compared with those of controls. However the serpentine yields
expressed in ug ml^{-1} culture were high relative to control values
(Figure 2). These serpentine levels were reflected in the autofluorescent properties of these cell lines when visualised under
short-wave UV light (Table 2). It is possible that our results
reflect an adaptation to MZ medium in addition to an acquired
resistance to 5MT.

While having a comparable growth profile to that of the control cell
line, the resistant cell line MV57 selected on M3 medium accumulated
much lower levels of serpentine (see Table 3).

FIGURE 3.   Effects of feeding indole alkaloid precursors
to M3-growth cell suspension of C. roseus.

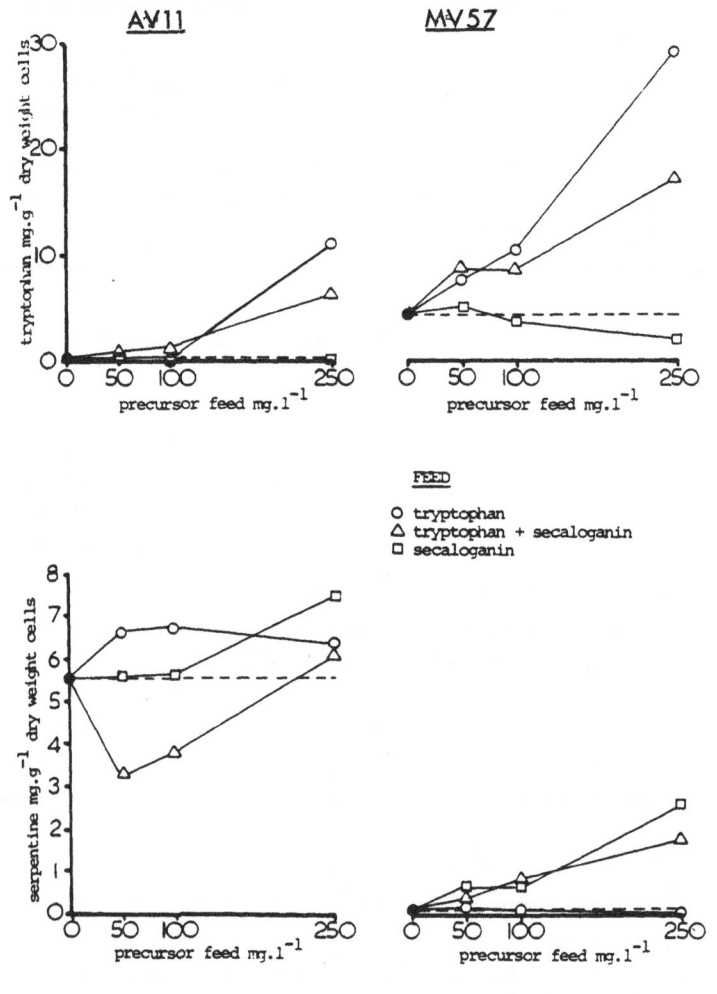

FEEDING OF TRYPTOPHAN AND SECALOGANIN TO RESISTANT AND CONTROL CELL LINES

Precursor additions were made on the first day of sub-culture, and cells harvested after 7 days. In all cases, the addition of secaloganin alone produced a reduction in the intracellular tryptophan level and an increase in the level of serpentine indicating a possible stimulation of the condensation reaction to strictosidine. The addition of tryptophan alone to both resistant cell lines (MV57 and MV14/5MT(2)) resulted in almost complete suppression of serpentine accumulation. Figure 3 shows the response of the M3 cell lines to feeding. Trends were similar for MZ cell lines.

CONCLUSIONS

1. Selection for cell lines with elevated intracellular tryptophan levels will not necessarily result in high alkaloid cultures. However, the high tryptophan cell lines described here may be mixtures of several cell types which could be separated by single cell cloning. Selection for 5MT resistance in suspension cultures could form the basis for a selection programme for high alkaloid yielding cell lines, and has in fact been exploited with apparent success by Deus-Neumann and Zenk (1984).

2. Feeding of precursors can have a positive effect upon alkaloid level in some cell lines, e.g. in MV57 where a 37 fold increase in serpentine level was produced by a secaloganin feed. This indicates a possible limitation on alkaloid yield at the point of secaloganin synthesis in these cell lines. An investigation of the effect of secaloganin precursors on alkaloid yield could help to pinpoint a bottleneck on this pathway.

3. Primary precursor supply is not necessarily a major limiting factor with respect to alkaloid biosynthesis in C. roseus cell cultures. Other mechanisms which may operate are: (a) restrictions on the flux of intermediates at a later stage in the pathway. Noe & Berlin (1985) have suggested that the conversion of tryptophan to tryptamine by the decarboxylase enzyme may be a rate-limiting step in the indole alkaloid pathway. (b) Rapid turnover of alkaloid "end-products." If a significant proportion of cells lack the correct physiological status for the accumulation of alkaloids in the vacuole, then alternative fates for synthesized alkaloids are possible. The first is secretion of alkaloids to the medium, a phenomenon claimed for C. roseus only in immobilised cells (Rosevear and Lambe, 1983). A second alternative is the intracellular degradation of the alkaloids in the cytoplasm before they reach cytotoxic levels. (c) High demand by other biosynthetic pathways for precursor compounds, or "end-product" repression by the indole alkaloid or other pathways.

ACKNOWLEDGEMENT

My thanks go to Mrs. H. Woodhead and Mr N. Rhodes for performing product analyses.

REFERENCES

Deus-Neumann, B. & Zenk, M.H. (1984). Instability of indole alkaloid production in C. roseus cell suspension cultures. Planta Medica 50, 427-431.

Doller, G. (1978). Influence of the medium on the production of serpentine by suspension cultures of C. roseus (L) G. Don. In: Production of Natural Compounds by Cell Culture Methods. ed. Alfermann, W. & Reinhard, E. Gesellschaft fur Strahlen und Unweltforschung mbH pp.109-117. Munchen.

Fowler, M.W. (1983). Commercial applications and economic aspects of mass plant cell culture. In: Plant Biotechnology. ed. Mantell, S.H. & Smith, H. pp.3-37. Cambridge University Press.

Gamborg, O.L., Miller, R.A. & Ojima, K. (1968). Nutrient requirements of suspension cultures of soybean root cells. Exp. Cell Res. 50; 155-158.

Noe, W. & Berlin, J. (1985). Induction of de novo synthesis of tryptophan decarboxylase in cell suspensions of C. roseus. Planta 166, 500-504.

Rosevear, A. & Lambe, C.A. (1983). Immobilised plant and animal cells. Top. Enz. Ferment. Biotechnol. 7, 13-37.

Schallenberg, J. & Berlin, J. (1979). 5-methyl tryptophan resistant cells of C. roseus. Z. Naturforsch. 34c, 541-545.

Scott, A.I., Mizukami, H. & Leo, S-L. (1979). Characterisation of a 5-methyl tryptophan resistant strain of C. roseus cultured. cells.

Stafford, A., Smith, L. & Fowler, M.W. (1985). Regulation of product synthesis in cell cultures of C. roseus (L) G. Don. Plant Cell Tissue Organ Culture 4, 83-94.

Zenk, M.H., El-Shagi, H., Arens, H., Stockigt, J., Weiler, E.W. & Deus, B. (1977). Formation of the indole alkaloids serpentine and ajmalicine in cell suspension cultures of C. roseus. In: Plant Tissue Culture and its Biotechnological Applications. ed. Barz, W., Reinhard, E. & Zenk, M.H. Berlin, Springer-Verlag.

34. LONG TERM STABILITY OF ALKALOID PRODUCTIVITY IN CELL SUSPENSION CULTURES OF CATHARANTHUS ROSEUS.

P. Morris

Wolfson Institute of Biotechnology,
University of Sheffield,
Sheffield.
S10 2TN. U.K.

INTRODUCTION
Over the past ten years the levels of some secondary metabolites found in plant cell cultures has increased substantially, mainly as a result of optimization of culture conditions for product biosynthesis and by selection of high yielding cell lines (see Yamada & Hashimoto, 1984). Long term maintenance of product accumulation in these high yielding cultures is now essential for their exploitation (Deus-Neumann & Zenk, 1984).

While non-selective subculture has generally been found to cause a decrease in product accumulation for certain metabolites (Dougall et al, 1980), some stable cell lines have been obtained without selection (Zenk et al, 1975, 1977a), although in many cases only the stability of a class of compound has been investigated. Stable cell lines have also been obtained by repeated cloning. However the production of berberine in repeatedly cloned "stable" cell lines of Coptis japonica (Sato & Yamada, 1984), ajmalicine & serpentine in cell lines of Catharanthus roseus (Deus-Neumann & Zenk, 1984), rosmarinic acid in Anchusa officinalis (Ellis, 1982) and nicotine in selected cell lines of tobacco (Ogino et al, 1978) all fluctuated during repeated non-selective subculture. Ellis (1982) however was able to show greater stability in some of his lower yielding cell lines. The long term instability of product production in plant cell cultures is compounded by the lack of a general method for long term preservation of high yielding lines, as routinely used for preserving strains of micro-organisms. Thus plant cell cultures are maintained by regular repeated subculture, with or without intermitant selection of high yielding sub-lines.

Sato & Yamada (1984) suggest that variations in product yields during this type of long term maintenance of cultures may be due to factors in the subculture methodologies, whereas Deus-Neumann & Zenk (1984) suggest that instability in chromosome numbers and an unconscious selection for these cells may explain the decline in product production during long term subculture.

VARIATION IN ALKALOID PRODUCTION AND CELL GROWTH
We have recently described a cell culture system for C.roseus in which high levels of growth associated alkaloid accumulation occurred (Morris, 1986). This is in contrast to the

system first described by Zenk <u>et al</u>. (1977b) and subsequently used by
other workers, where separate growth and production conditions are
required. Using this system of culture, Deus-Neumann & Zenk (1984)
have recently shown that alkaloid production over an 8 year period in
both selected and non-selected cell lines of <u>C.roseus</u> was unstable. In
our unselected cell lines, sustained growth and capacity for alkaloid
accumulation occurred over repeated subculture. The stability of
alkaloid productivity compared to a two-stage system and the factors
affecting product yield are considered in this article.

The variation in alkaloid yields in a single cell line (C87) of
<u>C.roseus</u> subcultured for two years under growth associated alkaloid
production conditions is shown in Fig. 1. Considerable instability in
both growth rate and serpentine yield were found, with a direct
relationship between the rate of dry wt. accumulation and the
serpentine yield. Average yields of 0.2-0.3% serpentine were however
maintained over the 2 year period. During this period of repeated
subculture, occasional analysis of cells during a growth cycle showed
that maximum serpentine yields always occurred at 20-22 days, hence
the variability in yield shown in Fig. 1 is unlikely to be due to
variations in the time to reach peak serpentine levels.

Fig. 1.   Growth and alkaloid production of <u>Catharanthus
roseus</u> cell line C87 during a two year period of serial
subculture on M3 medium. Cells were subcultured at 14 day
intervals at a dilution of 1:5. Growth and alkaloid
content of 20 day old cells. AGR = average growth rate
mg/l/day dry wt. accumulation (▲), serpentine (O),
ajmalicine (■).

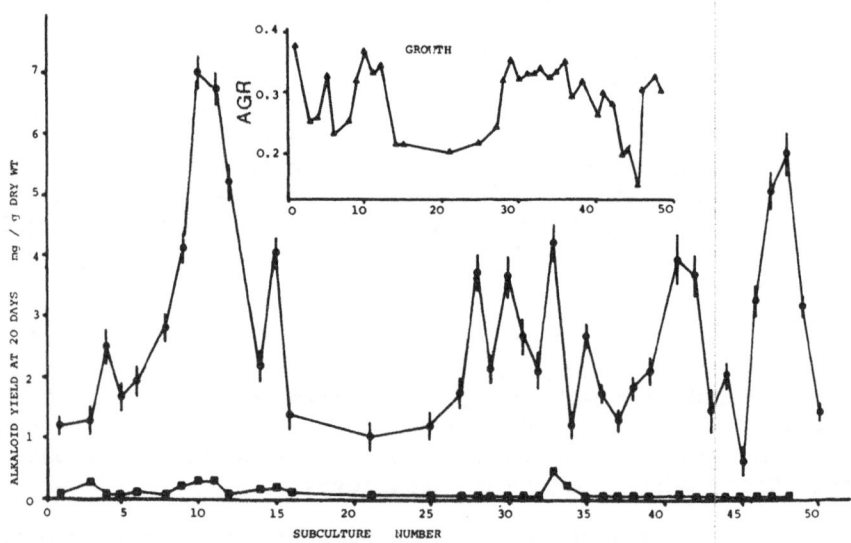

However, as ajmalicine peaked at 14 days and then declined (Morris, 1986) the level of ajmalicine in the cells at 20 days was always low compared to peak levels at 14 days.   Similar results as in Fig. 1 have been obtained for 2 other cell lines which contain different proportions and levels of ajmalicine and serpentine (data not shown).

FACTORS CONTRIBUTING TO VARIATION IN PRODUCT YIELD
Several factors other than inherent genetic instability may be responsible for the observed variation in alkaloid yields and growth rates.   These are mainly concerned with the reproducability of the subculture routine and with the long term maintenance of stable culture conditions.

Inoculation density
The density at which cells are inoculated into fresh medium at each subculture has been found to affect both the growth rate, the rate of alkaloid accumulation and the final alkaloid yield. Decreasing the inoculation density increased the specific growth rate (dry wt. accumulation) and decreased both the rate of synthesis and final yield of serpentine in cells transferred from growth medium to production medium Z (Zenk et al., 1977b) (Fig. 2).

Fig. 2.   Effect of inoculation density on growth (dry. wt. accumulation) and alkaloid accumulation in C. roseus cells cultured on Zenks production media.   Max serpentine yield (■), Max ajmalicine yield (O), growth rate (▲), rate serpentine accumulation (□).

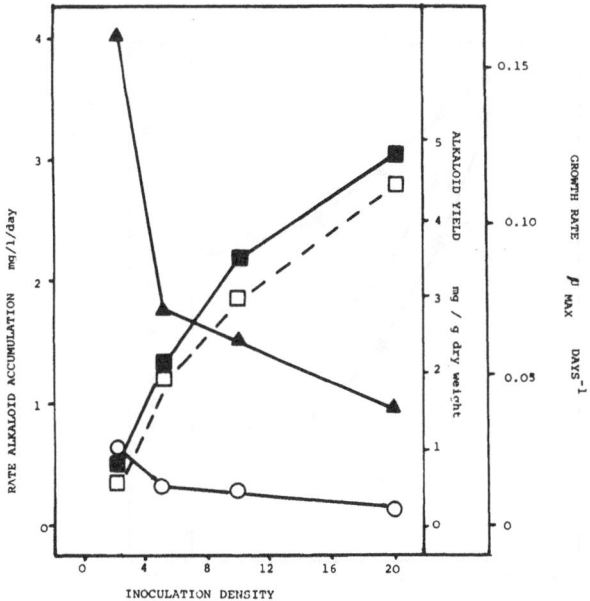

Growth and alkaloid production was therefore sensitive to inoculation density and hence this should be controlled at each subculture. Subculture by volume can give rise to such variations in inoculation density if growth rates vary between subcultures. In Fig. 1 inoculation densities over the 50 subcultures varied from 1.5 - 2.5 g. dry wt./l. with a mean of 2.00 ± 0.07 g. dry wt/l, hence variations in inoculation density in this case are unlikely to account for the observed variation in alkaloid yields and growth rates.

Subculture period

Stafford et al. (1985) have shown that both growth and alkaloid productivity in cells transferred to Zenks production medium were sensitive to the physiological age of the inoculated cells. Fig. 3 (reconstructed from the data in Stafford et al., 1985) shows that as the age of the cells at transfer was increased the growth rate, rate of serpentine synthesis and final serpentine yield declined. Thus variations in the physiological age of the cells at subculture arising either from alterations in culture cycle length or from variations in the growth rate during the preceding culture period can give rise to oscillations in growth rate and alkaloid yields. While the cell line used here (Fig. 1) was subcultured strictly at 14 days, because variations in growth rates between subcultures occurred, the cells may not always have been of similar physiological ages at each subculture.

Fig. 3.   Effect of culture age on growth and alkaloid accumulation on Zenks production medium, Growth rate (▲), serpentine yield (O), rate serpentine accumulation (■).

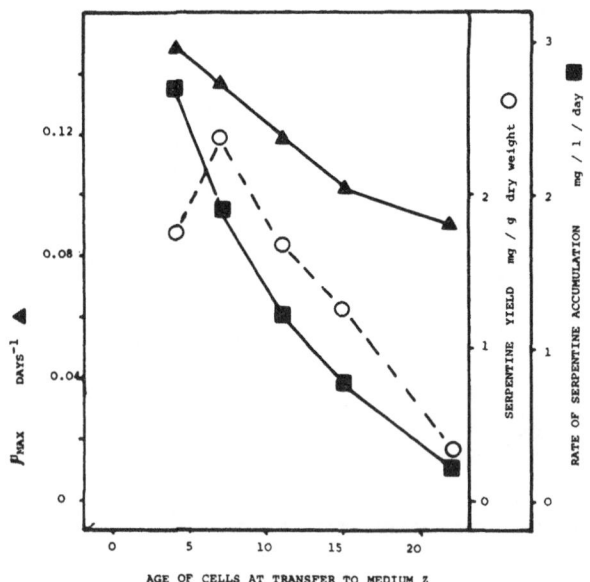

AGE OF CELLS AT TRANSFER TO MEDIUM Z

Culture temperature
        Growth and alkaloid production in the C87 cell line
growing on M3 medium was found to be very sensitive to culture
temperature (Fig. 4).   Small seasonal variations in the average
temperature between subcultures would therefore result both in
oscillations in growth rate and in alkaloid yields.

Analysis of culture temperature records over the 2 year period showed
that the average culture temperature over the whole period was
24-26°C.   While the variation in temperature within a single growth
period was ± 1°C, the maximum and minimum average temperatures between
growth periods was 27°C (in Summer) and 23°C (in Winter).   Thus cells
were grown over a range of temperatures from 23± 1°C to 27± 1°C, a
difference of 4°C.   Fig. 1 suggests that this difference markedly
affects both the growth rate and the yield of alkaloids.

CONCLUSIONS
        Several cell lines of C.roseus cultured under growth
associated alkaloid production conditions for 2 years maintained their
capacity for alkaloid accumulation, although large oscillations in
growth rate and alkaloid yields were observed.   Some evidence has been
presented which indicates that the variability observed both in growth
rates and in maximum alkaloid yields at each subculture may be due to
inadequate control of sub-culture methodologies and environmental
culture conditions.   It is easy to see how the cells can enter into
such oscillations if one considers that an alteration in culture
conditions such as temperature, culture cycle, length, or inoculation
density at one subculture can result in a change in the growth rate

Fig. 4.   Effect of culture temperature on grwoth and
alkaloid accumulation in the C87 cell line on M3 medium.
Growth rate (O), Max serpentine yield (▲), Max ajmalicine
yield (△), rate serpentine accumulation (■).

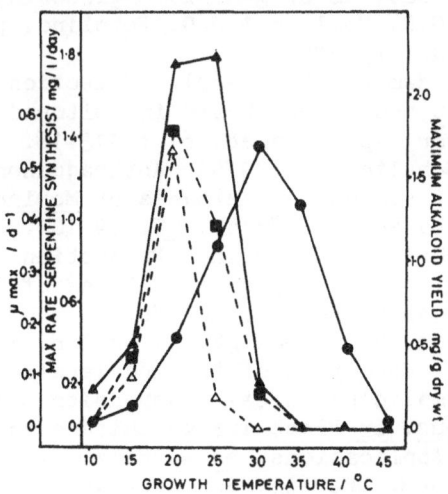

such that at the next subculture cells of a different physiological
stage of development are subcultured. While this may explain some of
the variations in yield observed in many culture systems, it cannot
explain why in some systems the average yield declines with
subculture. However if the oscillations in culture conditions are
large, this may result in the selection of cells which have lost the
capacity for product accumulation. Possible reasons for this loss are
discussed by Stafford 1986 (this volume 28).

REFERENCES

Deus-Neumann, B. & Zenk, M.H. (1984). Instability of indole alkaloid
        production in Catharanthus roseus cell suspension
        cultures. Planta Medica, 50, 427-31.

Dougall, D.K., Johnson, J.M. & Whitten, G.H. (1980). A clonal
        analysis of anthocyanin accumulation by cell cultures of
        wild carrot. Planta, 149, 292-97.

Ellis, B. (1982). Cell to cell variability in secondary metabolite
        production within cultured plant cell populations. In
        Plant Tissue Culture, 1982. Ed. A. Fujiwara. pp. 395-96.
        Tokyo: Abe Photo Printing.

Morris, P. (1986) Regulation of product synthesis in cell cultures of
        Catharanthus roseus II Comparison of production media.
        Planta Medica (in press).

Ogino, T., Hiroaka, N., Tabata, M. (1978). Selection of high
        nicotine-producing cell lines of tobacco callus by single-
        cell cloning. Phytochem. 17 : 1907-10.

Sato, F. & Yamada, Y. (1984). High berberine-producing cultures of
        Coptis japonica cells. Phytochem., 23, 281-85.

Stafford, A., Smith, L. & Fowler, M.W. (1985). Regulation of product
        synthesis in cell cultures of Catharanthus roseus : Growth
        related indole alkaloid accumulation in batch culture.
        Plant Cell Tissue & Organ Culture, 4, 83-94.

Yamada, Y. & Hashimoto, T. (1984). Secondary products in tissue
        culture. In Applications of genetic engineering to crop
        improvement. ed. G.B. Collins & J.G. Petolino. pp. 561-99.
        Dordrecht: Martinus Nijhoff.

Yamamoto, Y., Mizuguchi, R. & Yamada, Y. (1982). Selection of a high
        and stable pigment producing strain in cultured Euphorbia
        milli cells. Theor. Appl. Genet. 61 : 113-16.

Zenk, M.H., El. Shagi, H. & Schulte, U. (1975) Anthraquinone
        production by cell suspension cultures of Morinda
        citrifolia. Planta Medica. Suppl. 75, 79-101.

Zenk, M.H., El Shagi, H. & Ulbrich, B. (1977a) Production of
        rosmarinic acid by cell suspension cultures of Coleus
        blumei. Naturwissenschaften 64, 585-86.

Zenk, M.H., El Shagi, H., Arens, H., Stockigt, J. Weiler, E.W. &
        Deus, B. (1977b). Formation of the indole alkaloids
        serpentine and ajmalicine in cell suspension cultures of
        Catharanthus roseus. In Plant Tissue Culture and its
        Biotechnological Applications. ed W. Barz, E. Reinhard &
        M.H. Zenk. pp. 27-43. Berlin: Springer-Verlag.